SHIFTING THE EARTH

SHIFTING THE EARTH

The Mathematical Quest to Understand the Motion of the Universe

ARTHUR MAZER

A JOHN WILEY & SONS, INC., PUBLICATION

Published by John Wiley & Sons, Inc., Hoboken, New Jersey
Published simultaneously in Canada

For general information on our other products and services or for technical support, please contact our Customer Care Department within the United States at (800) 762-2974, outside the United States at (317) 572-3993 or fax (317) 572-4002.

Wiley also publishes its books in a variety of electronic formats. Some content that appears in print may not be available in electronic formats. For more information about Wiley products, visit our web site at www.wiley.com.

Library of Congress Cataloging-in-Publication Data:

Mazer, Arthur, 1958–
   Shifting the Earth : the mathematical quest to understand the motion of
the universe / Arthur Mazer.
      p. cm.
   Includes bibliographical references and index.
   ISBN 978-1-118-02427-0 (pbk.)
   1. Kepler's laws. 2. Motion. 3. Celestial mechanics. I. Title.
   QB355.3.M39 2012
   521'.3–dc23
                                    2011013592

Printed in the United States of America

oBook ISBN: 978-1-118-10829-1
ePDF ISBN: 978-1-118-10832-1
ePub ISBN: 978-1-118-10831-4

10 9 8 7 6 5 4 3 2 1

CONTENTS

PREFACE

There is a comic of a skinny little kid who, while standing on a stool so that he can look at himself in a mirror, views the image of a ripping, six-packed, Shwartznegger-esque champ. It's all too true to life and, like the best of comics, begs a question. Suppose that the same skinny kid looks at the mirror 35 years later when he is a pudgy, balding man. What is the image that he will see? Will he continue to see the champ, or will he have come to accept who he is, but in doing so see more possibilities for himself than the little boy could have imagined? This question is the topic of many novels and movies. It is intriguing because there is no predestined path for the boy to follow and there is no definitive outcome.

Shifting the Earth relates the story of how humanity collectively dispossessed itself of its geocentric fabrication and accepted a less prestigious position in the universe. It was not preordained that humanity would toss out its self-image, pack up what remained, and move to a new heliocentric earth. Psychological barriers resisted the move. Vested interests battled to bound humanity on a geocentric earth, and while they did not prevail, they put up a good fight. One can even imagine a scenario where they might have prevailed. The mathematicians and astronomers who contributed to the accomplishment did not live in a vacuum and were very much affected by their cultural environment. *Shifting the Earth* describes of the interplay between the forces tugging at the contributors in different directions and how the contributors ultimately created a prevailing force of their own.

The question of a heliocentric versus geocentric structure was indisputably resolved in the seventeenth century, a century steeped in the controversies of the Counter-Reformation. The accomplishment was the conduit to the Enlightenment and the subsequent scientific–industrial revolution that has transformed our lives. The accomplishment has received short shrift within the historical literature where philosophy lies at the center of historical evolution. *Shifting the Earth* holds the perspective that since Newton proposed his laws of motion with the explicit intent of determining a planet's pathway, the scientific revolution has been the main driver of history. Philosophy has become a backseat passenger that must either adapt to the discoveries of the scientists, or be forcibly removed by another philosophy that wants to get on board. Starting with the ancient Greeks, *Shifting the Earth* examines this transition as it unfolds.

Since the investigation into the universal order began, mathematics has guided the investigators and has also been the language through which they convey their results. This is true for geocentric theories as well as the development of the heliocentric theory. Ptolemy's universe rests firmly on Euclid's mathematical foundation, which Ptolemy skillfully exploits to describe heaven's trajectories. Fifteen centuries later, Isaac Newton's universe rests firmly on Newtonian calculus and axioms of motion. In the interim, mathematics wielded in the hands of mathematicians and astronomers was the prodding rod shepherding humanity from a geocentric to a heliocentric world. Accordingly, mathematics is central to the story and occupies a central role in *Shifting the Earth*. Each chapter includes a mathematical treatment of significant contributions to both the geocentric and heliocentric theories.

This book encompasses two areas, history and mathematics, which our culture normally segregates. There is no delineated boundary between these areas. History subsumes mathematics, and mathematics influences history, so the segregation is as unnatural as separating water from broth. Both constitute the chicken soup, and with the same sentiment, I uphold that history and mathematics both constitute this story.

As a preview, let me provide an offering that highlights the interplay between the history and the mathematics. The preeminent European scientist of his day, Johannes Kepler, using methods that Archimedes developed two millennia before Kepler, asserts conservation of angular momentum in integral form, and uses this assertion to hunt down the ellipse. With his discovery of the ellipse as the planetary pathway, Kepler becomes Europe's most formidable advocate of a heliocentric system. Kepler is a celebrity throughout Europe, and the Jesuit community in Vienna woos Kepler, an excommunicated Lutheran, with the hope of converting him to Catholicism. During this same time, the Jesuits in Rome were leading an Inquisition against Galileo, charging him with heresy on the basis of his arguments in favor of a heliocentric universe.

The situation is pregnant with both mathematical and historical questions. What mathematics did Archimedes develop two millennia before Kepler that assisted Kepler with his discovery? If the mathematics had been in place for so long, why did it take two from the time of Archimedes to discover the true heliocentric system? Why did the Lutherans excommunicate Kepler? Why did the Jesuits woo Kepler, while at the same time persecute Galileo? Answers to these questions require a historical as well as mathematical response. Our story must address all of these questions, or the story will be incomplete.

Concerning the mathematics, I had a few choices to make. A first decision was to determine which material to include. The combined works of Copernicus, Kepler, Galileo, and Newton are well over 2000 pages. Add to that the works of Eudoxus, Aristarchus, Archimedes, Apollonius, Hipparchus, and Ptolemy, and one can conceive of over 3000 pages of material. Clearly, this had to be pared down. I applied subjective judgment weighing two objectives, historical significance, and direct significance, to the completion of a heliocentric theory. The result is that I've sheared the mathematical hedge into a shape that pleases me, recognizing that others would do so differently.

My subjective judgment is guided by the manner in which the discovery of the ellipse as the planetary pathway unfolds. Looking backward through time, Kepler discovered the pathway for the planet Mars using Tycho Brahe's observations. Both Copernicus and Ptolemy also present their models of Mars' motion, and there is a chain that links all three works. The presentation focuses on the planet Mars so that the links in the chain are made visible.

Just as there is no delineation between mathematics and history, there is no distinct historical line that marks the beginning or the end of this story. I chose to begin with Eudoxus, a contemporary of Plato and Aristotle who gives the first known systematic, mathematical model of the motion of the heavens in the ancient Western world. One could begin at an earlier time, looking back to the Babylonians, or a later time, with Ptolemy. The former was excluded because of a blur in our knowledge of the influence of pre-Eudoxian astronomy on Eudoxus. On the mathematical front, while Eudoxus' model is not a visible link in the Ptolemaic–Copernican–Keplerian chain, it initiates Western mathematical analysis of the universe's motions and as such it is that chain's invisible anchor. On the historical front, it was the Athenian culture that fostered not only Eudoxus but also Ptolemy. To delve into Ptolemy without giving some historical context is akin to a read of Abraham Lincoln's signature Gettysburg Address without an awareness of the issues causing the Civil War. I begin with Eudoxus and Athens because they leave their signature on all that follows.

As for the story's endpoint, the decision was more difficult. Had I written this book before the twentieth century, the distinct historical line marking the end would have been unmistakable. Newtonian reasoning was the final assault that deflated the geocentric myth and launched Europe into the Enlightenment. So successful were Newton's laws and so powerful were the mathematical tools he bequeathed that another myth displaced the geocentric myth; all would be understood by Newtonian reasoning. The twentieth century revealed the existence of a distinct historical Newtonian line but also that the universe is more than the physical manifestation of Newtonian reasoning. We now know that the universe behaves in ways that completely challenge our notions of what is reasonable. This realization is inherent in our story where our balding man looks at himself in the mirror and then outwardly toward the universe and sees previously unimagined possibilities in both. Einstein was the first to expose a bizarre universe and convince a mystified world of its reality. Einstein is where I choose to end the story.

While I believe that mathematics is central to the story, popular sentiment may disagree. Showing a flexible disposition, I accommodate popular sentiment by writing in a style that allows readers to engage the mathematics to their level of comfort. If you do not find the mathematics engaging, don't fret, skip over the mathematics and enjoy the narrative.

Let me finish this preface by addressing the reader who is interested in following some or all of the mathematical content. A decision point in presenting mathematical material concerns the presentation of works that predate modern mathematical concepts and notation. I had to decide whether to derive results using only the tools available to the original authors or make full use of modern mathematics. Clarity of exposition guided my decision. So that their bosses in the State Department can understand their reports, a China specialist writes in English. Similarly, I convey results using modern concepts and notation that are familiar to the present-day reader.

Another decision that is inherently bound with the choice of material, addresses the mathematical prerequisites. My efforts are aimed at reaching as broad an audience as possible, and that means keeping the prerequisites at a minimum. In all chapters except Chapters 2 and 4, I endeavor to present material at its simplest level. This means that whenever a conflict between presenting elementary mathematics or more technically advanced mathematics arises, I choose the path of an elementary exposition even if it is the less succinct. For example, I give a mathematical presentation of Newton's laws without the use of calculus. Indeed, with the exception of Chapters 2 and 4, the mathematical material is accessible to a high school student. Okay, I confess, I couldn't resist giving a calculus-based proof of the ellipse. With the exceptions of Chapter 2, Chapter 4, and the calculus-based section of Newtonian mechanics in Chapter 8, all of the material is accessible to a high school student.

Concerning Chapter 2, this material is somewhat tangential to the remainder of the book. Chapter 2 presents Eudoxus' universe. Eudoxus' work is all the more remarkable in that he accomplished his task prior to the development of foundational mathematics that I needed to describe it. With college-level mathematics, the model becomes tractable. Without college-level mathematics, I couldn't imagine how one gives a mathematical description, let alone how one solves the problem. I had two choices, not include the work or use higher-level mathematics, and for reasons described above, I chose the latter.

Chapter 4 presents work from Apollonius' *On Conics*. As with Eudoxus, this work is tangential to the remaining material and far ahead of its time. While this material could be presented at an elementary level without the use of calculus, in this one instance I chose to be succinct. Aside from succinctness, the use of calculus allows for an exploration into concepts that Apollonius hints at, but are now fully developed. The mathematical material of Chapters 2 and 4 is independent of the material in the rest the book. The reader without the prerequisites may skip these sections and follow everything else.

On a final note, I was quite pleased by the accessibility of special relativity. One can give a rather complete account of the mechanical side of special relativity and remain fully bounded within a standard high school mathematics curriculum. In fact, key findings of Einstein, those that assault our common sense, are more readily accessible than the fully intuitive works of Newton. It is my hope that *Shifting the Earth* is an agent of the assault.

ACKNOWLEDGMENTS

By and large, this book is not the result of my individual effort, but the circumstances in which I find myself; indulge my wish to give details. There is a nasty side of human character that takes pleasure in causing jealousy. While I like to think myself above it, I'm not. My family feeds my nasty side, for anyone who knows my wife, Lijuan, and children, Julius and Amelia, would most certainly be jealous of me. How could I write a book and hold down a job without my family behind me all the way? Lijuan gives me a pass on much of the honey-do list; our bushes perpetually look like they've had a bad-hair day. Julius and Amelia root me on and give motivation. Despite their full awareness that the question "Dad, how's your book going?" will likely lead to an enthusiastic response detailing the adversity that Kepler overcame in discovering the ellipse, a story they've been abused by more than any human deserves, they ask anyway. Then when thoughts of avoiding my writing and spending the evening watching The Simpsons™ surface, the mood to write is restored.

During the writing of this book, I had a chance to catch up with an elementary school friend, Joel Sher. I'll spare the embarrassment of the number of decades that have passed since we last met, but despite the time lapse, Joel welcomed me into his home. He gave me exclusive use of the second floor of his house for a week. It came at a critical point when I really needed the time and solitude to get the job done. Joel is as I remembered him, a good friend who is always willing to help out.

It is difficult to find Kepler's *New Astronomy*. I offer my thanks to Beena Morar, who not only located a copy, but also made sure the copy was exclusively available to me throughout the writing of this book.

Also, I would like to thank Professor Roger Cooke from the Mathematics Department at The University of Vermont. The mathematics editor at Wiley,

Susanne Steitz-Filler, forwarded my proposal to Professor Cooke for him to review. Professor Cooke's careful line-by-line audit of the proposal was not what one would want from an IRS auditor, but exactly what one hopes for in a reviewer. His spotting of errors allowed me to pull my pants up from my ankles before going out in public. I became quite awed by Professor Cooke's knowledge. Whereas I read translations of the original Greek and Latin writings, Professor Cooke reads the original works. Not only did he point out a gaffe in a mathematical analysis; he also noted my poor theological interpretation of Ptolemy that resulted from a poor translation. His own translation and explanation of the original work influenced me to rewrite my own work. There were some disagreements between myself and Professor Cooke. If my pants are not fully buckled, I have to take responsibility, for Professor Cooke gave warning.

Finally, I would like to thank the people at Wiley and Susanne Steitz-Filler in particular. I have found my interactions with Wiley very professional and a pleasure. Susanne Steitz-Filler's support for this project is most gratifying. Without Susanne, this book would not be.

CHAPTER 1

PERFECTION

I am the circle. Within a plane, I am the set of points equidistant from a designated center. I am the perfect shape, and I boast a list of properties that no other shape can lay claim to.

Every point of my composition is equal. Segments containing midpoints and corners form the triangle, square, and every other polygon. An ellipse has two points closest to its center and another two points farthest from its center. A crescent has two vertices as well as midpoints to each of its arcs. Indeed, every shape other than mine contains points of special character, altering their status among other points within the assembly. Only I am composed of points that are all truly equal; the equality defines beauty along with perfection.

As all my points are equal, I am the only planar shape that remains unchanged by any rotation of any angle. Rotate a square by a multiple of other than 90 degrees, and it is obvious that you have altered its orientation. But rotate me through any angle, and you will perceive no change in my perfect and beautiful configuration.

I form the optimal boundary of an enclosure because the area enclosed by me is the largest area that can be enclosed within a boundary of fixed length. A triangle, square, ellipse, crescent, or any other shape with a perimeter that is the same length as my circumference, encloses a smaller area than mine. Perfection is optimal as well as beautiful.

Shifting the Earth: The Mathematical Quest to Understand the Motion of the Universe,
First Edition. Arthur Mazer.
© 2011 John Wiley & Sons, Inc. Published 2011 by John Wiley & Sons, Inc.

Every line through my center forms a line of reflective symmetry about my center. Place a mirror perpendicular to any line about my center, and the semicircle in front of the mirror is self-complementary so that its image with itself forms a perfect circle. Attempt this with an equilateral triangle, and unless the mirror is perfectly placed on one of only three lines of symmetry, the component in front of the mirror is not self-complementary. An ellipse has but two lines of symmetry through its center; an isosceles triangle and crescent have only one; and a scalene triangle, like most arbitrary shapes, has none. This infinite set of symmetries expresses itself only through my perfect shape in which every point is equal.

Inscribe an equilateral triangle within me, and every line of symmetry of the triangle is also a line of symmetry of mine. This is true for a square, an octagon, a dodecagon, a heptagon, a myriagon, or any other regular polygon, one with sides of equal length. Perfection dominates the imperfect as I dominate the polygons.

My dominance of the polygons also expresses itself in the attempt of a set of regular polygons to reach perfection. Order the regular polygons by the number of sides. Then, as one indefinitely climbs up the hierarchy, the shape of the polygons approaches my perfection. Indeed, by going far enough along the hierarchy to a target polygon, the points of all the subsequent polygons can be made as close as any arbitrarily small distance to my points. While they come close to me, only an insignificant fraction of the points settle on my perfect frame. I dominate the polygons. As they strive to reach me, they strive for perfection, but can never attain it.

My perfection inspires humans in all their endeavors. For three millennia engineers have used me to transport materials across the land whether by horse-powered cart, human-powered bicycles, steam-powered locomotive, or diesel-powered 18-wheelers. Circular gears drive mechanical devices; any other shape would cause uneven wear on the equipment. Electric power generators rotate through a circle yielding controllable voltage, current, and power output.

My symmetry inspires scientists to search for symmetry in nature. They have discovered the cyclic nature of time and stamped its daily rhythms on a circular clock. In the Northern Hemisphere, the North Star is a fixed center about which all other stars rotate along a circular pathway. Toss a stone into a lake, and a scientist confirms that the waves ripple across the lake's surface in concentric circles. I provide the pattern for the eye of a hurricane and the rainbow.

Just as mathematicians have discovered an impressive list of properties that I possess, artists and architects pay homage to my beauty. The artist adorns figures of religious admiration with a halo. Light enters a house of warship through a circular stained-glass window.

As I set the standard of equality, I am the foundation of many religious and political philosophies. Monotheism places all humankind equally about God; men are the points of a circle with God as their center. Both communism and democracy strive for the equality of citizens, one through an equal distribution of goods, the other through representative government via universal plebiscite in which every citizen has an equal vote in an electoral process.

I am a universal ideal pursued by engineers, scientists, mathematicians, artists, and philosophers. They pursue me because of my perfection. But like the polygons, humankind's pursuit is in vain for perfection is impossible to attain and so easy to destroy. A bicycle wheel is never in perfect true. Lay it on its side, and it doesn't evenly rest on its surface. Even with the naked eye, one can perceive a blip as it rotates. By adjusting the tension in the spokes, one can reduce a blip, but never eliminate it, and during the course of adjustment, a new blip always arises. Once close to true, a small bump in the road perturbs the wheel yet farther from perfection.

Human philosophical and social efforts toward a perfect circle of equality meet with similar road bumps. A system of privilege blanketed the democratic ideals of ancient Athens as citizenship was limited to a small select group. Others had limited or no rights, while some of the others were no more than the property of their slave-possessing owners. Over 2000 years after Athens' zenith, an assembly of men in the city of Philadelphia founded an independent nation vested on the circle of equality with God at its center as evidenced by the words of America's Declaration of Independence:

> We hold these truths to be self-evident, that all men are created equal, that they are endowed by their Creator with certain unalienable Rights, that among these are Life, Liberty and the pursuit of Happiness.

Echoing Athens, many of the very signers of this esteemed document left Philadelphia and returned to their estates where slaves who enriched America's founders were not among the equally created men, but were property.

It is not surprising that my perfect ideal cannot be achieved by human beings who are endowed with such imperfect character; human beings' character imperfections leach into their engineering, arts, philosophies, and institutions. So these very humans, with their imperfect character, have looked to nature to find me. Their first impressions of me as a true physical entity reflect their naivety and the immaturity of their science. Neither the stars' apparent path, the concentric waves, nor the rainbow's arc achieve my perfection. Just as a bump on the road perturbs the wheel away from true, nature introduces a wobble on the earth's axis perturbing the apparent path of the stars, and ripples on a lake perturb the concentric waves, and nonuniformity of the atmosphere perturbs the rainbow's arc. Not only am I too perfect for humankind; I am also too perfect for nature. I exist not as a fact, but as an ideal.

It is through the investigation into planetary motion that human beings revealed my true status as Utopian and not material. Humans unveiled the ellipse and replaced my perfection with its form. The scientists and mathematicians who contributed to the unveiling of the ellipse did so not in a vacuum, but with preconceived notions seeded in their instinct as well as those that society stamped on them. It was a monumental achievement to overcome humanity's obsession with perfection and find truth.

While scientists and mathematicians led the investigation, they were not the only participants. Because of the ramifications of this discovery on philosophy, religion, religious institutions, and the relation between the state and religious institutions, the highest levels of the ecclesiastic community weighed in on the debate attempting to influence its outcome. In doing so, they exposed themselves, and all except the most fervent noted that not only was the pathway of the heavens imperfect, but the institutions guiding humanity's moral precepts were themselves following an imperfect path. With the discovery of the ellipse, humans evicted me from their view of nature and their perception of their own self and institutions.

I am an ideal worth pursuing, but in pursuing me humanity has discovered a far more diverse universe with a far more interesting set of possibilities than I have to offer. Humanity has also established institutions that, however imperfect, are far more dynamic than those that claimed perfection. *Shifting the Earth* tells the story of how this unfolded.

CHAPTER 2

PERFECTIONISTS

A TOUR OF ATHENS

In 338 B.C.E., with the submission of Athens to Philip of Macedon (382–336 B.C.E.), the city-states of Greece forever lost their status as independent entities. The subsequent two millennia and two plus centuries saw Greece as a vassal territory to successive empires: Macedonian, Roman, Byzantine, and Turkish. A common historical view buttressed in school curricula is with their loss of independence; Greek and in particular Athenian culture waned from perfection. The post-Philippic Greeks themselves took a nostalgic view of their history as evidenced by latter scholarship focusing on investigations into their idyllic past as opposed to fresh creation.

It is not surprising that our public education follows Greek attitudes. Ancient Greece is where scholarship coalesced into distinct disciplines. The first European historians, philosophers, artists, poets, and playwrights were all Greek. Their output—including Homer's *Iliad* and *Odyssey*, Aristophanes' (446–386 B.C.E.) plays, the statues of forgotten artists, and the architecture of the Parthenon, not to mention the ancient philosophers, Socrates (469–399 B.C.E.), Plato (429–347 B.C.E.), and Aristotle (384–322 B.C.E.) and the philosophic schools that they inspired—stuns the admirer. Athens is also where the democratic ideal was born as citizens openly debated and voted on public policy. Prior to Philip, citizen armies defended the state out of devotion to an ideal, while after Philip, the empires raised armies of booty-seeking mercenaries. The historical narrative

Shifting the Earth: The Mathematical Quest to Understand the Motion of the Universe,
First Edition. Arthur Mazer.
© 2011 John Wiley & Sons, Inc. Published 2011 by John Wiley & Sons, Inc.

informs us that post-Philippic Greek cultural offerings, like the votive offerings of the Athenians to the Gods so that the Gods might bestow victory over Philip, fell short.

We seek perfection, but everyday confront an imperfect world. Only a denier who has gone past the point of delusion would describe our world as perfect. If not here and now in our present, then perhaps perfection existed in the past. Our wish to discover perfection is our exposed flank through which we contact the past and create legends.

That the culture that Philip inherited was far from perfect was as obvious to the Greeks of their period as the imperfection in our world is to us. They have left a record of indignities so vulgar that we dare not expose them to our schoolchildren; better that our children should think that our cultural progenitors were perfect. But our objective here is truth, and so we bring some counterbalance to the legend.

The Astynomoi of Athens was a civic organization overseeing several functions, including disposal of defecation, removal of the dead, and policing prostitution. If a counterbalance to the cultural legend of Athens exists, here's where one can find it. Concerning the former, removal of defecation, sanitation required human labor to perform this task. One might wonder who among the population were the designated. Given that the slave population was considerable and that this vocation is undoubtedly on the list of history's five worst vocations, it is a near certainty that slave labor disposed of the waste.

One can imagine that slave owners would dip into the public purse by offering up their human property for this task. While this was not documented, indications of profiteering at public expense most definitely were. Accusations of treasonous acceptance of bribes from the enemy during wartime are part of the historical record.

Under certain circumstances the Astynomoi was unable to satisfactorily perform its duty. Pericles' (495–429 B.C.E.) strategy against Sparta during the Peloponnesian War was to amass all Athenians within the city walls as opposed to engaging the superior Spartan infantry in the open. (Athens would engage Sparta only on the seas using its superior navy.) As the Spartans wrought destruction on all property beyond the walls, a swell of humanity several times the normal population sardined itself within Athens. Rather than being disposed of beyond the city gates, human waste remained with its Athenian producers inside the walls. The Spartans did not need a victory in the field to win the war—unsanitary conditions alongside the overcrowding caused an epidemic, reducing the population with great efficiency and defeating the Athenians. Pericles, the man who rallied the Athenians toward war when face-saving diplomatic options were available, fell victim to the plague. Somehow we envision a pillar of the democratic ideal to die a more dignified death, and yet, by dragging the Greek city-states into an utterly meaningless internecine war, perhaps the death was just.

As earlier, noted another responsibility of the Astynomoi was to ensure removal of the dead. On accepting his punishment and poisoning himself with hemlock, should Socrates' body have been left unattended, the Astynomoi would have been notified. What misdeed caused Socrates' fellow citizens to bring charges against him, and what crime did the unimposing septuagenarian commit that caused Socrates' jury of peers to sentence him to death? It seems that the elderly philosopher was a smart aleck. He had a way of humiliating others in debate. Apparently Socrates humiliated some of the more than equal among equals, and their personal pride trumped free speech. Those with influence could and did drum up charges. Socrates' student, Plato, attended to Socrates as Socrates executed his own sentence by drinking a concoction of hemlock. Presumably Plato arranged for a dignified burial, so there was no need to notify the Astynomoi and have them remove the body. Plato should have notified the Astynomoi about the other casualty, the democratic ideal in which free speech is paramount.

Speaking of the dead, prostitutes, both male and female, would accompany them through the night as the cemetery also proffered resting grounds for the homeless. Perhaps a prostitute and client would finalize their dealings in the cemetery, creating the image of the circle of life with its beginning and ending at the same point. As awful as one imagines this life, one wonders whether, among the women, Athens held out anything better.

Women were possessions of men who desired to possess them in one of two forms. Foremost, to establish himself within the community, a respected man desired a childbearing wife. The Athenians were extreme in their precautions against a tryst. The wife was not to be alluring to any man, lest she attract attention. She was not to know anything of the world of men, so there was no possibility of her carrying on a conversation that another man might find stimulating. She was in a cultural hermetic seal. To match her cultural isolation, the Athenians placed their wives in physical isolation as well, essentially imprisoning them. With the exception of a few holidays, the wife was not allowed outdoors; she spent most of her life within her home's walls.

The dull housebound wife did not provide the attractive lure that many men desire, so the prostitute filled that role. How did a man recognize a prostitute? There was a simple code for determining who could render this service, any women outdoors by definition was an available prostitute. But not only women on the street were prostitutes, there were whorehouses where daughters of slaves who, after being sold to a whorehouse proprietor as a commodity, would offer their service as a commodity, enriching the pimping proprietor. There was also an additional category of women who lived in an amorphous zone between housewife and prostitute. These women were self-educated lovers who could not only satisfy a man's physical lust but also be a partner in conversation and ideas. Returning to the streetwalker and the Astynomoi, their entwinement was strictly administrative. The Astynomoi enforced a government-set fee for a prostitute's services.

Our tour of Athens alongside the Astynomoi is jolting and admittedly not balanced. Perhaps things were not so bad. Certainly there were husbands and wives with genuine affection for one another, fulfilled in their relationship. Some households may have allowed their girls to have an education or at least did not interfere with a precocious intelligent girl who was quick to pick things up on her own. There may have been owners and slaves who developed an understanding of one another that bordered on mutual respect. Chest thumpers did not unilaterally prevail in leading citizens to internecine warfare, and when the Persians thrust war on the Greeks, the Athenian minority endowed with democratic rights proved to be more than up to the task. Furthermore, the public discourse in Athens was not, unlike some of our modern-day news programs, 24/7 venom and there is strong evidence of freedom of speech; Socrates' fate was the outlier, not the norm. Nevertheless, social attitudes and institutions gave rise to much abuse. Charles Dickens would not have had any problem spotting the hypocrisies in a society striving for perfection through a democratic ideal, while protecting a broad set of privileges that apply to a minority. In fact, Athens had its own Dickensian moral watchdogs in the form of playwrights and philosophers so that the only Athenian believing that Athens had achieved perfection was a denier who had gone past the point of delusion.

PLATO'S CHALLENGE

If we, with a 360-degree view of the imperfections of our times, look for ideals in the past, where did the Athenian similarly aware of society's shortcomings look for perfection? In this book we are concerned with two refuges for perfection, mathematics and the heavens. These were not exclusive domains. Mathematics was and continues to be the channel through which we theorize about the heavens. As such, the heavens have amply fertilized the field of mathematics.

Plato posed the following challenge to his students and fellow scholars: "What are the uniform and ordered movements by the assumption of which the apparent motion of the planets can be accounted for?" While seemingly an innocuous challenge, little did Plato know the influence that his challenge would have on humankind's development. Here we begin with the first response by Eudoxus (410–355 B.C.E.), whose solution was brilliant.

The constraint that Plato places on the challenge, motions must be uniform and ordered, is of interest and contains the essence of Plato's challenge. To understand the constraint, and to initiate an explanation of Eudoxus' solution, consider the motion of the heavens absent the planets, the sun, and the moon. In this universe, while the earth is cold and lonely, there is a simple response to Plato's challenge. The earth is, of course, at the center of this universe, and a universal sphere contains the stars. The axis of the universal sphere is aligned with the earth's north–south axis, and the sphere rotates uniformly about its axis in a counterclockwise direction when viewed from the earth's Northern Hemisphere. In this universe all the heavens rotate in perfect circular motion about the universal north–south axis. There would be no sun to distinguish a complete day, so one might associate time with the circular motion of the stars. Each star would complete one full circle about the universal axis in around 4 minutes less than the current day (see Figure 2.1). It is manifest that both the

Figure 2.1. Rotation of stars. A camera pointed at Polaris shows the apparent rotation.

circle and heavens are perfect, and so it is natural to find the one in the other while further extending the circle's perfection into the measurement of time.

The Greek word *planes* ($\pi\lambda\alpha\nu\eta\varsigma$) means wanderer. The Greeks bestowed the planets with this title because unlike the stars, they do not execute uniform circular motion about the universal axis. Instead, like Shakespeare's fairies in *Midsummer Night's Dream*, they dart about, causing mischief in the heavens. To some they may have been a delight, but to others they were a blemish, a blemish because they upset the perfection of the heavens. We examine Mars to describe the path of the planets as they dance about the stars. Figure 2.2 illustrates the pathway of Mars through the constellations as viewed by an Earthbound observer. Mars does not rotate on the universal sphere as do the other stars, but follows another path. There is a general trend to Mars' drift; however, without warning Mars at times reverses direction and executes a loop, just to show he's his own man. This renegade behavior is known as *retrograde motion*.

There are two ways to interpret Mars' independence, Mars, indeed, follows his own rules, so the heavens are not perfectly coordinated and chaos reigns in the heavens just as it does on Earth, or the heavens are perfect and humans are ignorant of their orchestration. Plato axiomatically believes in the perfection of the heavens, and his constraint that the motions must be uniform and ordered is a challenge to discover its synthesis.

For Eudoxus, the circle is perfect and is the basis of any perfectly orchestrated motion. He rises to Plato's challenge with brilliance by demonstrating that the

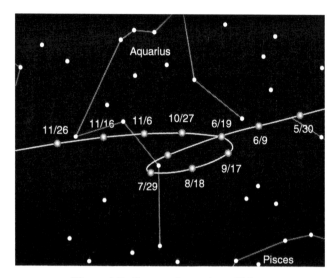

Figure 2.2. Retrograde motion of Mars.

circle does, indeed, regulate the heavens, including the planets, sun, and moon; and so, as Plato foresaw, perfection, not chaos, reigns supreme in the heavens. Eudoxus' solution is influential; his success inspires future mathematicians. But equally, perhaps due to the lure of perfection, Eudoxus collars the imagination of his scientific progeny for nearly two millennia. Apart from the philosophical implications, the solution is an intellectual feat that is comparable to those of modern-day Nobel laureates.

EUDOXUS' UNIVERSE

Eudoxus' universe contains not only the earth and the stars but also five planets, the sun, and the moon. As with the description of the lonely and cold-earthed universe, the earth lies at the universe's center and the rotating universal sphere governs the motion of the stars. A composition of concentric spheres tethered to the universal sphere and then inwardly to one another, each spinning in a perfect circle, steers the motion of the remaining bodies. Below, we present Eudoxus' model for the stars, sun, moon, and planets. We first provide visual illustrations and accompanying explanations. Afterward, a more technical approach demonstrates the ability of Eudoxus' system to produce retrograde motion.

The Stars

Consider that we project the earth's longitudinal and latitudinal coordinates onto the universal sphere. As Eudoxus' universe is geocentric and the earth is fixed, these coordinates are fixed while the heavens move through the coordinate

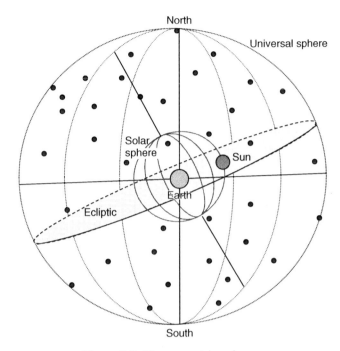

Figure 2.3. The geocentric universe.

system. A star seated on the universal sphere's axis is, like the earth and our modern-day North Star, fixed. All the other stars float through the earth's fixed longitudes at a uniform rate while their latitudes are steadfast. As already noted, the stars move along a circular path around the universal sphere's north–south axis (see Figure 2.3), completing slightly more than one cycle in a solar day.

The Sun

Within Eudoxus' geocentric universe, the sun varies in both longitude and latitude. The longitudinal motion of the sun is slightly slower than that of the stars. Whereas in one year's time, the stars rotate about the earth roughly 366 times, during the same annual period, the sun executes only slightly more than 365 orbits about the earth. The sun's latitude follows a sinusoidal pathway that varies across the seasons. Choosing an arbitrary starting time the latitude begins at 23.5 degrees north at the winter solstice, proceeds to zero degrees at the spring equinox, then continues on to 23.5 degrees south at the summer solstice. From then on the sun's latitude retraces its pathway back to zero degrees for the fall equinox and then returns to its initial point of 23.5 degrees north for the winter solstice.

The sun traces an interesting pathway across the heavens. Because of the offset between its longitudinal rotation and that of the stars, the sun promenades through the constellations, visiting each member of the zodiac at its allocated time. These constellations do not lie on an even stage; some are higher in the sky, and some are lower as dictated by the sun's oscillating latitude (see Figure 2.4).

Let the universal sphere envelope a smaller sphere and set the sun on the smaller sphere's equator (see Figure 2.3). Initially align the poles of the solar sphere with those of the universal sphere, and spin the solar sphere in opposition to the spin of the universal sphere so that it completes one cycle in one year's time. Next tilt the poles of the solar sphere so that its axis of rotation and that of the universal spheres are at an angle of 23.5 degrees and affix the axis of the solar sphere to the interior of the universal sphere. By affixing the solar sphere's axis to the universal sphere, the axis and hence the solar sphere subsume the spin of the universal sphere.

To an observer on earth, the sun's movement is a composition of the two circular motions. The spin of the universal sphere dominates the sun's longitudinal velocity giving rise to the daily motion and the spin of the solar sphere governs the sun's seasonal fluctuations in latitude. One point of note is that had we not tilted the axis of the solar sphere and kept it aligned with that of the universal sphere, the plane of rotation for the sun would have been in the earth's equatorial plane and there would be no variation in the sun's latitude. Instead, the sun rotates about a plane known as the ecliptic, giving rise to the sun's wavy path. In addition, the rotation of the solar sphere slightly offsets the longitudinal speed from that of the stars so that the sun completes one cycle about the earth in a day as opposed to slightly the slightly faster stars.

This is nearly, but not quite Eudoxus' model. Eudoxus incorporates another sphere with the belief that the motion was more complex than our standard model.

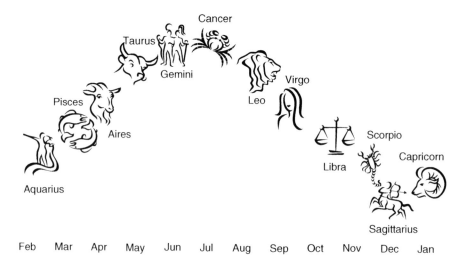

Figure 2.4. Motion of Sun through constellations.

The Moon

The Greeks observed that the (earth's) moon, like the sun, changes in both longitude and latitude. There is the near-daily cycle of moonrise and moonset. Additionally, like the sun, the moon does not sit in a fixed position with the stars. In comparison with the sun, which saunters through the zodiac once per year, the moon zips around the zodiac 13 times annually. As the moon's orbital plane does not lie in the earth's equatorial plane, like the sun, the moon's latitude fluctuates. There are two components to the latitude fluctuations, one that follows the moon's orbital cycle with reference to the stars and another due to the precession of the moon's orbital plane. The cycle of the precession is approximately 18.5 years. We first describe Eudoxus' solution as if there were no precession and afterward add another ring to complete Eudoxus' model.

To account for the orbital cycle in latitude, initially align a lunar sphere's poles with those of the universal sphere and place the moon on the sphere's equator. Spin the sphere in opposition to the spin of the universal sphere at a rate of 13 cycles per year. Next, tilt the axis of the lunar sphere to an angle of 18.4 degrees with that of the universal sphere and affix the axis to the universal sphere. As with the sun, the rotation of the universal sphere dominates the daily motion of the moon with some adjustment from the lunar sphere. The motion of the lunar sphere regulates the variations in the moon's latitude.

The ancient Greeks were aware of a slight precession in the moon's rotational plane. This is rather awe-inspiring and says a lot about the Greeks' devotion to their studies of the heavens. To account for the precession, Eudoxus inserts an additional sphere between the universal sphere and the lunar sphere. This additional ring rotates once every 18.5 years, is affixed to the surface of the universal sphere, and is aligned with the solar sphere. The lunar sphere is then affixed not to the universal sphere but to the precessional sphere, and is tilted back toward the universal axis at an angle of 5.1 degrees to that of the precessional sphere.

The Planets

Eudoxus knew of five planets. The motion of the planets have two or three spheres tethered to the universal sphere depending on whether they are an inner planet, Mercury and Venus, or an outer planet, Mars, Saturn, and Jupiter. As with the sun and the moon, the universal sphere dominates the daily motion of the stars while the inner spheres provide a minor correction. The sphere immediately tethered to the universal sphere spins at a speed so that the planets dance through the constellations at their observed speed. Eudoxus aligns the axis of this sphere with that of the solar sphere so that its equator lies within the ecliptic. For the inner planets, the innermost sphere holds the planet along its equator and imputes a precession of the planet's orbital plane. For the outer planets, Eudoxus structured the inner two spheres in a manner that creates retrograde motion. The planet sits on the equator of the innermost sphere.

By generating retrograde motion through this final construction, Eudoxus demonstrates his mastery of geometry. Below, we discuss some details of the construction. The presentation fully exploits mathematics that were not available to Eudoxus, including trigonometry, Cartesian coordinates, analytic geometry, linear algebra, and Euler angles. With these tools, not fully developed until two millennia after Eudoxus, the solution is manageable. Before continuing, take a moment to imagine the orbit of a point along the equator of a spinning sphere fixed on another spinning sphere that is then again fixed on another spinning sphere, which is finally tethered to an outermost spinning sphere. Imagine, without our modern mathematical tools, structuring the correct alignment of the axes and the speed of the spins to create a desired motion, such as retrograde motion. As you imagine this, Eudoxus earns your respect.[1]

The Whole

Eudoxus' heavens, the universal sphere and all other spheres perfectly centered around the earth, fit together as a whole through a system ordered according to distance from the earth. The center of the universe is Earth. On the closest sphere sits the moon. The next occupied sphere hosts Mercury, followed by Venus, the sun, Mars, Jupiter, Saturn, and finally the universal sphere. Aristotle, memorialized this universe in his writings. Later in time, the Church would deify Aristotle and in the process, canonize Eudoxus' universe.

THE MATH

Below we present the mathematics of Eudoxus' solar orbit and the orbit of a planet with retrograde motion. To begin, let us describe the construction of the orbits as an algorithm. Once this verbal description is available, we can associate mathematical operations with each step. The algorithm applies to each body separately.

Preliminary Step.   Suspend the spheres concentrically so that they are all centered on the earth and so that their axes of rotation are all aligned along the north–south axis of the earth. Seat the planet or sun upon the equator of the innermost sphere.

Step 1.   Among the stationary spheres, spin the innermost one at its appropriate speed and direction.

Step 2.   Tilt the axis of the sphere from step 1 at an appropriate angle, and affix the axis to the immediate sphere that envelopes it.

Step 3.   Return to step 1 until all the spheres are spinning.

---

[1]As an aid, a nice visual that displays the retrograde motion of a planet along three concentric circles is available on the website http://hsci.cas.ou.edu/images/applets/hippopede.html.

Two operations are of interest: (1) spinning a sphere with its axis aligned along the north–south direction and (2) tilting and affixing the sphere. These operations proceed from the innermost sphere outward toward the universal sphere.

To describe these operations mathematically, some notation is necessary. We consider three-dimensional Euclidean space in which (x,y,z) form the standard orthogonal coordinate system. Center the earth at the origin of the system and point the z axis directly north. The z axis is the axis of rotation for the universal sphere. Alongside Euclidean space, we append a time variable t, with the solar year as its unit measurement. A vector $(x(t), y(t), z(t))^T$ gives the (x,y,z) position of the sun or planet at time t.

Each sphere has a rate and direction of rotation designated by the real number ω_j. A positive value of ω_j indicates spin in a clockwise direction as viewed by an individual on the z axis looking down on the universal sphere from the outside. The subscript j identifies the associated sphere from the innermost to the universal sphere. With this notation, ω_1 is the angular velocity of the innermost sphere, ω_2 is the angular velocity of its immediate outer neighbor, ω_3 is the angular velocity of the third most inner sphere, and so forth. If there are m spheres, then ω_m is the angular velocity of the universal sphere.

Aside from the angular velocity, each sphere, excluding the universal sphere, has a tilt angle with respect to the axis of its immediate outer neighbor. We designate the tilt angle by θ_j, whereas with the angular velocity, the subscript j identifies the associated sphere.

With this notation, we are ready to mathematically describe the algorithm. The preliminary step sets the position of the body prior to performing any operations. The body maintains a constant distance from the earth that we arbitrarily set at one. This requires that the innermost sphere's radius be one. The preliminary step initially seats the body at the position $(0, 0, 1)^T$ on the innermost sphere; note that this position lies on the sphere's equator.

Step 1 imputes a rotation about the z axis through the angle $\omega_j t$, the angle is the product of the angular velocity and time. A vector executes the desired rotation when operated on by the following matrix.

$$R_{\omega_j}(t) = \begin{bmatrix} \cos(\omega_j t) & -\sin(\omega_j t) & 0 \\ \sin(\omega_j t) & \cos(\omega_j t) & 0 \\ 0 & 0 & 1 \end{bmatrix}$$

For example, applying the matrix to the vector $(1, 0, 3)^T$ yields the following result:

$$R_{\omega_j}(t) \begin{bmatrix} 1 \\ 0 \\ 3 \end{bmatrix} = \begin{bmatrix} \cos(\omega_j t) & -\sin(\omega_j t) & 0 \\ \sin(\omega_j t) & \cos(\omega_j t) & 0 \\ 0 & 0 & 1 \end{bmatrix} \begin{bmatrix} 1 \\ 0 \\ 3 \end{bmatrix} = \begin{bmatrix} \cos(\omega_j t) \\ \sin(\omega_j t) \\ 3 \end{bmatrix}$$

The (x,y) components of the vector $(1, 0, 3)^T$ are rotated to the values $(\cos(\omega_j t), \sin(\omega_j t))$, while the z component remains fixed at the value 3.

Step 2 tilts the sphere at a prescribed angle θ_j. A tilt is also a rotation that we perform in the (y, z) plane leaving the x axis fixed. The rotation matrix is given as follows.

$$T_{\theta_j} = \begin{bmatrix} 1 & 0 & 0 \\ 0 & \cos(\theta_j) & -\sin(\theta_j) \\ 0 & \sin(\theta_j) & \cos(\theta_j) \end{bmatrix}$$

For example, applying the matrix to the vector $(3, 0, 1)^T$ yields the following result:

$$T_{\theta_j} \begin{bmatrix} 3 \\ 0 \\ 1 \end{bmatrix} = \begin{bmatrix} 1 & 0 & 0 \\ 0 & \cos(\theta_j) & -\sin(\theta_j) \\ 0 & \sin(\theta_j) & \cos(\theta_j) \end{bmatrix} \begin{bmatrix} 3 \\ 0 \\ 1 \end{bmatrix} = \begin{bmatrix} 3 \\ -\sin(\theta_j) \\ \cos(\theta_j) \end{bmatrix}$$

The x value of the vector remains fixed, while the (y,z) components are tilted through the angle θ_j.

We are now in a position to mathematically describe the motion of the sun as follows:

$$\begin{bmatrix} x(t) \\ y(t) \\ z(t) \end{bmatrix} = R_{\omega_2}(t) T_{\theta_1} R_{\omega_1}(t) \begin{bmatrix} 1 \\ 0 \\ 0 \end{bmatrix} = \begin{bmatrix} \cos(\omega_2 t) & -\sin(\omega_2 t) & 0 \\ \sin(\omega_2 t) & \cos(\omega_2 t) & 0 \\ 0 & 0 & 1 \end{bmatrix}$$

$$\times \begin{bmatrix} 1 & 0 & 0 \\ 0 & \cos(\theta_1) & -\sin(\theta_1) \\ 0 & \sin(\theta_1) & \cos(\theta_1) \end{bmatrix} \begin{bmatrix} \cos(\omega_1 t) & \sin(\omega_1 t) & 0 \\ \sin(\omega_1 t) & \cos(\omega_1 t) & 0 \\ 0 & 0 & 1 \end{bmatrix} \begin{bmatrix} 1 \\ 0 \\ 0 \end{bmatrix}$$

We can march through the algorithm by reading the expression from right to left, reflecting the fact that the first operation on the initial vector is the far-right matrix, the second operation is the next matrix in the center, and the final operation is the far-left matrix.

Preliminary Step.   The sun's preliminary position is set at $(1,0,0)$.

Step 1.   Since the solar ring completes one revolution annually, the value ω_1 is -2π. The matrix,

$$\begin{bmatrix} \cos(-2\pi t) & -\sin(-2\pi t) & 0 \\ \sin(-2\pi t) & \cos(-2\pi t) & 0 \\ 0 & 0 & 1 \end{bmatrix},$$

spins the initial vector in a counterclockwise direction.

Step 2. The value θ_1 is equal to the angle of the ecliptic with the equatorial plane. The matrix,

$$\begin{bmatrix} 1 & 0 & 0 \\ 0 & \cos\left(2\pi\dfrac{23.5}{360}\right) & -\sin\left(2\pi\dfrac{23.5}{360}\right) \\ 0 & \sin\left(2\pi\dfrac{23.5}{360}\right) & \cos\left(2\pi\dfrac{23.5}{360}\right) \end{bmatrix},$$

tilts the solar ring through an angle of 23.5 degrees aligning the sun's orbit with the ecliptic.

Return to Step 1. Since the stars circle about the universal axis 366 times per year, the value ω_2 is 732π. The matrix,

$$\begin{bmatrix} \cos(732\pi t) & -\sin(732\pi t) & 0 \\ \sin(732\pi t) & \cos(732\pi t) & 0 \\ 0 & 0 & 1 \end{bmatrix},$$

rotates the universal sphere clockwise. We are done.

Performing the matrix multiplications and simplification gives the following result for the sun's trajectory:

$$\begin{bmatrix} x(t) \\ y(t) \\ z(t) \end{bmatrix} = \begin{bmatrix} \cos(\omega_2 t)\cos(\omega_1 t) - \cos(\theta_1)\sin(\omega_2 t)\sin(\omega_1 t) \\ \sin(\omega_2 t)\cos(\omega_1 t) - \cos(\theta_1)\cos(\omega_2 t)\sin(\omega_1 t) \\ -\sin(\theta_1)\sin(\omega_2 t) \end{bmatrix}$$

There are a few notable features in this solution. As expressed by the z value, the sun oscillates up to a maximum value of $\sin(\theta_1)$ and a minimum value of $-\sin(\theta_1)$. The way everything has been set, the initial time represents the fall equinox, where the ecliptic and the equatorial plane intersect. The sun first attains its minimum and maximum values at $t = \frac{1}{4}$, the winter solstice, and $t = \frac{3}{4}$, the summer solstice, and annually after that. The maximum angle of latitude is θ_1, which, as previously noted, is the angle between the ecliptic and the equatorial plane.

Because the rates of rotation, ω_1 and ω_2, are both integral multiples of 2π, the values $(x(t), y(t), z(t))$ are cyclic with an annual period. Indeed, the sun returns to its initial position at $(1,0,0)$ after one year and then repeats its orbit.

After this solar warmup exercise, we are ready to chase the planets and capture retrograde motion. The rotate, tilt, rotate, tilt ... algorithm applied to four spheres expresses itself through the following mathematical equation:

$$\begin{bmatrix} x(t) \\ y(t) \\ z(t) \end{bmatrix} = R_{\omega_4}(t)T_{\theta_3}R_{\omega_3}(t)T_{\theta_2}R_{\omega_2}(t)T_{\theta_1}R_{\omega_1}(t)\begin{bmatrix} 1 \\ 0 \\ 0 \end{bmatrix}$$

Here we are concerned primarily with the first and second spheres as these spheres are most responsible for the retrograde motion. Set the angular velocities of the two innermost spheres equal in magnitude but opposite in direction to one another: $\omega_2 = -\omega_1$. We examine this case for an arbitrary tilt angle θ_1.

Evaluating the expression

$$R_{-\omega_1}(t)T_{\theta_1}R_{\omega_1}(t)\begin{bmatrix}1\\0\\0\end{bmatrix}$$

gives the following result:

$$\begin{bmatrix}\tilde{x}(t)\\\tilde{y}(t)\\\tilde{z}(t)\end{bmatrix} = R_{-\omega_1}(t)T_{\theta_1}R_{\omega_1}(t)\begin{bmatrix}1\\0\\0\end{bmatrix}$$
$$= \begin{bmatrix}\cos^2(\omega_1 t) + \cos(\theta_1)\sin^2(\omega_1 t)\\(\cos(\theta_1) - 1)\sin(\omega_1 t)\cos(\omega_1 t)\\\sin(\theta_1)\sin(\omega_1 t)\end{bmatrix} \quad (2.1)$$

Our first problem is to describe the path of the planet absent the influence of the third sphere and universal sphere. (The tildas emphasize that this is not a complete description of the path since not all spheres are considered.) First let us describe the relation between $\tilde{x}(t)$ and $\tilde{y}(t)$. Using trigonometric identities, it is possible to solve for both $\sin(\omega_1 t)$ and $\cos(\omega_1 t)$ in terms of $\tilde{x}$. On substituting the solutions into the equation for $\tilde{y}(t)$, the relation between $\tilde{x}$ and $\tilde{y}$ emerges. The steps are as follows:

$$\tilde{x} = \cos^2(\omega_1 t) + \cos(\theta_1)\sin^2(\omega_1 t)$$
$$= 1 - \sin^2(\omega_1 t) + \cos(\theta_1)\sin^2(\omega_1 t)$$
$$= 1 + (\cos(\theta_1) - 1)\sin^2(\omega_1 t)$$

Solving for $\sin^2(\omega_1 t)$ yields the following equality:

$$\sin^2(\omega_1 t) = \frac{\tilde{x} - 1}{\cos(\theta_1) - 1} \quad (2.2)$$

A similar calculation in which one uses the Pythagorean identity to substitute for $\sin^2(\omega_1 t)$ yields the following result:

$$\cos^2(\omega_1 t) = \frac{\tilde{x} - \cos(\theta_1)}{1 - \cos(\theta_1)} \quad (2.3)$$

Substituting expressions (2.2) and (2.3) into the equation for $\tilde{y}(t)$ [i.e., (2.1)] and simplifying gives the following relation between $\tilde{x}$ and $\tilde{y}$:

$$\tilde{y} = (\cos(\theta_1) - 1)\sin(\omega_1 t)\cos(\omega_1 t)$$

$$\tilde{y}^2 = (\cos(\theta_1) - 1)^2 \sin^2(\omega_1 t)\cos^2(\omega_1 t)$$

$$= (\cos(\theta_1) - 1)^2 \left(\frac{\tilde{x} - 1}{\cos(\theta_1) - 1}\right)\left(\frac{\tilde{x} - \cos(\theta_1)}{1 - \cos(\theta_1)}\right)$$

$$= (1 - \tilde{x})(\tilde{x} - \cos(\theta_1))$$

$$= -\tilde{x}^2 + (1 + \cos(\theta_1))\tilde{x} - \cos(\theta_1)$$

$$= -\left(\tilde{x} - \frac{1 + \cos(\theta_1)}{2}\right)^2 + \left(\frac{1 + \cos(\theta_1)}{2}\right)^2 - \cos(\theta_1)$$

$$= -\left(\tilde{x} - \frac{1 + \cos(\theta_1)}{2}\right)^2 + \left(\frac{1 - \cos(\theta_1)}{2}\right)^2$$

In two dimensions the path of $\tilde{x}(t)$ and $\tilde{y}(t)$ is the circle centered at $(\tilde{x}, \tilde{y}) = \left(\frac{1+\cos(\theta_1)}{2}, 0\right)$ with radius $\frac{1-\cos(\theta_1)}{2}$:

$$\left(\tilde{x} - \frac{1 + \cos(\theta_1)}{2}\right)^2 + \tilde{y}^2 = \left(\frac{1 - \cos(\theta_1)}{2}\right)^2 \qquad (2.4)$$

Returning to three-dimensional space, the result obtained above necessitates that the planet's orbit confine itself to the surface of the cylinder whose cross section at any constant value of z is the circle of equation (2.4). Also note that in three dimensions, because the inner sphere's center always remains fixed and the inner sphere's radius is one, the planet meanders about the sphere $\tilde{x}^2 + \tilde{y}^2 + \tilde{z}^2 = 1$. The shape of the planet's path, absent the influence of the third sphere and the universal sphere, is then the intersection of these two surfaces (see Figure 2.5).

The resulting intersection appears like a figure eight that has been stamped on a sphere. A further analysis of the path given by equation (2.1) reveals that the planet begins at the cross of the figure eight and skates about the upper loop (assuming that θ_1 ranges between 0 and 180 degrees). After returning to the cross, the planet skates about the lower loop. Eudoxus calls the figure a hippopede, meaning horse fetter.

Before continuing, note the influence of the tilt angle θ_1 as it ranges from 0 to 180 degrees. When the tilt angle is 0 degree, the two spheres are aligned and spinning in opposite directions. The planet never moves but remains at its initial point. One can see that the circle of equation (2.4) shrinks to a point at which the corresponding cylinder shrinks to its central axis and intersects the sphere tangentially at the point (1,0,0). As we increase the tilt angle toward 90 degrees, the radius of the cylinder increases, while its surface remains tangent to the sphere at the point (1,0,0). This increases the size of the figure eight, and

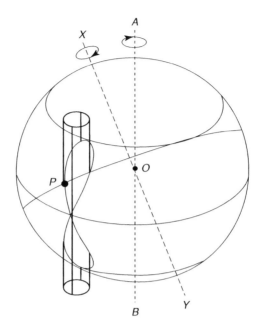

Figure 2.5. The hippopede.

the figure reaches its greatest vertical extent when the angle is 90 degrees and the loops touch the poles. Beyond 90 degrees the loops begin to approach one another, sliding toward the equator around the poles until at 180 degrees they meet up with one another and the planet spins around the equatorial plane at the rate $2\omega_1$.

Proceeding to step 2 of the algorithm, we tilt the figure eight and affix it to the third sphere. Let the tilt angle θ_2 be 90 degrees so that the figure eight lies on its side, becoming an infinity sign. Applying the tilt yields the following result:

$$
\begin{bmatrix} \tilde{x}(t) \\ \tilde{y}(t) \\ \tilde{z}(t) \end{bmatrix} = T_{\theta_2} R_{-\omega_1}(t) T_{\theta_1} R_{\omega_1}(t) \begin{bmatrix} 1 \\ 0 \\ 0 \end{bmatrix}
$$

$$
= T_{\theta_2} \begin{bmatrix} \cos^2(\omega_1 t) + \cos(\theta_1)\sin^2(\omega_1 t) \\ (\cos(\theta_1) - 1)\sin(\omega_1 t)\cos(\omega_1 t) \\ \sin(\theta_1)\sin(\omega_1 t) \end{bmatrix}
$$

$$
= \begin{bmatrix} 1 & 0 & 0 \\ 0 & 0 & -1 \\ 0 & 1 & -1 \end{bmatrix} \begin{bmatrix} \cos^2(\omega_1 t) + \cos(\theta_1)\sin^2(\omega_1 t) \\ (\cos(\theta_1) - 1)\sin(\omega_1 t)\cos(\omega_1 t) \\ \sin(\theta_1)\sin(\omega_1 t) \end{bmatrix}
$$

$$
= \begin{bmatrix} \cos^2(\omega_1 t) + \cos(\theta_1)\sin^2(\omega_1 t) \\ -\sin(\theta_1)\sin(\omega_1 t) \\ (\cos(\theta_1) - 1)\sin(\omega_1 t)\cos(\omega_1 t) \end{bmatrix}
$$

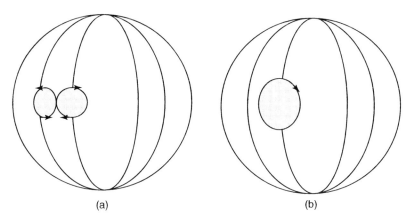

Figure 2.6. Comparison of motion along hippopede (a) and circle (b).

After this tilt, the planet skates around the resting figure eight traversing back and forth somewhat along the universal equator (see Figure 2.6a). The next step is to impute a rotation that governs the drift of the planet through the zodiac. It is the combination of the uniform drift along with the skate around the figure eight that creates retrograde motion. As an aid toward visualizing what occurs, let's consider a simpler construction with a similar feature.

For the simplified construct, assume that the planet lies at the intersection of a cylinder whose central axis is the x axis and a sphere. The intersection is a circle as displayed in Figure 2.6b, which contrasts this construct with that of Eudoxus. As displayed in the figure, the planet rotates uniformly about the intersecting circle, causing it to change direction, just as Eudoxus' skating planet. Next let's rotate this planet about the heavens. To make a mechanical device that replicates the motion, build a track around a great circle of the universal sphere and place a wheel connected to Earth with an axle on the track. Somewhat closer to Earth, on the axle, attach a pointer that rotates with the wheel. As the wheel rotates uniformly about its track, the tip of the pointer gives the location of the planet. As Figure 2.7 illustrates, while following a trend through the constellations, the orbit loops about itself in retrograde motion. Eudoxus' skating planet, skating back and forth along its figure eight as it trends through the zodiac, produces retrograde motion in a similar manner.

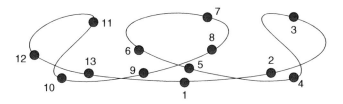

Figure 2.7. Simplified model yielding three retrograde loops every trip around the zodiac.

Returning to Eudoxus' construction, we next spin the third sphere. The third sphere governs the rate at which the planet floats through the constellations, requiring that the angular velocity ω_3 give the radian measurement of the planet's cycle through the universal sphere in one year's time. (Recall that the sun cycles through the zodiac one time per year, so the angular velocity of its sphere is 2π.) Mars skates through the zodiac around 1.06 times every 2 years or gets a little over halfway through in one year. The associated angular velocity ω_3 is about -1.06π. Applying the spin of the third sphere to the planet results in the following trajectory:

$$
\begin{bmatrix} \tilde{x}(t) \\ \tilde{y}(t) \\ \tilde{z}(t) \end{bmatrix} = R_{\omega_3} T_{\theta_2} R_{-\omega_1}(t) T_{\theta_1} R_{\omega_1}(t) \begin{bmatrix} 1 \\ 0 \\ 0 \end{bmatrix}
$$

$$
= R_{\omega_3} \begin{bmatrix} \cos^2(\omega_1 t) + \cos(\theta_1)\sin^2(\omega_1 t) \\ -\sin(\theta_1)\sin(\omega_1 t) \\ (\cos(\theta_1) - 1)\sin(\omega_1 t)\cos(\omega_1 t) \end{bmatrix}
$$

$$
= \begin{bmatrix} \cos(\omega_3 t) & \sin(\omega_3 t) & 0 \\ -\sin(\omega_3 t) & \cos(\omega_3 t) & 0 \\ 0 & 0 & 1 \end{bmatrix}
$$

$$
\times \begin{bmatrix} \cos^2(\omega_1 t) + \cos(\theta_1)\sin^2(\omega_1 t) \\ -\sin(\theta_1)\sin(\omega_1 t) \\ (\cos(\theta_1) - 1)\sin(\omega_1 t)\cos(\omega_1 t) \end{bmatrix}
$$

$$
= \begin{bmatrix} \left(\cos^2(\omega_1 t) + \cos(\theta_1)\sin^2(\omega_1 t)\right)\cos(\omega_3 t) \\ \quad - \sin(\theta_1)\sin(\omega_1 t)\sin(\omega_3 t) \\ -\left(\cos^2(\omega_1 t) + \cos(\theta_1)\sin^2(\omega_1 t)\right)\sin(\omega_3 t) \\ \quad + \sin(\theta_1)\sin(\omega_1 t)\cos(\omega_3 t) \\ (\cos(\theta_1) - 1)\sin(\omega_1 t)\cos(\omega_1 t) \end{bmatrix}
$$

As the $\tilde{z}(t)$ component of the trajectory indicates, the planet oscillates about the equatorial plane $\tilde{z} = 0$. In reality, the planets hug the ecliptic. A tilt of the third sphere toward the ecliptic ($\theta_3 = 23.5$ degrees) reorients the planet:

$$
\begin{bmatrix} \tilde{x}(t) \\ \tilde{y}(t) \\ \tilde{z}(t) \end{bmatrix} = T_{\theta_3} R_{\omega_3} T_{\theta_2} R_{-\omega_1}(t) T_{\theta_1} R_{\omega_1}(t) \begin{bmatrix} 1 \\ 0 \\ 0 \end{bmatrix}
$$

$$
= T_{\theta_3} \begin{bmatrix} \left(\cos^2(\omega_1 t) + \cos(\theta_1)\sin^2(\omega_1 t)\right)\cos(\omega_3 t) \\ \quad - \sin(\theta_1)\sin(\omega_1 t)\sin(\omega_3 t) \\ -\left(\cos^2(\omega_1 t) + \cos(\theta_1)\sin^2(\omega_1 t)\right)\sin(\omega_3 t) \\ \quad + \sin(\theta_1)\sin(\omega_1 t)\cos(\omega_3 t) \\ (\cos(\theta_1) - 1)\sin(\omega_1 t)\cos(\omega_1 t) \end{bmatrix}
$$

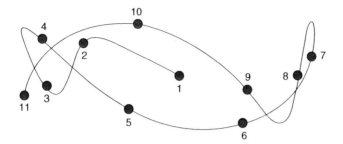

Figure 2.8. Eudoxus' planet with retrograde orbit.

$$T_{\theta_3} = \begin{bmatrix} 1 & 0 & 0 \\ 0 & \cos(\theta_3) & -\sin(\theta_3) \\ 0 & \sin(\theta_3) & \cos(\theta_3) \end{bmatrix}$$

The results for $(\tilde{x}(t), \tilde{y}(t), \tilde{z}(t))$ are as follows:

$$\tilde{x}(t) = \left(\cos^2(\omega_1 t) + \cos(\theta_1) \sin^2(\omega_1 t)\right) \cos(\omega_3 t)$$
$$- \sin(\theta_1) \sin(\omega_1 t) \sin(\omega_3 t)$$

$$\tilde{y}(t) = \left[-\left(\cos^2(\omega_1 t) + \cos(\theta_1) \sin^2(\omega_1 t)\right) \sin(\omega_3 t)\right.$$
$$+ \sin(\theta_1) \sin(\omega_1 t) \cos(\omega_3 t)] \cos(\theta_3)$$
$$- (\cos(\theta_1) - 1) \sin(\omega_1 t) \cos(\omega_1 t) \sin(\theta_3)$$

$$\tilde{z}(t) = \left[-\left(\cos^2(\omega_1 t) + \cos(\theta_1) \sin^2(\omega_1 t)\right) \sin(\omega_3 t)\right.$$
$$+ \sin(\theta_1) \sin(\omega_1 t) \cos(\omega_3 t)] \sin(\theta_3)$$
$$+ (\cos(\theta_1) - 1) \sin(\omega_1 t) \cos(\omega_1 t) \cos(\theta_3)$$

These equations encode retrograde motion through the zodiac. Figure 2.8 is a graph of the orbit. The animation at http://hsci.cas.ou.edu/images/applets/hippo-pede.html uses these equations. There is only one more operation left to complete the description: to spin the universal sphere to endow the planet with its daily motion. We leave Eudoxus' model incomplete and allow the reader seeking perfection to remove the tildas and complete it.

CHAPTER 3

THE ICONOCLAST

INSANITY

In 326 B.C.E. Alexander the Great (356–323 B.C.E.) stood on the western shore of the Beas River beholding the east. His victories from Turkey south to Egypt, east through Arabia, and further east through Persia, Afghanistan, Pakistan, across the Indus, and on to the limits of the known world did not quell his desire for more conquest. Alexander assembled his Macedonian staff and outlined his plan to penetrate the Indian subcontinent toward the Ganges. The minds of those veterans who had followed Alexander for more than 8 years, among both the Macedonian officer corps and the mixed Macedonian, Greek, and Persian foot soldiers, were not with Alexander further east, but with their homelands far beyond the western horizon. The war had become a quagmire that the soldiers could not escape, not because of an implacable enemy but because of an implacable leader. After accomplishing the objective of their expedition, the conquest of Persia, long ago and faithfully following their leader on and on and on, these men longed to return to their homes and families. The high-ranking general Coenus delivered the message to Alexander. How do you tell an egomaniac determined to expand his domain to every stone on Earth that his army will no longer follow him? Coenus was direct and forceful; the message was unmistakable.

Shifting the Earth: The Mathematical Quest to Understand the Motion of the Universe,
First Edition. Arthur Mazer.
© 2011 John Wiley & Sons, Inc. Published 2011 by John Wiley & Sons, Inc.

While Alexander's own identity was Macedonian, he was educated as a Greek; Aristotle was Alexander's tutor. Like all Macedonians, Alexander payed homage to the Greek Gods and attributed his success to them. Now he and his army needed to know whether the Gods would continue to bless Alexander's venture. Folklore states that with his soldiers looking on, Alexander stood upon the Beas and released an eagle. The eagle took flight and then landed on the western bank of the Beas. After consulting with his diviners, Alexander concluded that the Gods guided the eagle toward the west, directing Alexander and his army homeward. Of course, if the eagle had crossed the river to the east, the Gods' message would have been that the Gods side with the Indian defenders of their native land and so Alexander should return home toward the west. The omens would have pointed west either way because Alexander knew that his 15,000-mile trail of conquest had come to an end and this ceremony was all a face-saving ruse. With acid bitterness Alexander agreed to what he viewed as a retreat.

Fighting men were only a part of the returning entourage. As it was Alexander's policy to perpetuate a mobile army through generations, often soldiers' families accompanied Alexander. Bitterness guided Alexander's choice of a return route. Rather than return along a known road, Alexander turned south to the unknown. Alexander built a navy, and a flotilla carried supplies south along the Indus River. Once reaching the Indian Ocean, the coordinated naval–land return was to continue with half of Alexander's men in the navy and the other half alongside the women and children in the army. When the Makran mountains inserted themselves between the Indian Ocean and Alexander's army, the smooth return became a death march. The army cut through the Gedrosian desert along a mountainous path. On the western flank of the desert, having marched through hundred-plus-degree weather, having faced a parched landscape from which it was impossible to squeeze the amount of water required to support the army; having been deprived of supplies from the navy, only one-quarter of those entering the Gedrosian exited. The greatest casualties were among women and children. The Phoenicians, Egyptians, Persians, and inhabitants of the subcontinent never dealt a defeat on Alexander's army the likes of which Alexander dealt on his own men. Was this retribution for their retreat?

Once through the Gedrosian, Alexander became restless. The management of his vast empire was of no interest as his lust was singular, conquest, and the slim pickings along the return route yielded little satisfaction.[1] In addition to his repressed ambition, Alexander's personal life suffered when his friend, confidant, and lover, General Hephaestion, who had shared in Alexander's victories from the very first, died. Alexander vented his rage on the Cosseans, who had the misfortune to be within proximity. After the slaughter, Alexander would drink himself unconscious for days. Perhaps it was while in a drunken stupor that the idea of self-deification resurfaced. Alexander's mother had encouraged a belief that his true father was the God Heracles. After Hephaestion's death, Alexander

---

[1]This is not to say there was no fun. Along the way he unleashed his warriors and claimed some token villages.

ordered the construction of temples where his subjects were to pay homage to him as the God he perhaps believed himself to be. Soon afterward, as if the Gods themselves mocked his efforts, Alexander proved his mortality and died. The event occurred in Babylonia as Alexander was preparing an army to traverse north Africa.

Efforts by his inner circle to have Alexander settle his affairs and arrange for succession failed. Alexander's response to a query concerning his successor was that he would be succeeded by the strongest—and so the games began. A reading of the deconstruction of Alexander's empire into tangled subempires takes one into a pit of depravity. It is a limbo competition of unscrupulous behavior in which the competitors performed with abominable deft, each sinking lower and lower into the pit. Victors who died of natural causes left the competition to their offspring. A spectacle of homicides, alliances of convenience, convenient breaking of alliances, incestuous marriages, and warfare continued unabated over two and a half centuries.

Often the killing was domestic as relations through political marriage followed by competition among offspring enclosed the limbo jamboree into an all-in-the-family event. There was fratricide, infanticide (nip a potential adversary in the bud), uxoricide, sororicide, patricide, matricide, afunculcide, and other familial homicides for which words have not been invented. Perhaps the most famous of the family slayings occurred when, after a forced marriage required for succession to the Ptolemaic Empire, Ptolemy XI murdered his wife, who also happened to be his mother-in-law (some claim his mother), Cleopatra Berenice.

This sordid lesson in human intrigue is not subtle. Like watching the aftermath of a gruesome car accident, it is sickening, but we still slow down to catch a view with riveted fascination. In this case, by focusing on the savagery, we miss the truly amazing post-Alexandrian development. More unbelievable than the savagery is that in this environment, the best of Greek culture spread throughout the contested expanse and took firm root. The Greeks hated the Macedonians and viewed them as foreign usurpers. It is likely that a Greek living in what we now call the Hellenistic age would have recoiled at the description of the post-Alexandrian empires as Hellenic. For the Greek, the empires were most certainly Macedonian, and yet, the term *Hellenistic age* is appropriate. The mercenary Greeks who killed for profit established themselves in the conquered territories and invited colonizers from their homeland. A constellation of Greek cities suddenly appeared throughout the conquered areas. While the Macedonians firmly held the sword, the Greeks romanced the intellect.

By all accounts, locals accepted the Greek offerings and held them in high esteem. As evidence of the influence that Greek culture had on local culture, among the Jewish population, Hellenization had become so commonplace that traditionalists viewed Greek culture as a threat to Jewish survival. Civil strife between traditionalists and Hellenized Jews caused the Macedonian Seleucids to intervene with an overly heavy hand and tip local sentiment in favor of the traditionalists. The Jewish holiday Hanuka celebrates the victory of a Jewish band of guerrilla warriors against a Seleucid army.

Contemporaneous with their fratricidal warring, the Macedonians encouraged Greek culture and competed with one another to be known as centers of scholarship. In Athens, the Lyceum that Plato established after Socrates' death continued to promote scholarship. The Attilid kingdom, shrewdly culled from its Seleucid origins, established a library at Pergamon. The library amassed a considerable number of scrolls, each painstakingly drafted on its papyrus surface letter at a time by a scribe, and became a center for scholars. Damascus, the capital of the Seleucid Empire, also provided public support for scholarship. However, the preeminent center of intellect was the Library at Alexandria. There has since never been an institution that dominated scholarship in the manner of the Library at Alexandria, and there will, most likely, never be another one.

Very few individuals can claim to have lived a life as interesting as Ptolemy, a childhood friend of Alexander who was a trusted general. In addition to being among the few contenders for succession to Alexander's empire to die of natural causes (at age 81), Ptolemy, through deft political maneuvering and convenient alliances, participated in the early retirement of many other contenders; founded the Ptolemaic dynasty; was an embezzler, embezzling political currency in the form of Alexander's dead body; had some difficulties among his four wives—two political unions, one for sex, and one apparently out of love; and founded the Library at Alexander. It is the latter achievement that redeems Ptolemy and for which we in fact owe him a debt. (Concerning marital difficulties, Ptolemy's scorned wife, Eurenice, absented herself from Ptolemy and delivered herself to one of his political foes, Poliorcetes, bearing Poliorcetes a daughter. Protecting his inheritance, Ptolemy Philadelphius, the son of Ptolemy's beloved wife, Berenice, murdered one of his half brothers, a son of Eurenice.)

Hellenic Alexandria was among the most cosmopolitan cities in the world. A fertile mix of cultures—Egyptian, Jewish, Macedonian, and Greek—that Ptolemy guided in an enlightened and tolerant spirit, spawned a creative atmosphere. At the Library of Alexandria, Ptolemy offered generous support for a cadre of intelligentsia, including lifetime tenure for academic achievement. A quip from the mathematician Euclid (circa 300 B.C.E.) to his slave attests to the generosity of the employment contract; one could afford a slave only with a high salary. The terms attracted the most able intellects throughout the Hellenic kingdoms. This was their guilded sanctuary in an otherwise harsh world.

As noted in Chapter 2, there is a nostalgia associated with pre-Philippic Athenian culture promoting a sentiment that the culture of the Hellenistic age was somewhat inferior. In the case of mathematics, this sentiment is dead wrong. The mathematicians associated with Alexandria, building on the foundations of their predecessors, furthered mathematical knowledge well beyond the Athenians. Their work endures to this day and forms the foundation for modern mathematics. The history books often understate their influence and relegate them to a junior role beneath the disciplines of philosophy and the arts. Not only historians misunderstand mathematicians, when it comes to a public conversation, the terms "dweeb," "geek," "nerd," "freak," "dork," and "weirdo," dress the mathematician in a cloak of insanity. This view holds that mathematical research is esoteric and unrelated to any practical application. Furthermore, only one maybe not wholly insane, but certainly endowed with a healthy spirit, of craziness would invest one's life in the field. Let's address the field from the mathematician's perspective.

Above, we refer to a quip from Euclid indicating that his compensation was generous. The full quip was a response to a student who challenged Euclid to demonstrate any use for geometry. Euclid instructed his slave to "Give him two pence so that he may make a profit of what he learns."

Interpreting Euclid's response requires an understanding of Euclid's accomplishment. Euclid systematized the axiomatic–deductive process that underlies mathematics. The beginnings of such an undertaking were already apparent in the wisecracking Socrates. Socrates' method for deconstructing his target was to seek his counterpart's opinion on some issue. He would then seek the counterpart's opinion on another issue and yet another. It all seemed innocent, and, indeed, at first the counterpart may have felt honored; here is the wise Socrates seeking out my opinions on all sorts of matters.

In the world of the nonnerds, the nongeeks, the fully sane, compromises must be made to accommodate actual circumstances. Opinions formulated and presented implicitly reflect these compromises. Socrates would highlight the compromises and explicitly expose them as contradictions. The counterpart's belief system imploded while Socrates enjoyed his victory at the victim's expense. Socrates sought a belief system without any contradictions, the perfect system. In fact, he proposed the following: (1) it is better to be the victim of harm than to cause harm, (2) it is not possible to harm a just person, and (3) justice is always more beneficial than injustice. We do not analyze this system. Suffice it to say that Socrates' counterparts may have seen it of limited use and that its limits indicate the difficulty of creating a perfect, contradiction-free philosophy that allows one to address the complex problems of one's daily life.

Plato's challenge to find a perfect system of governance for the motions of the heavens extends Socrates' quest to the universe. Here, the challenge shifts from a social context to a scientific one, and with good argument, one can claim that Eudoxus had more success in defining a system than did Socrates. But it is in the mathematical realm (Eudoxus also has influence here), where the quest for a perfectly consistent system has met with its greatest success. Prior to Euclid, there were efforts to set mathematics firmly on an axiomatic–deductive foundation, but Euclid's works surpass his predecessors in scope and technicality. Euclid, in his work *The Elements*, after proposing five basic axioms, deduces perfect, contradiction-free theorems that number in the hundreds. *The Elements* germinated a body of mathematics that is alive and continues to expand today.

So, what did Euclid really say to his student? Let us take license, a great deal of license, and imagine that he embellishes further:

I am showing you perfection and you are unable to see it's value? Do you only attach value to physical acquistion? Go into the world where chaos, confusion and contradiction are supreme. Go into the world where ambition trumps reason. Live among those who war with their countrymen, deceive their neighbors, and murder their relatives. Here then are two pents. Take it and by doing so choose to leave my world and enter the world where they call me a dweeb, nerd, a geek, a dork, and a weirdo. But in doing so, you choose insanity, for only a crazy man would choose to forego my world where one can find solace and reside where lunacy prevails.

A DISTURBING INSIGHT

Let us return to astronomy and the pathway of the heavens. The Greeks applied axiomatic–deductive reasoning to the sciences. Science, being a human endeavor, is not immune to either prejudice or ego. Implicit within the works of the ancient Greeks is a belief that truth and perfection cohabit the same space. Find one and you reveal the other. But prejudice intertwined with ego taints the view of perfection. For the Greeks, perfection dictated that circles govern the motion of the heavens, and ego demanded that the earth occupies the critical defining point of each circle, their common center. This was the axiomatic truth.

Aristarchus (310–230 B.C.E.), a contemporary of Euclid, was a most unusual man. Although he was a first-rate mathematician and astronomer, what lifts him above his contemporaries is his ability to overcome prejudice and separate ego from his work. Whereas others did not question the truth that the earth lies at the center of the universe, Aristarchus did. Against his cultural inheritance, he rejected the geocentric universe and replaced it with a heliocentric system in which the earth moves around the sun. Among his peers, nobody was moved. Their minds remained fixed just as the earth was fixed at the center of their geocentric world. In fact, on hearing of Aristarchus' theory, Cleanthes, a citizen of Athens, became so enraged that he demanded that charges of sedition be prepared against Aristarchus. There is no evidence of a trial. Nevertheless, this initial response provides a leading indicator of the sacredness with which the geocentric universe was held for the next two millennia.

The manuscript in which Aristarchus presents his heliocentric theory has been lost. It is only through secondary sources that we are aware of Aristarchus' accomplishment. However, there is a surviving manuscript, *On the Sizes and Distances of the Sun and Moon*, that could well be Aristarchus' point of departure from geocentrism. Using a measurement taken during a lunar eclipse, Aristarchus presents the conclusion that the sun is a much larger body than the earth. Perhaps it was this realization that caused Aristarchus to question the status quo, as to why the smaller body should be at the center. Below is a mathematical treatment of Aristarchus' method for determining the relative sizes of the sun, the earth, and the moon. As with our presentation of Eudoxus' model, we use all the mathematical tools available to us, but not available to the original author.

ARISTARCHUS AND *On the Sizes and Distances of the Sun and Moon*

The apparent size of the moon relative to the sun is evident during a full solar eclipse when the moon dominates as the sun cowers behind the moon with no

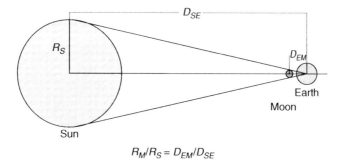

$$R_M/R_S = D_{EM}/D_{SE}$$

Figure 3.1. Similar triangles during a solar eclipse.

room to spare. But full domination lasts a mere instant before the sun reappears as the stronger body. Were the moon somewhat larger, it could linger in its glory for more than an instant; were the moon smaller, it would never fully dominate the sun. But, as Figure 3.1 illustrates, the apparent sizes of the sun and the moon are nearly the same, so the moon tastes authority for only an instant before ceding it back to the sun. Ratios on the similar triangles that are illustrated in Figure 3.1 result in

$$\frac{R_M}{R_S} = \frac{D_{EM}}{D_{SE}} \tag{3.1}$$

where R_S is the radius of the sun, R_M is the radius of the moon, D_{SE} is the distance between the sun and the earth, and D_{EM} is the distance between the earth and the moon.

Aristarchus recognized that when there is a half-moon, it is possible to determine the right-hand side (RHS) of equation (3.1). Figure 3.2 illustrates that during a half-moon the earth, the moon, and the sun form a right triangle with the moon at the apex of the right angle. The sine of the corresponding angle with the sun at its apex is precisely the desired ratio, $\sin(\theta) = \frac{D_{EM}}{D_{SE}}$. With a good measurement of the angle θ and some knowledge of trigonometry, it is possible to determine the ratio of the actual, radii, $\frac{R_M}{R_S}$.

Since a solar eclipse reveals the ratio of the moon's and sun's radii, might one not expect by symmetry that a lunar eclipse would reveal the ratio of the

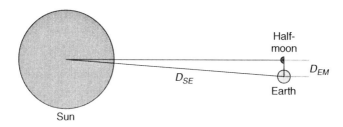

Figure 3.2. Formation of right triangle during half-moon.

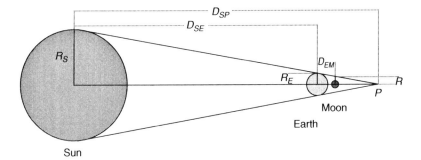

Figure 3.3. Lunar eclipse. Similar triangles used by Aristarchus.

earth's and sun's radii? It does, but the mathematics is less obvious. Aristarchus' method attests to his ability.

Figure 3.3 displays the alignment of the sun, the earth, and the moon during an eclipse. The earth casts a conical shadow that extends to the point P. The base of the shadow is the circle on the earth's surface that forms the boundary between day and night. We denote the distance from the center of the sun to the point P by D_{SP}. There are three similar triangles with bases respectively given by the radius of the sun R_S, the radius of the earth R_E, and one with a base along the chord of the moon's path through the earth's shadow R. Aristarchus demonstrates his perfectionist nature by taking care to account for the distance between the moon's center and the chord between the moon's entry and exit points through the earth's shadow. The distance, which we denote ε, assumes a circular lunar pathway (see Figure 3.4).

Ratios of the similar sides of the aforementioned three triangles yield the following equalities:

$$\frac{R_S}{D_{SP}} = \frac{R_E}{D_{SP} - D_{SE}} = \frac{R}{D_{SP} - D_{SE} - D_{EM} + \varepsilon} \tag{3.2}$$

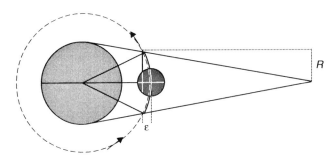

Figure 3.4. Closeup of the Moon–Earth pathway during a lunar eclipse, indicating distance ε.

Simplifying the first of these equalities results in the following:

$$\frac{R_E}{R_S} = \frac{(D_{SP} - D_{SE})}{D_{SP}} = 1 - \frac{D_{SE}}{D_{SP}} \tag{3.3}$$

Similarly, simplification of the first and last expressions of equation (3.2) using equation (3.3) results in the following:

$$\frac{R}{R_S} = \frac{D_{SP} - D_{SE} - D_{EM} + \varepsilon}{D_{SP}}$$

$$= \frac{R_E}{R_S} - \frac{D_{EM}}{D_{SP}} \tag{3.4}$$

For Aristarchus, the distance D_{SP} is problematic because there is no way to directly measure it. Let us solve for D_{SP} in equation (3.4), place the result into equation (3.3), and simplify:

$$\frac{R}{R_S} = \frac{R_E}{R_S} - \frac{D_{EM} - \varepsilon}{D_{SP}}$$

$$D_{SP} = \frac{(D_{EM} - \varepsilon) R_S}{R_E - R}$$

$$\frac{R_E}{R_S} = 1 - \frac{D_{SE}}{D_{SP}}$$

$$= 1 - \frac{D_{SE} (R_E - R)}{(D_{EM} - \varepsilon) R_S} \tag{3.5}$$

$$= 1 - \frac{D_{SE} R_E}{(D_{EM} - \varepsilon) R_S} + \frac{D_{SE} R}{(D_{EM} - \varepsilon) R_S}$$

$$\times \left(1 + \frac{D_{SE}}{D_{EM} - \varepsilon}\right) \frac{R_E}{R_S} = 1 + \frac{D_{SE} R}{(D_{EM} - \varepsilon) R_S}$$

$$\frac{R_E}{R_S} = \frac{1 + \dfrac{D_{SE} R}{(D_{EM} - \varepsilon) R_S}}{1 + \dfrac{D_{SE}}{D_{EM} - \varepsilon}}$$

Aristarchus expresses the quantity R in terms of the moon's radius R_M so that $R = \lambda R_M$. It is the quantity λ that Aristarchus must determine, and this is the quantity revealed by measurements taken during an eclipse. With the substitution $R = \lambda R_M$, the equation for the ratio $\frac{R_E}{R_S}$ becomes the following:

$$\frac{R_E}{R_S} = \frac{1 + \dfrac{D_{SE} \lambda R_M}{(D_{EM} - \varepsilon) R_S}}{1 + \dfrac{D_{SE}}{D_{EM} - \varepsilon}} \tag{3.6}$$

Substituting equation (3.1), $\frac{R_M}{R_S} = \frac{D_{EM}}{D_{SE}}$, into equation (3.6) and simplifying yields the following:

$$
\begin{aligned}
\frac{R_E}{R_S} &= \frac{1 + \dfrac{\lambda D_{EM}}{D_{EM} - \varepsilon}}{1 + \dfrac{D_{SE}}{D_{EM} - \varepsilon}} \\[2ex]
&= \frac{(1 + \lambda) D_{EM} - \varepsilon}{D_{EM} + D_{SE} - \varepsilon} \\[2ex]
&= \frac{D_{EM} + D_{SE} - \varepsilon + \lambda D_{EM} - D_{SE}}{D_{EM} + D_{SE} - \varepsilon} \\[2ex]
&= 1 + \frac{\lambda - \dfrac{D_{SE}}{D_{EM}}}{1 + \dfrac{D_{SE} - \varepsilon}{D_{EM}}} \\[2ex]
&= 1 + \frac{\lambda - \dfrac{1}{\sin(\theta)}}{1 + \dfrac{1}{\sin(\theta)} - \dfrac{\varepsilon}{D_{EM}}}
\end{aligned}
\tag{3.7}
$$

Aristarchus needs only three measurements to determine the ratio between the earth's and the sun's radii, the value λ, the angle θ, and the ratio $\frac{\varepsilon}{D_{EM}}$. An analysis of the influence of these three numbers leaves one perplexed by Aristarchus' final result. The casual observer notes that the angle θ is very close to zero, yielding a tiny value for the term $\sin(\theta)$. The term's reciprocal, $\frac{1}{\sin(\theta)}$, is both large and sensitive to the the measurement for θ. In fact, this is the most critical of all three measurements. By contrast, the term $\frac{\varepsilon}{D_{EM}}$ adjusts the result from a linear pathway to a circular one. (In Figure 3.4 we set ε to zero so that the moon's path lies along the line segment.) The casual observer notes that the moon's trajectory as it passes through the earth's shadow is nearly a line, so this correction must be small. Indeed, it is insignificant in comparison with the other terms. Yet, Aristarchus uses a very poor estimate for θ, and rather than ignoring $\frac{\varepsilon}{D_{EM}}$, provides a sophisticated geometric analysis to ensure that he mathematically accounts for the curvature of the moon's pathway as it passes through the earth's shadow. This approach leads one to believe that Aristarchus is more enthralled by the mathematical challenge than worried about the precision of his result.

Aristarchus sets the angle θ to 3 degrees. The novice can ascertain that this is far too large, and certainly Aristarchus was aware of the overestimate. How can one explain his choice? Perhaps Aristarchus knew the result would be controversial. Perhaps he knew of a better angle, but chose to use a conservative estimate. He demonstrates that even if one were to make this very conservative assumption, the result would be that the sun is very much larger than the earth.

Placing in a more accurate, smaller value for θ only results in an even larger sun. Indeed, the reader with knowledge of limits and calculus can show that the ratio $\frac{R_E}{R_S}$ approaches zero as the angle θ approaches zero, so the relative size of the sun's radius increases as θ becomes smaller.

We may leap 23 centuries ahead of Aristarchus, pull out our twentieth-century calculators, and find that $\sin(3°) = 0.052336$. In 300 B.C.E. calculating the sine of angles to any degree of accuracy was not at all trivial—in fact trigonometry had not yet existed as a discipline. Before following Aristarchus any further, let us take full advantage of modern technology, make the reasonable approximation that the moon's trajectory through the shadow is a line, and get a numerical result for equation (3.7). This simplifies the problem and provides preparation for Aristarchus' more complex analysis. We still need a value for λ. Aristarchus estimates that at the very moment when the moon completely enters the earth's shadow, the moon's journey through the shadow is half complete. This, along with our straight-path assumption, places the value of lambda at a little less than 2; we'll take 2 (see Figure 3.5). Recall that the straight-path assumption also sets the value ε to zero. All of the parameters for equation (3.7) are ready, so we get out our calculators and start punching:

$$\frac{R_E}{R_S} = 1 + \frac{\lambda - \dfrac{1}{\sin(\theta)}}{1 + \dfrac{1}{\sin(\theta)} - \dfrac{\varepsilon}{D_{EM}}}$$

$$= 1 + \frac{2 - \dfrac{1}{\sin(3°)}}{1 + \dfrac{1}{\sin(3°)}}$$

$$= 0.149199377$$

Taking the reciprocal gives the result that the sun's radius is around 6.7 times larger than that of the earth. We now know this to be very inaccurate as the sun's radius is in fact on the order of 109 times that of the earth. Aristarchus' method is not the source of the error; his geometry is impeccable. Rather, it is his poor choice of θ that causes this poor result. The actual value for θ is nearer to 10 minutes, a quantity 18 times smaller than Aristarchus' value. With this choice,

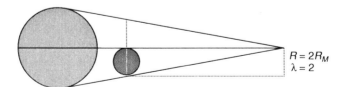

$R = 2R_M$
$\lambda = 2$

Figure 3.5. Schematic representation of straight-path assumption.

the calculator displays the ratio $\frac{R_E}{R_S} = 0.008701323$, yielding a solar radius that is 114 times that of the earth.

Let us return to Aristarchus, consult his estimates of the trigonometric values, and account for the moon's curved path. Using rather sophisticated geometric arguments, Aristarchus proves that $\frac{1}{20} < \sin(3°) < \frac{1}{18}$. Aristarchus is aware of the limitations of his mathematics and adjusts his objective accordingly. As he cannot precisely determine $\sin(3°)$ but can only place bounds on the value, he does not establish the actual ratio $\left(\frac{R_E}{R_S}\right)$ but only places bounds on it.

First, it is necessary to determine the bounds for the value λ, which depends on a measurement during a lunar eclipse. As noted above, Aristarchus estimates that at the very moment that the moon completely enters the earth's shadow, the moon's journey through the shadow is half complete. Figure 3.6 illustrates the geometry where one observes the following relationships:

$$D_{EM} = R_M \cot\left(\frac{\alpha}{2}\right)$$

$$R = D_{EM} \sin(\alpha)$$

$$= R_M \cot\left(\frac{\alpha}{2}\right) \sin(\alpha)$$

Recall that $\lambda R_M = R$ gives the following result:

$$\lambda = \cot\left(\frac{\alpha}{2}\right) \sin(\alpha)$$

The angle α is the angle that the moon subtends. As with θ above, Aristarchus overestimates this angle setting it to two degrees. Using a calculator, one arrives at the value $\lambda = 1.999390827$. Demonstrating his geometric acumen, Aristarchus bounds λ between the following numbers: $\frac{88}{45} < \lambda < 2$.

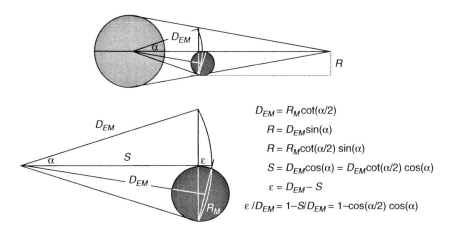

Figure 3.6. Geometric representation for determination of lambda and epsilon.

The final estimate necessary for Aristarchus to complete his analysis is the ratio $\frac{\varepsilon}{D_{EM}}$. Once again, Figure 3.6 illustrates the following relationships:

$$D_{EM} - \varepsilon = S\cos(\alpha)$$

$$S = D_{EM}\cos\left(\frac{\alpha}{2}\right)$$

$$D_{EM} - \varepsilon = D_{EM}\cos\left(\frac{\alpha}{2}\right)\cos(\alpha)$$

After simplification, the following relationship results:

$$\frac{\varepsilon}{D_{EM}} = 1 - \cos\left(\frac{\alpha}{2}\right)\cos(\alpha)$$

Returning once more to the calculator with α set to 2 degrees yields the value $\frac{\varepsilon}{D_{EM}} = 0.000761385$. Summoning his geometric skills, Aristarchus finds that $0 < \frac{\varepsilon}{D_{EM}} < \frac{1}{675} = 0.00148148$.

We make one final visit to the calculator to obtain the ratio of radii with all of Aristarchus' parameters present:

$$\frac{R_E}{R_S} = 1 + \frac{\lambda - \dfrac{1}{\sin(\theta)}}{1 + \dfrac{1}{\sin(\theta)} - \dfrac{\varepsilon}{D_{EM}}}$$

$$= 1 + \frac{1.999390827 - \dfrac{1}{0.052336}}{1 + \dfrac{1}{0.052336} - 0.000761385}$$

$$= 0.149136862$$

Differences between this result and that of the straight-line approximation are immaterial with the sun's radius close to 6.7 times that of the earth's.

Finally, we use Aristarchus' approximations to place bounds on the result. For the lower bound of the ratio, use the lower bound of λ $\left(\frac{88}{45}\right)$, the lower bound of $\sin(3^\circ)$ $\left(\frac{1}{20}\right)$, and the upper bound of $\frac{\varepsilon}{D_{EM}}$ $\left(\frac{1}{675}\right)$. Conversely, for the upper bound of the ratio, use the upper bound of λ (2), the upper bound of $\sin(3^\circ)$ $\left(\frac{1}{18}\right)$, and the lower bound of $\frac{\varepsilon}{D_{EM}}$ (0). The results are as follows:

$$1 + \frac{\lambda_L - \dfrac{1}{\sin(\theta)_L}}{1 + \dfrac{1}{\sin(\theta)_L} - \dfrac{\varepsilon}{D_{EM\,U}}} < \frac{R_E}{R_S} < 1 + \frac{\lambda_U - \dfrac{1}{\sin(\theta)_U}}{1 + \dfrac{1}{\sin(\theta)_U} - \dfrac{\varepsilon}{D_{EM\,L}}}$$

$$1 + \frac{\dfrac{88}{45} - 20}{1 + 20 - \dfrac{1}{675}} < \frac{R_E}{R_S} < 1 + \frac{2 - 18}{1 + 18}$$

$$\frac{997}{7087} < \frac{R_E}{R_S} < \frac{3}{19}$$

$$0.140680119 < \frac{R_E}{R_S} < 0.157894737$$

According to Aristarchus' analysis, the sun's radius is between 6.3 and 7.1 times that of the earth. Beyond any doubt, Aristarchus demonstrates that we do not inhabit the largest sphere in the neighborhood. Later, he would take aim at another egocentric myth: that we are at the center of the universe.

CHAPTER 4

INSTIGATORS

A HEAVY HAND

On an autumn morning in 213 B.C., the consul Marcellus had launched his flotilla against Syracuse. He had every reason to believe that by sunset, his Roman legions would be in control of the Greek city on the western coast of Sicily. After all, the tactics employed to breach the defenses had become procedural, and the flow of events as predictable as a religious sacrificial ceremony. First, the boats would row to the city wall and secure themselves so that they could support siege ladders and towers. Eventually the enemy's resistance would give way to the sheer magnitude of Roman wall hoppers and a vanguard would breach the wall. Once the wall was breached, a portion could be secured, providing access to tens of thousands of legionnaires. Romans had a policy of giving no quarters to the enemy, and once within the city walls, the legionnaires had license to give expression to every savage and utterly despicable aspect of the human character. There was no line separating the acceptable from the unacceptable; theft, rape, murder, and destruction of property were not only acceptable but also often encouraged. Marcellus had recently overseen such an operation against Syracuse's sister city, Leontoni. But Archimedes (287–212 B.C.) was a citizen of Syracuse, not Leontoni.

Marcellus' legionnaires must have been in a particularly ornery mood. Most were drawn from those lucky to have survived a humiliating defeat at the hands

Shifting the Earth: The Mathematical Quest to Understand the Motion of the Universe,
First Edition. Arthur Mazer.
© 2011 John Wiley & Sons, Inc. Published 2011 by John Wiley & Sons, Inc.

of Hannibal, the Carthaginian terror who ransacked the Italian peninsula at will. Hannibal crossed over the Alps 5 years earlier with an army of infantry, warrior elephants, and most importantly a Numidian cavalry that provided Hannibal with unmatched mobility on the battlefield. The Romans had engaged Hannibal three times, always with the same result. Hannibal's cavalry would outflank the Romans, encircle, and defeat them. The defeat at Cannae was the most devastating. The Romans assembled an army of 86,000 to face Hannibal's 50,000. Even though the numbers overwhelmingly favored Rome, the flexibility of Hannibal's deployment proved superior to Rome's static tactics. Unconfirmed battle casualties are put at 50,000 for the Romans against 6000 for the Carthaginians.

The remnants of Rome's army withdrew and regrouped. After it became apparent that Hannibal would not assault Rome, the Romans left him free to roam about the Italian peninsula and sent a contingent of survivors with Marcellus to put down Carthaginian-inspired uprisings on Sicily. On the unfortunate island, the legionnaires would extract their revenge.

As Marcellus' ships rowed toward the walls, he might have been somewhat taken aback by the distance to which the Syracusan catapults could hurl their payload. No matter, he might have thought, if the Syracusans were silly enough to calibrate their catapults for such a distance, all the better. The vessels would soon be closer to the wall where the catapults, calibrated for great distances, would be ineffective. As the payloads continued to harass Marcellus' marines unabated regardless of their distance, Marcellus may have been further taken aback by the losses he incurred. Nevertheless, he had more than enough vessels and manpower to reach and compromise the wall. From his vantage point, Marcellus might have wondered why the boats stalled as they neared the wall. Could he see the barrage of darts and missiles that thwarted their advance? However, Marcellus unmistakably saw what was in store for any boat that did make it to the wall. A claw came down from above and grappled the boat. Then the claw seemingly without effort hoisted the several-ton quintereme and its crew out of the water, dangling the boat vertically on a rope. After lifting the boat to sufficient height and shaking it about, the claw released the boat so that Marcellus might see it smash into pieces as it struck rocky bottom. A land assault similarly failed; the attack went awry as the Romans could not penetrate the gauntlet of catapults, darts, missiles, and the claw. The designer of these defenses was the brilliant mathematician Archimedes. Rather than face Archimedes, Marcellus withdrew and set siege to Syracuse. The siege would last $2\frac{1}{2}$ years.

It is a remarkable coincidence of history that a mathematical genius whose influence is comparable to Newton's should have a role in the Punic Wars. Because of this role, like Hannibal and Marcellus, Archimedes receives attention from the historian.[1] In fact, regardless of the Punic Wars, Archimedes is more than worthy of the historian's attention, his merit reaches beyond that of both Hannibal and Marcellus, and this book aims to demonstrate why this is so. Yet,

---

[1] The attention follows the trail of the absurd. One myth has Archimedes focusing an array of mirrors onto oncoming Roman ships, thus incinerating them at sea. The Discovery Channel's *Mythbusters* incinerates this myth with aplomb.

the lives of all three of these men illustrate the completeness and finality with which the Roman empire stamped its authority. This is germane to our story, so we will follow their fates.

After $2\frac{1}{2}$ years under a siege that left the Syracusans in a weakened state, Marcellus launched a final and successful assault. This completed the Roman reconquest of Sicily, and Marcellus returned to Rome as a hero. The celebration included an ostentatious display of the wealth looted from Syracuse; certainly this was Marcellus' most shining moment. But then, the old adage that no good deed goes unpunished found application. The Roman senate confirmed on Marcellus the command of a legion to hunt down Hannibal.

Marcellus held his opponent in high regard and ensured that the Roman legions under his command maintained their distance from the Carthaginians. This continued for 2 years, resulting in occasional skirmishes. Hannibal bested Marcellus in every encounter, proving the wisdom behind the Roman's timidity. Then, during the summer of 208 B.C. the two armies chanced on one another. It was a now-or-never moment. In preparation for a conclusive battle the next day, Hannibal sent a contingent of his Numidian cavalry to post themselves on a strategic knoll. Hannibal was not alone in his preparations or his grasp of the topography. When the Numidians arrived, Marcellus was on the knoll with a contingent of his men. Marcellus died in the ensuing brush. Having lost one of their consuls, the Romans broke camp and departed. The now-or-never dice had been tossed and fate had answered never. Indeed, the Romans did not challenge Hannibal on Italian soil for another 7 years. All in all, Hannibal had maintained an army on the Italian peninsula for 17 years, from 218 through 201 B.C., without losing a battle. It was not Italy where a decision would be forced, but Africa.

Rome's answer to Hannibal was the very capable consul Scipio. Following Hannibal's lead, Scipio took the battle to the enemy. At age 25, he overran Hannibal's Spanish capital, Cartagena, and after a major victory against Hasdrubal, Hannibal's brother ensconced the Romans as the predominant force in Iberia. On his return to Rome, Scipio successfully lobbied to lead an army against Carthage. In North Africa, he would mirror Hannibal's disruption of Italy, but with more success.

Scipio's first order of business was to obtain a cavalry of his own. Exploiting a rift in Numidian politics, Scipio allied himself with a worthy partner, Masinissa, gaining a cavalry that was more than a match for the Carthaginian faction. With his cavalry, superbly trained army, and unmatched tactical mind, Scipio terrorized the Carthaginians; they cried for Hannibal to return from Italy.

The day before the battle, the two men met. Perhaps sensing that he was disadvantaged, Hannibal offered peace terms, but to no avail. Each man was an exceptional tactician. Each man had penetrating insight and could read the minds of their enemy. Each man had shown the capacity to surprise their enemy through brilliant maneuvers. On that day as they sized up one another; they recognized their twin. Each knew that he had no surprise to throw at the other. On the next day when the two finest military men of their day controlled a collision of over 80,000 men, it was a static, hold-that-line slugfest. When the carnage ended, the

Romans held the field. The second Punic War had ended and Carthage was a vassal state.

Hannibal never quit. He left Carthage for Damascus and became an adviser to Antigonus, the Seleucid king. Within Antigonus' court until his death, Hannibal unsuccessfully lobbied for an army to take to Italy. As with Alexander, history associates greatness with the name. How is it that both these men who caused such widespread suffering for no purpose other than self-aggrandizement are viewed with such respect? No answer is offered, nor will this be explored, but it is something to ponder.

As for Carthage, it became a model vassal territory. The citizens adapted to Roman authority, promoted their economy, and paid their taxes to Rome. But Roman enmity toward Carthage never subsided. In 149 B.C. the enmity culminated in an unprovoked siege of Carthage led by Scipio's grandson through adoption. In 146 B.C., after a brutal campaign fought by a truly gallant citizen's army, the Carthaginians capitulated. The Romans enslaved a sizable portion of the already depleted population, scattered the remainder into the countryside, took their booty, and burned the city to the ground, where it left no mark on the landscape. Equally, the Phoenician culture of Carthage leaves no mark on the cultural landscape of North Africa.

Returning to Archimedes, he, too, was the target of Roman indignation. Legend states that during Marcellus' final and successful assault on Syracuse, a soldier found the septuagenarian drawing geometric figures in a sandbox. Archimedes responded to the intruder with a shout: "Don't disturb my circles." The soldier promptly thrust his spear into Archimedes, killing him. Marcellus gave Archimedes an honorable burial. We will see that as the Roman empire aged and expanded, it showed less tolerance for cultural diversity and, once siding with the Christian philosophies, buried Hellenism just as it had Carthage and Archimedes. Unlike Carthage, Archimedes' works would resurface 1800 years after his death and 1100 years after the fall of Rome.

PATRIARCHS

Although born in Syracuse, Archimedes received his education in Alexandria, where he most likely immersed himself in Euclid's *Elements*. Certainly his unequaled ability would have attracted the attention of the intellectual cadre and undoubtedly, Archimedes could have remained at Alexandria should he have so chosen. Instead, he chose to return to Syracuse for his ill-fated rendezvous with Marcellus' army. In the intervening years, Archimedes left for posterity works that leave the modern-day reader dumbfounded, for the rest of humanity took nearly two millennia to catch up with him.

In his treatise *On the Sphere and the Cylinder*, Archimedes determines the volume of the sphere using integral calculus. Indeed, his proof would be familiar to every modern-day calculus student as it is a topic in every standard calculus book: volumes of revolution. Showing familiarity with the fundamental theorem of calculus, Archimedes also determines the surface area of the sphere.

In his treatise *On Floating Bodies*, Archimedes, once again using limiting methods of integral calculus, determines the center of gravity of paraboloids. This goes beyond developing brilliant mathematical methods; it is a bold thrust into mathematical physics. After presenting the principle of buoyancy, Archimedes uses the center of gravity to determine configurations of stability for a floating paraboloid. Here, not only do we see methods of integral calculus; we see the understanding of balance of forces, equilibrium, stability, and instability; concepts that all encroach on Newton's (1643–1727) and Euler's (1707–1783) domains.[2]

In his treatise, *On Spirals*, Archimedes investigates the motion of a point forced to move along a spiral pathway. In a manner that once again illustrates Archimedes' awareness of the fundamental theorem of calculus, Archimedes relates the velocity of the point with the area it sweeps out. Eighteen hundred years later, Kepler would have a similar Archimedean moment with the discovery of his area sweeping law of planetary motion. Archimedes also parses the point's motion along two directions, the radial and the angular directions and analysis them independently, Galileo (1564–1642) and Emmy Noether (1882–1935) took note.[3]

There are so many novelties in these and other beautiful works that Archimedes achieves what Alexander only dreamed of. If not himself a God, Archimedes came closer to them than any mortal. But, returning to the topic of the motion of the heavens, Archimedes shows that after all he is only human. Despite the force of Aristarchus' arguments, which the intellectually superior Archimedes was capable of grasping, Archimedes could not shake the weight of his culture and outright rejected Aristarchus' heliocentric proposal. He was not alone, the other towering intellect of his day, Apollonius (262–190 B.C.), stood alongside Archimedes on the geocentric world leaving Aristarchus all alone to fly about the universe on his heliocentric planet.

Like Archimedes, Apollonius left his home to study at Alexandria, but then Apollonius chose to remain. Among the powers of the day, the Ptolemaic kingdom was the most prosperous. The kingdom was unresponsive to Hannibal's efforts at forming a concerted alliance against Rome, instead going about the business of making money and supporting its prestigious library. Apollonius was the preeminent mathematician and astronomer in Alexandria. His intellectual gravitas was of such magnitude that the program he outlined for describing the motions of the heavens is the program that guided researchers over the next several centuries.

---

[2] Archimedes' results are the first analysis of forces on a static object. This is a necessary precursor to understanding how forces affect motion. Newton takes the dynamic leap as described in Chapter 9. A stability analysis determines whether a static object in balance by the forces acting on it is stable to a minor perturbation in the forces. Euler was the first to perform an analysis using Newton's equations. Archimedes, nearly 2000 years earlier, had devised the analysis for his paraboloids.

[3] Galileo parsed the motion of a projectile into two directions: the vertical motion, which is influenced by gravity; and horizontal motion, which is not. This allowed Galileo to solve for each component of motion independently. Emmy Noether generalized the concept, giving conditions when such a parsing is possible and finding equations for the independent components.

THE IRONY

The deficiencies with Eudoxus' spherically layered universe became apparent. While the model replicates retrograde motion, it is nearly impossible to match the motion with actual observations. A different approach was necessary, and Apollonius met the challenge. While maintaining Eudoxus' axiomatic principles of a geocentric universe and governance of the heavens through circular motions, he redefined the composition of circular motions so that they could be more easily matched with observation. The proposal known as *epicyclic motion* posits that embedded on a planet's primary circular motion about the earth are further circular motions that cause retrograde. It conjures up the image of a mechanical universe with the heavens rotating about the earth on a network of interlocking circular gears.

Apollonius did not set about specifying the gear sizes, ratios, and speeds that regulate planetary motion. He left this to future astronomers who, as noted above, most enthusiastically took on the project (this is the topic of Chapter 5). We know the outcome; for centuries this effort substantiated the false axioms of its foundation, namely, geocentrism and circular motion of the heavens. However, once free of the axioms, the mechanical universe ceased to exist, rendering Apollonius' epicycles irrelevant. Apollonius, however, was not irrelevant in the dismantling of the universe that he had helped build.

Apollonius' best-known work is his eight-chapter manuscript *On Conics*. With this treatise Apollonius seizes the position of humankind's leading authority on conic sections, including the ellipse. While giving no physical application for the ellipse or any of the other conic sections, Apollonius generates theorems of uncanny originality and proves them, leaving the reader in stunned amazement. The work is that of an artist with unmatched imagination. It is to be admired for its own beauty, but serves no other purpose. Then, 1800 years after Apollonius wrote his treatise, Apollonius' ellipse appears in the formidable mind of Kepler. Like Archimedes' claw, it comes out of nowhere, it is entirely unanticipated, and it shatters the geocentric universe of Eudoxus, Aristotle, Archimedes, and Apollonius. Apollonius' work *On Conics* is also fundamental to René Descartes' (1596–1650) development of analytic geometry, a powerful unification of geometry and algebra that changed the way mathematics was performed and communicated. This facilitated Newton's development of calculus and its use in mathematically demonstrating the ellipse as the orbital pathway. While guiding his successors down the wrong road, Apollonius more than redeems himself by unintentionally creating the very tools that illuminate the heliocentric universe.

Building on the works of Archimedes and Apollonius, Descartes, Newton, and Leibniz refuted the Counter-Reformation and marshaled Europe into the Enlightenment. Hannibal is dead, and Carthage lies beneath the sands. Marcellus, Scipio, and the Roman empire came and went, leaving nothing of relevance to our world. But Archimedes and Apollonius endure at the very foundations of modern science.

FUN

Apollonius sent versions of the first three chapters of *On Conics* to a colleague in Pergamon. After the death of this colleague, he sent the remainder to another individual. Accompanying each chapter is an introductory letter, the contents of which are unremarkable. Each letter, in a style similar to a bulleted power-point presentation, lists the contents of its accompanying chapter. There are additionally some remarks describing shortcomings with previous works and other remarks stating Apollonius' own contributions. The letters are more remarkable for what they do not contain; there is no trace of any utility for these works, and Apollonius offers no motivation beyond completing the works of others and generating new theorems on conic sections.

Below, we give a modern rendition of results from Chapter 5; these are considered the most novel and difficult results in all of *On Conics*. In Chapter 5, Apollonius poses the following problem: Given a conic section and any point in the plane of the conic section, how many normal lines are there from the conic section to the point? Normal lines are distinguished by the property that they are at right angles to the tangent line of the curve at the point where the normal line and curve intersect. Figure 4.1 illustrates the problem along with the concept of a normal line. There two normal lines to the ellipse pass through the point (x_0, y_0). In general, there may be more than two normal lines passing through a designated point. Apollonius' problem is to characterize the number of normal lines given the location of a point.

Before getting into the math, let us dwell a bit more on the motivation for this problem. Apollonius gives none, and there is certainly no apparent use for the solution. In his biography of Marcellus, the Greek historian Plutarch describes Archimedes' character. A reading of Plutarch's character sketch gives Plutarch's

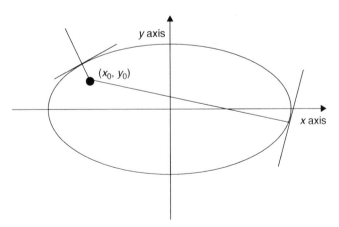

Figure 4.1. Schematic representation of normal-line concept showing orthogonal lines from a point to an ellipse.

views of a mathematician-philosopher's social role; he depicts the ideal Greek intellectual:

> Yet Archimedes possessed so high a spirit, so profound a soul, and such treasures of scientific knowledge, that though these inventions had now obtained him the renown of more than human sagacity, he yet would not deign to leave behind him any commentary or writing on such subjects; but, repudiating as sordid and ignoble the whole trade of engineering, and every sort of art that lends itself to mere use and profit, he placed his whole affection and ambition in those purer speculations where there can be no reference to the vulgar needs of life; studies, the superiority of which to all others is unquestioned, and in which the only doubt can be whether the beauty and grandeur of the subjects examined, of the precision and cogency of the methods and means of proof, most deserve our admiration. It is not possible to find in all geometry more difficult and intricate questions, or more simple and lucid explanations. Some ascribe this to his natural genius; while others think that incredible effort and toil produced these, to all appearances, easy and unlaboured results. No amount of investigation of yours would succeed in attaining the proof, and yet, once seen, you immediately believe you would have discovered it; by so smooth and so rapid a path he leads you to the conclusion required. And thus it ceases to be incredible that (as is commonly told of him) the charm of his familiar and domestic Siren made him forget his food and neglect his person, to that degree that when he was occasionally carried by absolute violence to bathe or have his body anointed, he used to trace geometrical figures in the ashes of the fire, and diagrams in the oil on his body, being in a state of entire preoccupation, and, in the truest sense, divine possession with his love and delight in science.

Plutarch lauds the investigator for his devotion to the discipline of mathematics and science. Plutarch's words indicate that the discipline merits study because it brings one in touch with truth, perfection, and beauty. Plutarch further suggests that any utility resulting from the study of the discipline sullies both the discipline and the investigator because seeking truth, perfection, and beauty is the ultimate objective and finding utility is at cross purposes. This is all consistent with Euclid's rebuke of his student who questioned the utility of geometry.

Was it his search for truth that motivated Apollonius to labor over the conics? Alternatively, was he seduced into seeking perfection and beauty? Perhaps these were contributing factors, but there is an overriding reason for doing math. It's fun. There is plenty of evidence that the childlike fun of mathematics escapes a proportion of the population, but a not so unsizable proportion knows firsthand the pleasure of problem solving. Anyone who has seen mathematicians at work would recognize the playground environment. It is not unlike watching jazz musicians, and certainly we recognize the joy of musicians as they express their creativity through music. As we seek a solution to Apollonius' problem, let us not worry about utility, truth, beauty, or perfection, but approach the problem perhaps as Apollonius did, with the sole intent of having fun.

APOLLONIUS' OSCULATING CIRCLE AND THE ELLIPSE

We restrict our investigation to the case of the ellipse. Begin with the equation
for an ellipse

$$\frac{x^2}{a^2} + \frac{y^2}{b^2} = 1 \tag{4.1}$$

where a is the major axis and b is the minor axis of the ellipse.

Let (x_0, y_0) be an arbitrary point in the plane of the ellipse. We seek an
equation for a normal line from the ellipse to the point. The equation for a
general line passing through the point (x_0, y_0) is $y - y_0 = m(x - x_0)$, where m
is the slope of the line. It is necessary to determine the slope for a normal line.
As Apollonius neatly demonstrates, it is possible to solve this without resorting
to calculus.[4] Here we fully employ the power of calculus. With the knowledge
that the gradient of a function points in the normal direction of the function's
constant level surfaces, we find the gradient of the function $f(x, y) = \frac{x^2}{a^2} + \frac{y^2}{b^2}$,
from which it is possible to determine the slope of the normal line:

$$\left(\frac{\partial f}{\partial x}, \frac{\partial f}{\partial y}\right) = \left(\frac{2x}{a^2}, \frac{2y}{b^2}\right)$$

[4]Select an arbitrary point on the ellipse, (x_e, y_e) and write a general equation for all lines passing
through that point:

$$y - y_e = m_t(x - x_e)$$

$$y = m_t(x - x_e) + y_e$$

Next, substitute for the y variable in (4.1) using the right-hand side (RHS) of this equation. The
result is a quadratic equation.

$$\frac{x^2}{a^2} + \frac{(m_t(x - x_e) + y_e)^2}{b^2} = 1$$

$$\left(\frac{1}{a^2} + \frac{m_t^2}{b^2}\right)x^2 + \left(\frac{2m_t(y_e - m_t x_e)}{b^2}\right)x + \frac{(y_e - m_t x_e)^2}{b^2} - 1 = 0$$

As the tangent line intersects this equation in one point, we look for a single solution to the quadratic
equation:

$$x = \frac{-\left(\frac{2m_t(y_e - m_t x_e)}{b^2}\right) \pm \sqrt{\left(\frac{2m_t(y_e - m_t x_e)}{b^2}\right)^2 - 4\left(\frac{1}{a^2} + \frac{m_t^2}{b^2}\right)\left(\frac{(y_e - m x_e)^2}{b^2} - 1\right)}}{2\left(\frac{1}{a^2} + \frac{m_t^2}{b^2}\right)}$$

The solution reduces to a single point when the expression under the radical vanishes. The reader
can check that this occurs when $m_t = -\frac{b^2 x_e}{a^2 y_e}$. The normal line is at right angles to the tangent line
and accordingly has slope $m = \frac{a^2 y_e}{b^2 x_e}$.

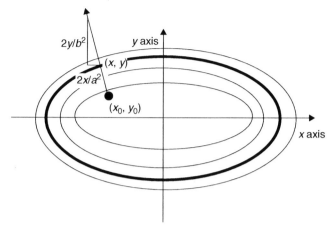

Figure 4.2. Gradient vector gives slope of normal line.

Then the slope of the normal line is given by the following equation (see Figure 4.2):

$$m = \frac{a^2 y}{b^2 x} \tag{4.2}$$

Placing the value of m into the equation of the normal line passing through the point (x_0, y_0) and simplifying results in an equation that describes a double branched hyperbola:

$$y - y_0 = m(x - x_0)$$

$$y - y_0 = \frac{a^2 y}{b^2 x}(x - x_0) \tag{4.3}$$

$$b^2 x(y - y_0) = a^2 y(x - x_0)$$

$$(a^2 - b^2)xy + b^2 y_0 x - a^2 x_0 y = 0$$

We can conclude that when the normal line to the point (x_0, y_0) intersects the ellipse of equation (4.1) at coordinates (x, y), it also intersects the double-branched hyperbola of equation (4.3) at the same coordinates (x, y). Several more conclusions are possible:

1. The point (x_0, y_0) determines the configuration of the hyperbola and hence the number of intersections with the ellipse.
2. We conclude that the number of intersections between the ellipse and the hyperbola gives the number of normal lines, which is the answer to Apollonius' problem. Figure 4.3 illustrates the possible ways in which the figures can intersect. We can see that at most there can be four normal lines to

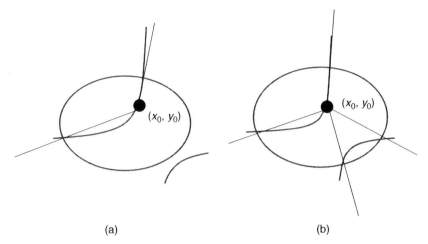

(a)   (b)

Figure 4.3. Intersection of hyperbola with ellipse: (a) two intersections, two normal lines; (b) four intersections, four normal lines.

a point as given by four points of intersection (Figure 4.3b). Noting that the minimum and maximum distances from the ellipse to a point are given by the length of the normal section, we can also conclude that there are at least two normals to a point, one associated with the minimum distance and one associated with the maximum distance. In fact, since one branch of the hyperbola passes through the origin and extends to infinity in two directions, there must be at least two points of intersection (Figure 4.3a).

3. Finally, we note that three normals to a point occur when one branch of the hyperbola is tangent to the ellipse, yielding a single point of intersection (Figure 4.4). This final conclusion is critical in our further analysis.

An algebraic analysis also yields some of the same conclusions. Let us find an equation for the intersections between the ellipse of equation (4.1) and the hyperbola of equation (4.3). Rearranging equation (4.3) and squaring results in the following equation:

$$(a^2 - b^2)xy + b^2 x y_0 - a^2 y x_0 = 0$$

$$(a^2 - b^2)xy - a^2 y x_0 = -b^2 x y_0$$

$$\left((a^2 - b^2)x - a^2 x_0\right) y = -b^2 x y_0$$

$$\left((a^2 - b^2)x - a^2 x_0\right)^2 y^2 = b^4 x^2 y_0^2$$

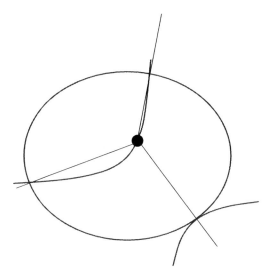

Figure 4.4. Intersection of hyperbola with ellipse: three intersections, three normal lines.

Next, solving for the value y^2 in equation (4.1) and substituting the result into this equation yields a quartic polynomial in x:

$$\frac{x^2}{a^2} + \frac{y^2}{b^2} = 1$$

$$y^2 = b^2 \left(1 - \frac{x^2}{a^2}\right)$$

$$\left((a^2 - b^2)x - a^2 x_0\right)^2 y^2 = b^4 x^2 y_0^2$$

$$\left((a^2 - b^2)x - a^2 x_0\right)^2 \left(1 - \frac{x^2}{a^2}\right) = b^2 x^2 y_0^2$$

$$\left((a^2 - b^2)x - a^2 x_0\right)^2 (a^2 - x^2) - a^2 b^2 x^2 y_0^2 = 0 \qquad (4.4)$$

Before continuing, we make the following observations. At most, there are four solutions to the quartic polynomial, indicating that any point has at most four normals. Evaluating the quartic polynomial at the points $x = \pm a$ and $x = 0$ yields the following:

$$\left((a^2 - b^2)a^2 - a^2 x_0\right)^2 (a^2 - a^2) - b^2 a^4 y_0^2 = -b^2 a^4 y_0^2$$

$$\left((a^2 - b^2)0 - a^2 x_0\right)^2 (a^2 - 0^2) - a^2 b^2 0^2 y_0^2 = a^6 x_0^2$$

Since the sign of the quartic changes between $x = -a$ and $x = 0$, there is at least one root between these values. By the same argument, there is at least one root between $x = 0$ and $x = a$. So there are at least two roots, equating to two normal lines to any point.[5] The algebraic analysis is consistent with the geometric analysis.

The first individual to solve the quartic polynomial was Lodovico Ferarri in the sixteenth century.[6] This solution was not available to Apollonius. Nevertheless, Apollonius went on to discover further relations that allow one to specify the number of normal lines to any given point. Below, we find the solution without explicitly solving the quartic polynomial.

It is of interest to find the set of points where there are three normals. It turns out that this set defines a closed curve; within the curve there are four normals, on the curve there are three normals, and outside the curve there are two normals. To determine the curve, let x_0 and y_0 be variables and seek a curve of the form $g(x_0, y_0) = 0$.

Note that equation (4.3) is linear in both x_0 and y_0. Using the observation that a branch of the hyperbola associated with a point having three normals is tangent to the ellipse (see Figure 4.4), one can obtain another relation between x_0 and y_0. At the tangent point the normal to the ellipse is aligned with that of the hyperbola. Using equation (4.3), we set $h(x, y) = (a^2 - b^2)xy + b^2xy_0 - a^2yx_0$, giving a family of hyperbolas. The gradient of $h(x, y)$ is aligned with the normal vector:

$$\left(\frac{\partial h}{\partial x}, \frac{\partial h}{\partial y} \right) = \left((a^2 - b^2)y + b^2y_0, (a^2 - b^2)x - a^2x_0 \right)$$

The slope of the normal line is $m = \frac{(a^2 - b^2)x - a^2x_0}{(a^2 - b^2)y + b^2y_0}$, which we can equate with the slope from the previous calculation, equation (4.2):

$$m = \frac{(a^2 - b^2)x - a^2x_0}{(a^2 - b^2)y + b^2y_0}$$

$$= \frac{a^2y}{b^2x} \tag{4.5}$$

Simplifying the expression yields another linear equation in x_0 and y_0:

$$\frac{(a^2 - b^2)x - a^2x_0}{(a^2 - b^2)y + b^2y_0} = \frac{a^2y}{b^2x}$$

$$\left((a^2 - b^2)x - a^2x_0 \right) b^2x = \left((a^2 - b^2)y + b^2y_0 \right) a^2y$$

---

[5] Because of the symmetry of the ellipse of equation (4.1), no spurious roots were introduced by squaring equation (4.2).

[6] An interesting story surrounds the discovery of the solution to the quartic polynomial. Ferarri and a compatriot, Tartaglia, went at one another over priority rights for discovery of the cubic. Their public argument culminated in a public mathematical duel. For details, see *The Ellipse* by the author [(Mazer 2010); see Bibliography at end of this book].

$$-a^2b^2(xx_0 + yy_0) = \frac{a^2 - b^2}{a^2b^2}(b^2x^2 - a^2y^2)$$

$$xx_0 + yy_0 = \frac{a^2 - b^2}{a^2b^2}(b^2x^2 - a^2y^2) \tag{4.6}$$

Equations (4.3) and (4.6) form a pair of linear equations in the variables x_0 and y_0. We next solve this system of equations for the variable x_0:

$$a^2yx_0 - b^2xy_0 = -(a^2 - b^2)xy$$

$$xx_0 + yy_0 = \frac{a^2 - b^2}{a^2b^2}(b^2x^2 - a^2y^2)$$

Multiplying the top equation by y and the bottom equation by b^2x and adding the two outcomes eliminates the y_0 variable and results in an equation in x_0:

$$(a^2y^2 + b^2x^2)x_0 = -(a^2 - b^2)xy^2 + \frac{a^2 - b^2}{a^2}(b^2x^3 - a^2xy^2)$$

$$(a^2y^2 + b^2x^2)x_0 = \frac{a^2 - b^2}{a^2}b^2x^3$$

Next, using the equation for the ellipse [eq. (4.1)], note that $a^2y^2 + b^2x^2 = a^2b^2$. Substituting this identity into the expression above results in a relationship between x and x_0:

$$(a^2y^2 + b^2x^2)x_0 = \frac{a^2 - b^2}{a^2}b^2x^3$$

$$a^2b^2x_0 = \frac{a^2 - b^2}{a^2}b^2x^3 \tag{4.7}$$

$$x_0 = \frac{a^2 - b^2}{a^4}x^3$$

Equation (4.7) has a nice geometric interpretation as follows. Assume that we start off with a point on the ellipse, (x, y) and ask ourselves at what value x_0 does the normal line to (x, y) intersect exactly two other normal lines. Equation (4.7) answers this question. With this relationship it is possible to find the curve of interest: $g(x_0, y_0)$.

One can make a similar calculation, or assert by symmetry the following equation between y and y_0:

$$y_0 = \frac{b^2 - a^2}{b^4}y^3 \tag{4.8}$$

From equations (4.7) and (4.8) one obtains the following:

$$\left(\frac{ax_0}{a^2 - b^2}\right)^{2/3} = \frac{x^2}{a^2}$$

$$\left(\frac{by_0}{a^2 - b^2}\right)^{2/3} = \frac{y^2}{b^2}$$

Adding these equations, using the equation for the ellipse, and simplifying yields the desired curve:

$$\left(\frac{ax_0}{a^2 - b^2}\right)^{2/3} + \left(\frac{by_0}{a^2 - b^2}\right)^{2/3} = \frac{x^2}{a^2} + \frac{y^2}{b^2}$$

$$\left(\frac{ax_0}{a^2 - b^2}\right)^{2/3} + \left(\frac{by_0}{a^2 - b^2}\right)^{2/3} = 1$$

$$(ax_0)^{2/3} + (by_0)^{2/3} = \left(a^2 - b^2\right)^{2/3} \qquad (4.9)$$

With one caveat that we address below, the points on the curve given by equation (4.9) are intersections of three normal lines from the ellipse. Figure 4.5 illustrates this curve. Apollonius demonstrates that within this curve, each point stands on four normal lines from the ellipse while outside this curve, each point stands on only two normal lines from the ellipse. The curve is known as the *evolute of the ellipse*. After addressing the caveat, we will return to the evolute.[7]

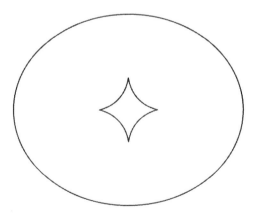

Figure 4.5. The evolute of the ellipse. The evolute consists of all points having three normal lines to the ellipse.

[7]Because ancient Greek mathematicians, who called themselves *geometers*, did not express functional relationships as we do today, Apollonius did not find an expression for the evolute that is equivalent

At this point, a judge would most certainly uphold an objection that the reader may wish to place forward. Twice in the derivation the author was less than careful concerning possible divisions by zero. This occurs when determining the slope value m once using the ellipse and once using the tangent hyperbola. The reader can verify that all the divisions by zero occur when either x and x_0 are zero or y and y_0 are zero; that is, the location of all illegal operations is on the x axis or the y axis. Using symmetry, one can deduce that any point on either axis stands on an even number of normals, precluding the possibility of standing on only three normals. One can see this for a point on the x axis as follows. If (x_0, y_0) lies on the x axis, then the point stands on two normals emanating from the points $(a, 0)$ and $(-a, 0)$. Suppose that there is a third normal emanating from the ellipse at the point (x_n, y_n). Then, by symmetry, another normal emanates from the point $(x_n, -y_n)$, precluding the possibility of only three normals. There can be only two or four normals. Figure 4.6 illustrates the argument; the point (x_0, y_0) lies at the corner of the evolute. On the corner point there are only two normals. A similar argument applies to points on the y axis. The outcome is that points on both the evolute and the coordinate axes stand on only two normal lines, not three. With this statement the caveat has been put to rest.

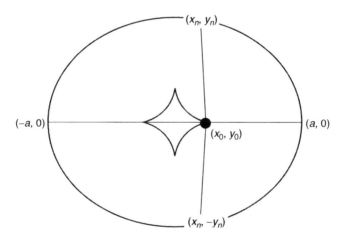

Figure 4.6. A point on the x axis cannot be on the evolute as well. By symmetry, there must be an even number of normals to a point on the x axis.

to equation (4.8). Nevertheless, he did find all the relationships necessary to derive the evolute. In particular, Apollonius writes relations similar to that of the ellipse [eq. (4.1)], the hyperbola [eq. (4.3)], and the slope of the tangent hyperbola [eq. (4.5)]. In his book, *A History of Greek Mathematics*, Heath uses Apollonius' relations to derive the evolute in its modern form (Heath 1921).

The Radius of Curvature

In the spirit of this chapter, just for fun, we now take an alternative path to the evolute. The path leads to a generalized evolute for any given curve, but the solution for the ellipse indicates a direction. For the ellipse, one may consider the point (x, y) at which the hyperbola of equation (4.3) becomes tangent as a limiting point at which two intersections of the receding hyperbola collide. The corresponding point on the evolute, namely, (x_0, y_0), is then the limiting point of intersection for two colliding normal lines (see Figure 4.7). Let us calculate the limiting point of intersection for a general curve given by $f(x, y) = 0$.

Before proceeding, some notation is necessary. Let $\mathbf{n}(x, y)$ be the unit normal vector to the curve having components $\mathbf{n} = (n_1, n_2)$. Note that both components are also functions: $n_1 = n_1(x, y)$ and $n_2 = n_2(x, y)$. Also since $\mathbf{n}$ is the unit normal, its length is one: $n_1^2 + n_2^2 = 1$. Below, we will indicate the direction in which $\mathbf{n}(x, y)$ points. Let (x, y) be on the curve $f(x, y) = 0$ and (x_a, y_a) also be on the curve with x_a close to x so that $x_a = x + \Delta$. Associated with both points (x, y) and (x_a, y_a) are normal lines that intersect at some point denoted by (x_0, y_0). With this notation the problem is to determine $\lim_{\Delta \to 0}(x_0, y_0)$ (see Figure 4.8).

The expression $(x + n_1(x, y)r, y + n_2(x, y)r)$ yields the normal line to (x, y) as the value r ranges over all real numbers. Similarly, the expression $(x_a + n_1(x_a, y_a)s, y_a + n_2(x_a, y_a)s)$ yields the normal line to (x_a, y_a) as the value s ranges over all real numbers. Because it is at the intersection of these two lines, the point (x_0, y_0) must satisfy the following set of equations for specified values of r and s:

$$\begin{aligned}
x_0 &= x + n_1(x, y)r = x_a + n_1(x_a, y_a)s \\
y_0 &= y + n_2(x, y)r = y_a + n_2(x_a, y_a)s
\end{aligned} \tag{4.10}$$

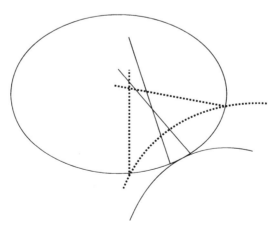

Figure 4.7. Intersections of normals associated with hyperbolas. As the hyperbolas recede and become tangent to the ellipse, the intersections approach a point on the evolute.

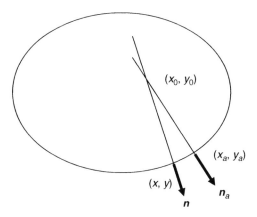

Figure 4.8. The limit of (x_0, y_0) as (x_a, y_a) approaches (x, y) is a point on the evolute.

These equations allow one to solve for the values of r and s that specify the point of intersection. The values r and s converge in the limit as $\Delta \to 0$. Accordingly, we focus on r:

$$n_1(x, y)r - n_1(x_a, y_a)s = x_a - x$$
$$n_2(x, y)r - n_2(x_a, y_a)s = y_a - y$$

Eliminating the s variable and solving for r yields the following equation:

$$(n_1(x, y)n_2(x_a, y_a) - n_2(x, y)n_1(x_a, y_a))\, r = n_2(x_a, y_a)(x_a - x)$$
$$-n_1(x_a, y_a)(y_a - y)$$

From this equation one obtains the value of r:

$$r = \frac{n_2(x_a, y_a)(x_a - x) - n_1(x_a, y_a)(y_a - y)}{n_1(x, y)n_2(x_a, y_a) - n_2(x, y)n_1(x_a, y_a)} \tag{4.11}$$

The objective is to determine $\lim_{\Delta \to 0} r$. Toward this end, the expressions are expanded to first order as a series in Δ. By assumption, the term $x_a - x = \Delta$. Next, let's determine the series expansion for the term y_a in terms of Δ, $y_a = y(x) + y'(x)\Delta + $ higher-order terms. To accomplish this, we assume that the equation $f(x, y) = 0$ allows one to express y as a differentiable function of x.[8] With this assumption, we note the following:

$$f(x_a, y_a) = f(x, y) + \frac{\partial f}{\partial x}\Delta + \frac{\partial f}{\partial y}\Delta_y + \text{higher-order terms}$$

---

[8]The necessary conditions are known from the implicit function theorem.

Both (x, y) and (x_a, y_a) lie on the curve yielding the following expression:

$$\frac{\partial f}{\partial x}\Delta + \frac{\partial f}{\partial y}\Delta_y + \text{higher-order terms} = 0$$

$$\frac{\partial f}{\partial x}\Delta + \frac{\partial f}{\partial y}\frac{dy}{dx}\Delta + \text{higher-order terms} = 0$$

$$\left(\frac{\partial f}{\partial x} + \frac{\partial f}{\partial y}\frac{dy}{dx}\right)\Delta + \text{higher-order terms} = 0$$

It follows that the coefficient $\frac{\partial f}{\partial x} + \frac{\partial f}{\partial y}\frac{dy}{dx}$ is identically zero, from which we calculate the derivative, $y'(x)$:

$$\frac{\partial f}{\partial x} + \frac{\partial f}{\partial y}\frac{dy}{dx} = 0$$

$$y'(x) = \frac{dy}{dx} = -\frac{\dfrac{\partial f}{\partial x}}{\dfrac{\partial f}{\partial y}}$$

Since the gradient and normal vectors are aligned, $\dfrac{\dfrac{\partial f}{\partial x}}{\dfrac{\partial f}{\partial y}} = \dfrac{n_1}{n_2}$ and we obtain the result for $y'(x)$.

$$y'(x) = -\frac{n_1}{n_2} \qquad (4.12)$$

Finally, we determine the expansion for the normal components, $n_1(x_a, y_a)$ and $n_2(x_a, y_a)$, in which the higher-order terms are ignored:

$$
\begin{aligned}
n_1(x_a, y_a) &= n_1(x, y) + \frac{\partial n_1}{\partial x}\Delta + \frac{\partial n_1}{\partial y}\Delta_y \\
&= n_1(x, y) + \frac{\partial n_1}{\partial x}\Delta + \frac{\partial n_1}{\partial y}\frac{dy}{dx}\Delta \\
&= n_1(x, y) + \frac{\partial n_1}{\partial x}\Delta - \frac{\partial n_1}{\partial y}\frac{n_1}{n_2}\Delta \qquad (4.13)\\
&= n_1(x, y) + \left(\frac{\partial n_1}{\partial x} - \frac{\partial n_1}{\partial y}\frac{n_1}{n_2}\right)\Delta
\end{aligned}
$$

Similarly, the expansion for $n_2(x_a, y_a)$ is the following:

$$n_2(x_a, y_a) = n_2(x, y) + \left(\frac{\partial n_2}{\partial x} - \frac{\partial n_2}{\partial y}\frac{n_1}{n_2}\right)\Delta \qquad (4.14)$$

Substituting the first-order expansions [eqs. (4.13) and (4.14)] into equation (4.11) results in the following expression for r:

$$r = \frac{\left[n_2 + \left(\dfrac{\partial n_2}{\partial x} - \dfrac{\partial n_2}{\partial y}\dfrac{n_1}{n_2}\right)\Delta\right]\Delta + \left[n_1 + \left(\dfrac{\partial n_1}{\partial x} - \dfrac{\partial n_1}{\partial y}\dfrac{n_1}{n_2}\right)\Delta\right]\dfrac{n_1}{n_2}\Delta}{n_1\left[n_2 + \left(\dfrac{\partial n_2}{\partial x} - \dfrac{\partial n_2}{\partial y}\dfrac{n_1}{n_2}\right)\Delta\right] - n_2\left[n_1 + \left(\dfrac{\partial n_1}{\partial x} - \dfrac{\partial n_1}{\partial y}\dfrac{n_1}{n_2}\right)\Delta\right]}$$

Eliminating all higher-order terms, simplifying, and taking the limit as $\Delta \to 0$ results in the desired expression:

$$\lim_{\Delta \to 0} r = \lim_{\Delta \to 0} \frac{\left[n_2 + \left(\dfrac{\partial n_2}{\partial x} - \dfrac{\partial n_2}{\partial y}\dfrac{n_1}{n_2}\right)\Delta\right]\Delta + \left[n_1 + \left(\dfrac{\partial n_1}{\partial x} - \dfrac{\partial n_1}{\partial y}\dfrac{n_1}{n_2}\right)\Delta\right]\dfrac{n_1}{n_2}\Delta}{n_1\left[n_2 + \left(\dfrac{\partial n_2}{\partial x} - \dfrac{\partial n_2}{\partial y}\dfrac{n_1}{n_2}\right)\Delta\right] - n_2\left[n_1 + \left(\dfrac{\partial n_1}{\partial x} - \dfrac{\partial n_1}{\partial y}\dfrac{n_1}{n_2}\right)\Delta\right]}$$

$$\lim_{\Delta \to 0} r = \lim_{\Delta \to 0} \frac{n_2\Delta + n_1\dfrac{n_1}{n_2}\Delta}{n_1\left(\dfrac{\partial n_2}{\partial x} - \dfrac{\partial n_2}{\partial y}\dfrac{n_1}{n_2}\right)\Delta - n_2\left(\dfrac{\partial n_1}{\partial x} - \dfrac{\partial n_1}{\partial y}\dfrac{n_1}{n_2}\right)\Delta}$$

$$\lim_{\Delta \to 0} r = \frac{n_2 + n_1\dfrac{n_1}{n_2}}{n_1\left(\dfrac{\partial n_2}{\partial x} - \dfrac{\partial n_2}{\partial y}\dfrac{n_1}{n_2}\right) - n_2\left(\dfrac{\partial n_1}{\partial x} - \dfrac{\partial n_1}{\partial y}\dfrac{n_1}{n_2}\right)}$$

$$\lim_{\Delta \to 0} r = \frac{n_2^2 + n_1^2}{n_1 n_2\left(\dfrac{\partial n_2}{\partial x} - \dfrac{\partial n_2}{\partial y}\dfrac{n_1}{n_2}\right) - n_2^2\left(\dfrac{\partial n_1}{\partial x} - \dfrac{\partial n_1}{\partial y}\dfrac{n_1}{n_2}\right)}$$

$$\lim_{\Delta \to 0} r = \frac{1}{n_1 n_2\left(\dfrac{\partial n_2}{\partial x} - \dfrac{\partial n_2}{\partial y}\dfrac{n_1}{n_2}\right) - n_2^2\left(\dfrac{\partial n_1}{\partial x} - \dfrac{\partial n_1}{\partial y}\dfrac{n_1}{n_2}\right)}$$

One final simplification provides the result:

$$\lim_{\Delta \to 0} r = \frac{1}{n_1 n_2\left(\dfrac{\partial n_2}{\partial x} + \dfrac{\partial n_1}{\partial y}\right) - \left(n_1^2\dfrac{\partial n_2}{\partial y} + n_2^2\dfrac{\partial n_1}{\partial x}\right)} \tag{4.15}$$

To complete the description of the limiting value, we take the direction of **n** in such a manner to ensure that the limit of equation (4.15) is positive. Once the limiting value of r is available, the limiting expressions for x_0 and y_0 are also available. Below we use the following notation for the limiting values:

$$\lim_{\Delta \to 0} r = r^* \quad \text{and} \quad \lim_{\Delta \to 0} (x_0, y_0) = (x_0^*, y_0^*)$$

By definition, the point (x_0^*, y_0^*) is on the evolute. With this notation, the following identities hold:

$$x_0^* = x + n_1 r^* \quad \text{and} \quad y_0^* = y + n_2 r^* \tag{4.16}$$

The Osculating Circle

We could return to the ellipse and use equations (4.16) to establish a relationship between x_0^* and y_0^*. But there is something else even more fun.

Note that r^* is a function of the point (x, y), $r^* = r^*(x, y)$, and the value r^* is the distance from the point (x, y) to the point (x_0^*, y_0^*). The distance r^* is known as the radius of curvature[9] at the point (x, y), and the point (x_0^*, y_0^*) is the center of curvature. Assume that the original curve, $f(x, y) = 0$, is a circle of radius R. Then, because all normal lines intersect at the center, r^* is the same value as R. Now return to the general case, $f(x, y) = 0$, any arbitrary curve. Because, at a given point (x, y), the general curve and a circle tangent to the curve at (x, y) with radius r^* both yield the same value for the radius of curvature, the curves are similar around the point (x, y). In fact, among all tangent circles, that of radius r^* is the circle that best approximates the original curve, as we next demonstrate.

First consider what is meant by the "best approximating circle to the original curve at a point (x, y)." Three choices specify any circle: two to designate its center, (x_c, y_c), and one for the radius, R. Let (x_a, y_a) be a point on the circle in the vicinity of (x, y), and express y_a as a function of x_a.[10] As above, it is possible to expand the functions in a Taylor series about the value x. A similar Taylor series about the value x is available using the original curve, $f(x, y) = 0$. As there are three parameter choices for the circle, it is possible to choose them so that the first three terms of each Taylor series formed from the circle match the first three terms from the Taylor series formed from the original curve. The circle that matches the first three terms is the best approximating circle at the point (x, y). Let us find this circle.

The following expression is the Taylor series of a function expanded about the value x:

$$y_a = y(x) + y'(x)\Delta + \tfrac{1}{2}y''(x)\Delta^2 + \text{higher-order terms} \tag{4.17}$$

The first three terms of two power series are equal when the value of y and its first two derivatives are the same at the given value x. Equation (4.16) gives the value of the first term, $y = y_0^* - r^* n_1$. Equation (4.12) provides the first-order term, $y'(x)\Delta = -\frac{n_1}{n_2}\Delta$; equivalently, $y'(x) = -\frac{n_1}{n_2}$.

Differentiating $y'(x)$ gives the second derivative:

$$y''(x) = -\frac{n_2\left(\dfrac{dn_1}{dx}\right) - n_1\left(\dfrac{dn_2}{dx}\right)}{n_2^2}$$

---

[9] The value $\kappa = \frac{1}{r^*}$ is known as the *curvature*.

[10] We assume that the conditions of the implicit function theorem hold so that the expression holds.

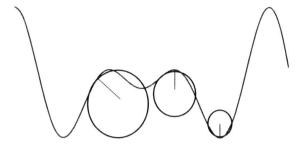

Figure 4.9. The osculating circle at different points on a curve.

Since y_a is considered as a function of x_a, the normal component n_1 can be considered as a function of x_a with expansion, $n_1(x_a) = n_1(x) + \frac{d}{dx}n_1(x)\Delta$ + higher-order terms. Equation (4.13) provides the first-order term of the expansion yielding the expression $\frac{d}{dx}n_1(x) = \left(\frac{\partial n_1}{\partial x} - \frac{\partial n_1}{\partial y}\frac{n_1}{n_2}\right)$. Similarly, from equation (4.14), $\frac{d}{dx}n_2(x) = \left(\frac{\partial n_2}{\partial x} - \frac{\partial n_2}{\partial y}\frac{n_1}{n_2}\right)$. Using these expressions and equation (4.15) results in the following expression for the second derivative of y:[11]

$$y''(x) = -\frac{n_2\left(\dfrac{\partial n_1}{\partial x} - \dfrac{\partial n_1}{\partial y}\dfrac{n_1}{n_2}\right) - n_1\left(\dfrac{\partial n_2}{\partial x} - \dfrac{\partial n_2}{\partial y}\dfrac{n_1}{n_2}\right)}{n_2^2}$$

$$= -\frac{n_2^2\left(\dfrac{\partial n_1}{\partial x} - \dfrac{\partial n_1}{\partial y}\dfrac{n_1}{n_2}\right) - n_2 n_1\left(\dfrac{\partial n_2}{\partial x} - \dfrac{\partial n_2}{\partial y}\dfrac{n_1}{n_2}\right)}{n_2^3}$$

$$= \frac{1}{r^* n_2^3} \tag{4.18}$$

For any curve, the three values $y(x)$, $y'(x)$, and $y''(x)$ depend only on the curve's normal components n_1 and n_2, the curve's radius of curvature r^*, and the curve's center of curvature (x_0^*, y_0^*). As such, the first three terms in a Taylor series of two curves sharing common normal components, radius of curvature, and center of curvature are identical. Figure 4.9 illustrates three circles tangent to three different points along a curve. The circles share a common normal with the curves at their points of intersection. Also, the center of each circle is the same as the curve's center of curvature: $(x_c, y_c) = (x_0^*, y_0^*)$. The circle is called the *osculating circle*.

The Evolute and the Osculating Circle

Associated with each point along a curve is an osculating circle and the osculating circle's center. The collection of all such centers forms another curve and this new

---

[11] The result for the second derivative is frequently written as $y''(x) = \kappa/n_2^3$, where κ is the curvature.

curve defines the evolute (see Figure 4.10). The preceding sections demonstrate that the points sitting on three normals from an ellipse are on the evolute of the ellipse. Our initial evaluation of the evolute didn't even apply the evolute's definition. It was an accidental outcome, and we weren't aware that the curve of equation (4.9) satisfies the properties of the evolute. Below, we calculate the evolute of the ellipse from its definition and show that the curve is the same curve resulting from the initial approach. The calculation simplifies equations (4.13) and (4.14) to obtain the expressions (4.7) and (4.8).

We begin by calculating the unit normals to the ellipse. We have already found normal vectors by taking the gradient of the function $f(x, y) = \frac{x^2}{a^2} + \frac{y^2}{b^2}$. The gradient is $\left(\frac{\partial f}{\partial x}, \frac{\partial f}{\partial y} \right) = \left(\frac{2x}{a^2}, \frac{2y}{b^2} \right)$. Normalizing the gradient and pointing the vectors inward yields the components of the unit normals:

$$n_1 = -\frac{x}{a^2 \sqrt{\dfrac{x^2}{a^4} + \dfrac{y^2}{b^4}}}$$

$$n_2 = -\frac{y}{b^2 \sqrt{\dfrac{x^2}{a^4} + \dfrac{y^2}{b^4}}}$$

Next, using equation (4.18), we obtain the following expression for the radius of curvature r^*:

$$r^* = \frac{1}{y''(x)n_2^3}$$

The derivative $y'(x) = -\frac{n_1}{n_2} = -\frac{b^2 x}{a^2 y}$. Differentiating this value results in the second derivative.

$$y''(x) = -\frac{b^2}{a^2} \left(\frac{y + \dfrac{b^2 x^2}{a^2 y}}{y^2} \right)$$

$$= -\frac{b^2}{a^4 y^3} \left(a^2 y^2 + b^2 x^2 \right)$$

$$= -\frac{b^2}{a^4 y^3} \left(a^2 b^2 \right) \qquad \text{from the equation of an ellipse}$$

$$= -\frac{b^4}{a^2 y^3}$$

Placing these expressions into equations (4.16) yields the desired expression:

$$x_0^* = x + r^* n_1$$

$$= x + \frac{n_1}{y''(x) n_2^3}$$

$$= x - \frac{x}{a^2 \sqrt{\dfrac{x^2}{a^4} + \dfrac{y^2}{b^4}}} \frac{a^2 y^3}{b^4} \frac{\left(b^2 \sqrt{\dfrac{x^2}{a^4} + \dfrac{y^2}{b^4}}\right)^3}{y^3}$$

$$= x - b^2 x \left(\frac{x^2}{a^4} + \frac{y^2}{b^4}\right)$$

$$= x - b^2 x \left(\frac{x^2}{a^4} + \frac{1}{b^2}\left(1 - \frac{x^2}{a^2}\right)\right)$$

$$= x - \frac{b^2 x^3}{a^4} - x + \frac{x^3}{a^4}$$

$$= \frac{a^2 - b^2}{a^4} x^3$$

A similar calculation beginning with $y_0^* = y + r^* n_2$ results in the equation $y_0^* = \frac{b^2 - a^2}{a^4} y^3$. Comparing this with equations (4.7) and (4.8), we find all the

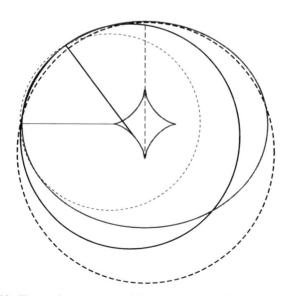

Figure 4.10. The evolute constructed from the centers of the osculating circles.

points (x_0, y_0) coincide with (x_0^*, y_0^*), affirming that the curve of expression 4.9, $(ax_0)^{2/3} + (by_0)^{2/3} = (a^2 - b^2)^{2/3}$ is, indeed, the evolute.

Figure 4.10 illustrates the construction of an ellipse's evolute using the family of osculating circles.

FUN'S HARVEST

An unanswered question is Apollonius' path to the evolute for the ellipse. Did he utilize the tangent hyperbola as in our initial approach, unaware of the osculating circle; discover and utilize the osculating as in our latter approach; or come to it through some other process? If Apollonius was aware of the osculating circle and its radius of curvature, he did not present its properties in his manuscript *On Conics*, but he most certainly motivated others to pursue it.

Christiaan Huygens (1629–1695) explicitly defines the radius of curvature and the osculating circle for for general curves. Leibniz applies his methods of calculus to determine the radius of curvature. Gauss (1777–1855), Riemann (1826–1866), and Ricci-Curbastro (1853–1925) create a general setting for non-Euclidean geometry in which the curvature is central. Then Einstein (1879–1955) incorporates curvature into his field equations of general relativity. Mathematics is fun, but not frivolous.

CHAPTER 5

RETROGRADE

THE IMPERIAL THEOCRACY

In 62 A.D.,[1] at the height of its power, the imperial city Rome suffered a devastating setback. The city housed over 1 million people, many ethnically Italian, but to be sure, there was an assortment of races from the conquered territories. Conquest brought slaves who begat children and created minority ethnic centers. Housing for the typical inhabitant, Roman or otherwise, was within a squalid apartment complex. There were thousands of such timber and mud brick structures packed in tight quarters, and when these fueled a great bonfire, leveling much of the city, someone had to take the blame.

The pagan emperor, Nero (37–68), considered his options. Connections to the imperial household protected the Jewish population, but not the Christians. These early Christians were dealt a lesson concerning the precarious position of a religious minority, one that their progeny would apply when they were in the position of the majority. Emperor Nero implicated the Christians and announced a spate of imperial anti-Christian decrees that set the tone for fierce persecution throughout the empire. Emperor Nero did this not with any particular enmity toward Christians, nor with any ideological fervor. It is doubtful that Nero ever actually met and spoke with any member of this group, so he had no chance

---

[1] After this point in the text, the abbreviation A.D. will be omitted for any *anno Domini* years mentioned (single years or multiyear ranges), for brevity.

Shifting the Earth: The Mathematical Quest to Understand the Motion of the Universe,
First Edition. Arthur Mazer.
© 2011 John Wiley & Sons, Inc. Published 2011 by John Wiley & Sons, Inc.

to develop any particular distaste toward them. Yet he corralled many innocent Christians into the Coliseum, where they delivered entertainment to the Romans and supper to the lions. His motivation was pure pragmatic politics.

Two and a half centuries after Nero's death, in his newly established imperial city Constantinople, Emperor Constantine (272–337) favored the Christian Church with state lands and exempted Christian clergy from taxes. He also opened up the imperial treasury to fund the construction of churches across the empire. Unlike Nero, Constantine was motivated by a sort of fervor; however, as with Nero, pragmatic politics loomed large. The Church had evolved into a well-organized hierarchy, one that Constantine could put to use as an agent of the empire. By coopting this hierarchy and acting as the protector of the Church, Constantine would maintain a loyal citizenry across a geographically and ethnically diverse empire. At the time of Constantine's baptism, within 2 months of his death, the majority of the empire's citizenry had not yet so bathed themselves. Constantine moved slowly but deliberately and pointed the empire toward a Christian theocracy.

By 390 Christians were far from a persecuted minority. In that year, the emperor, Theodosius (347–395), responded to the killing of a Roman sentry in the city of Thessalonica rather unwisely. It is unclear whether Theodosius ordered the local garrison to punish Thessalonica, or if he acquiesced, but the city received punishment in the style of the Roman Army. In a frenzy of slaughter perhaps 50,000 civilians perished.[2] Add to that the usual looting, rape, and burning that accompanied a good Roman Army disciplining, and the horror administered to the city enrages even the less sensitive among men. The outpouring of anger eroded Theodosius' political standing, and Theodosius had to act to restore his stature.

That Theodosius turned to Bishop Ambrose, the bishop of Milan, indicates a thorough transition in Christianity's status from the days of Nero. The number of baptized had become so large that the Church could sway public opinion. Clergymen, often appointees of the empire, held positions of political influence with judicial and administrative authority. The ambitious Bishop Ambrose used the occasion of Theodosius' public relations challenge to elicit even further influence. The image makeover was successful as a penitent Theodosius received God's forgiveness through Bishop Ambrose. But Bishop Ambrose subjected Theodosius to a political tax.

In 391 Theodosius illegalized the practice of any religion except Christianity. Although a Christian, Theodosius had previously shown himself to be tolerant. When Bishop Ambrose petitioned on behalf of a Christian horde that had illegally ransacked a Jewish synagogue, Theodosius berated the bishop and held to his view that the law applies uniformly to all. Theodosius had also courted the predominantly pagan Senate of Rome. Following his predecessors, Theodosius rooted his tolerance in pragmatism. So the most likely explanation

---

[2]Estimates range widely. Charles Freeman presents the estimate of 50,000 in his work. *The Closing of the Western Mind* (2005).

for Theodosius' reversal and most intolerant decree is that at that time it was pragmatic for him to make the decree.

Actual implementation of the law proved difficult and uneven as paganism was well entrenched. Nevertheless, zealous Christians did take up the cause. Christians ransacked pagan temples and Jewish synagogues, taking booty and installing it in their churches. As a pagan center of philosophy, Alexandria was hard-hit. In 392 a Christian mob assaulted the pagan temple at Serapeum. The assault included the burning of books housed in a library subsidiary to the well-known Library of Alexandria. There are credible claims that the mob reached the main library and torched the world's most complete collection of scrolls.

Regardless of whether the mob alighted the main library on the same day as its subsidiary, they did eventually snuff out the light of Alexandria. Succumbing to continued harassment, lecturers submitted to the inevitable and ceased lecturing. In 415, Christian vigilantes murdered the last remaining lecturer, a brave woman philosopher named Hypatia. With her death, lectures at the university ceased. While these activities were outside the scope of the law, far from being spontaneous occurrences, they were orchestrated by Bishop Cyril, who acted as an official of the empire. The Roman Empire, both its Byzantine and Roman halves, had transitioned to theocracies.

The use of the word *theocracy* is certainly controversial. We are brought up to think of Rome and Romans as the establishers of legal process, with the process residing in a balloon that floats on reason. This view points to the philosopher Cicero, a contemporary of Julius Caesar who embodies the rational legalistic spirit. (Caesar destroyed Cicero's republic, but then applied Cicero's laws to administer his empire.) There was a large gap between the Roman Empire of the Caesar dynasty and that of Theodosius. Whereas the Caesars had worshiped many Gods and tolerated many religions, Theodosius worshiped only one God and would tolerate only one religion. Whereas religion was peripheral to the Caesars' political apparatus, the Catholic Church was embedded into the politics of Theodosius as bishops were political appointees of the state. The balloon of reason had drifted toward a faith-based theocracy.

But there was something in the Roman culture that promoted the drift. The culture was a concoction mixing the ingredients of pragmatism with superstition to the point where they became entwined and inseparable. The Roman Empire could not have endured for seven centuries without a pragmatic perspective. Above, there are examples of the pragmatic politics of Nero, Constantine, and Theodosius. The pragmatic side of Roman culture also surfaces in its engineering feats, which we discuss later, but let us first consider superstition.

As a starting point we give superstition a definition. Let superstition be an instrument that actualizes social or individual delusion. It may be a ritual, it may be a belief, or it may be an emotion, so long as it causes delusion. Stated this way, superstition sounds like a ticket to the asylum. But what if that delusion promotes a supreme sense of confidence, direction, and purpose that, in turn, engenders poise? In early November 312 before the battle that would end civil war among Romans and decide who among the contenders would stand as the Augustus, emperor of Rome, Constantine saw a sign in the heavens. A cross illuminated the sky, indicating that the Christian God lent his full support to Constantine. Constantine's opponent, Maxentius, also received signs of victory from his pagan Gods. Prior to pursuing Constantine, Maxentius, sought out an oracle and asked who would be victorious. The response was that the true Augustus would win; Maxentius took this to mean himself. Both Maxentius and Constantine, confident in divine support, could now go about the business of overseeing final preparations for the battle. For both Constantine and Maxentius, the stakes were high; the victor would be the uncontested emperor while the vanquished had every reason to expect that his foe would execute him as a usurper. Constantine lived to fight further battles, while Maxentius escaped execution by drowning in the Tiber River as a bridge collapsed under the weight of his retreating army.

After the battle, Constantine accepted the Christian God. He accepted the Christian God as he placed his son Crispa on trial, found Crispa guilty of sedition, and sentenced Crispa to execution. He accepted the Christian God when, after reconsidering his son's execution, he concluded that Crispa was a victim of false rumors circulated by Constantine's more recent wife Fausta. Once found guilty of plotting against Crispa to promote her own son as heir, like Crispa before her, Constantine passed the sentence of execution on Fausta.[3] The Christian God that Constantine accepted was not a God of forgiveness and compassion. The Christian God was the God who proved that he could deliver victory on the battlefield. Insanity, delusion, superstition, flawless planning, glory, admired accomplishment, and wholly preventable disaster do not occupy distinct spaces in the human condition. They bounce about and embrace one another, sometimes contributing to the other's momentum and sometimes causing a change in course.

Aside from superstition, another cultural aspect of Rome was an antipathy toward mathematics and science. In its seven centuries of imperial glory, while producing legal philosophers, notable statespeople, great generals, artists, and accomplishing several rather astounding feats of engineering, the Romans produced not one notable mathematician or scientist. Our perfect shape, the circle, perfectly represents the entire contribution of Rome to science when one considers it to be the symbol for the number zero. It is a dismal record that is at odds with Rome's other accomplishments.

To reconcile this, let us return to Plutarch.[4] Plutarch was culturally Greek, and his praise of Archimedes reflects Greek attitudes. To the pragmatic Roman, such praise falls flat. Of what value is knowledge that yields no physical or political benefit? For seven centuries, the Romans stood alongside Euclid's questioning student in a perfectly continuous line. With his high praise for the purity of science, Plutarch unwittingly contributed to the perception that such knowledge was useless. The pagan Romans not only killed Archimedes; they jettisoned his works. When it was their turn, the Christian Romans not only ignored

---

[3]The specifics of Crispa's and Fausta's executions are uncertain. In other versions of the story, Crispa and Fausta are entangled in a love affair that went awry. The version given above follows Charles Freeman.
[4]Recall the quote of Plutarch in Chapter 4.

the knowledge of the scrolls in Alexandria; they torched them. In technology, the Romans introduced indoor plumbing and built their coliseums, bridges, aqueducts, and roads. One thousand years after the fall of Rome, Roman roads continued to be the main thoroughfares between eastern and western Europe and aqueducts continued to carry water to cities. The utility of such engineering was obvious, and this is where the Romans rewarded technical talent. Under the pagans, there were no rewards for Plutarch's Hellenic scientist. As there was not an established scientific cadre prior to the theocratization of the empire, there was virtually no chance of promoting science once theocracy established itself.

The Church assumed the role of Constantine's vision. It was the institution responsible for maintaining social order. It was a necessary function, and in its role, the Church assumed responsibilities that other institutions would not, such as constructing and maintaining hospitals as well as providing a safety net for the most vulnerable. It is also a naturally conservative role, and—because they were rational men—those in charge adopted conservative policies. After all, there is far less likelihood of social upheaval in an intellectually static society.

As pagan Romans morphed to Christian Romans, they brought with them a culture of superstition, a pragmatic sense of power politics, and an antipathy toward science. Superstition seeded a fascination with the afterlife, which, with their pragmatic sense of power politics, was applied to subjugating the population lest anyone who question Church doctrine go to hell. In turn, as it is easier to stamp authority on the unquestioning, ignorant mind than on the active, inquisitive mind, the pragmatic needs of power politics resonated with an antipathy toward science. In turn, antipathy toward science promoted superstition, for superstition germinates in creative minds with no scientific outlet. So it was this perfectly circular arrangement that wheeled Europe into the dark ages, setting a retrograde course that reversed the intellectual momentum of the Greeks.[5]

PTOLEMY'S UNIVERSE

In the preceding chapters, I claim that Hellenistic mathematicians sought truth, perfection, beauty, solace, and fun. The evidence for this claim is circumstantial as the mathematicians from Eudoxus to Apollonius write in a detached, "let's get down to business" manner. In their surviving manuscripts, there is little trace of wonderment and there are few words in praise of perfection or beauty. These are all technical, scholarly works to be interpreted by the reader. It is from the secondhand sources of the philosophical and historical community (Plato,

---

[5]Europe entered an intellectually barren era, not solely because of Church doctrine, but because it did not support liberal institutions that would counterbalance the Church's pragmatically conservative ideology. Concerning the Church's highly conservative views, their concerns that loosening their grip over intellectual endeavors would cause social upheaval were not unfounded. The upheavals of the industrial revolution, a revolution based on expansion of scientific knowledge, lead to overthrows of governments, rioting across Europe, two world wars, and ideological clashes that kept the world under the shadow of nuclear annihilation.

Aristotle, Plutarch), not the mathematicians themselves, that we find scraps of the mindset of the mathematicians. We place these scraps alongside our own prejudices, and a portrait emerges. This portrait is tenuous at best. Thus, when there is direct confirmation from the mathematicians, it is noteworthy, but loaded with traps.

In his work *The Almagest*, Ptolemy describes the motions of the heavens in great detail. It is a scientific work in the style of his predecessors. The translated version runs through roughly 470 pages in 13 books. The books are replete with mathematical tables, astronomical tables, astronomical observations, diagrams of geometric constructions, definitions, theorems, and geometric arguments. But then the English translation of the second sentence of Book 9, Section 2 reads as follows:

> Now, since our problem is to demonstrate, in the case of the five planets as in the case of the sun and the moon, all their apparent irregularities as produced by means of regular and circular motions (for these are proper to the nature of divine things which are strangers to disparities and disorders) the successful accomplishment of the aim as truly belonging to mathematical theory in philosophy is to be considered a great thing, very difficult and as yet unattained in a reasonable way by anyone.

I am unaware of nontechnical passages revealing philosophical thought in the available works of Aristarchus, Euclid, Archimedes, or Apollonius. Ptolemy's philosophical channel lies within parentheses, where Ptolemy indicates that the circle with regular motion is the nature of divine things. This one line corroborates the portrait that we have assembled as perfection, truth, and beauty, strangers to disparities, all reside within the divine, and it is the divine that Ptolemy pursues. Euclid's quest for solace in mathematics extends itself into Ptolemy's heavens, which being divine, are strangers to disorder.

Ptolemy represents the perspective of his mathematical heritage, one that has a similarity with the monotheists. Just as the monotheists devote themselves to one and only one God, there could be one and only one geometric symbol of perfection, the circle, and this must express itself in the heavens exclusive of every other geometric construct. Given the times that Ptolemy lived, the comparison with monotheism deserves a bit more scrutiny.[6] Another of his scarce philosophical statements is found right at the beginning of *The Almagest* in the second paragraph of Book 1:

> For given that all beings have their existence from matter and form and motion, and that none of these can be seen, but only thought, in its subject separately from the others, if one should seek out its simplicity the first cause of the first movement of the universe, he would find God invisible and unchanging. And the kind of science which seeks after Him is theological.

---

[6] In Ptolemy's days Alexandria had a significant Jewish population as well as a budding Christian population. Ptolemy was certainly exposed to monotheism.

The translation portrays Ptolemy as a monotheist.[7] There is mention of a single invisible, unchanging God, and the translation punctuates this belief with the singular "Him." This is in contrast to the multiple Gods of the pagan, Gods who present themselves in the forms of icons.

Beware, the monotheistic bent within Ptolemy rests on an interpretation of a translation. The errors compound. Another translation by G. J. Toomer is more glib, but more insightful:[8] "Now the first cause of the first motion of the universe, if one considers it simply, can be thought of as an invisible and motionless deity."

The second translation has more credibility than the first. The first uses capital letters to emphasize God and Him. These are the indicators that Ptolemy was a monotheist. As the ancient Greeks did not use capitalization, these are inventions of the translator and say more about the translator than they do about Ptolemy. There are no traces of monotheism within the second translation.

Let's interpret Ptolemy's views by his times. Ptolemy calls on a deity as the cause of the first movement, that is to say, the agent that keeps the heavens in motion along a well-designed pathway. There may be other Gods, some of great consequence, but there is only one that governs the motion of the heavens. This is consistent with Greek philosophy as Aristotle proposed a prime mover. Ptolemy continues to follow tradition with his discussion on the distinction between the theological and physical sciences:

> And the kind of science which seeks after the deity (prime mover)[9] is the theological; for such an act can only be thought as high above somewhere near the loftiest things of the universe and is absolutely part of some sensible things. But the kind of science which traces through the material and ever moving quality, and has to do with the white, the hot, the sweet, the soft, and such things would be called physical; and as such an essence since it is only to be found in corruptible things below the lunar sphere. And the kind of science which shows up quality with respect to forms and local motions, seeking figure, number, and magnitude, and also place, time, and similar things, would be defined as mathematical. For such an essence falls, as it were, between the other two not only because it can be conceived both through the senses and without the senses, but also because it is an accident in absolutely all beings both mortal and immortal, changing with those things that ever change, according to their inseparable form, and preserving unchangeable the changelessness of form in things eternal and of ethereal nature.

Here Ptolemy shows that he is an Aristotelian. Ptolemy attributes perfection to the heavens while corruption prevails on eárth (i.e., below Eudoxus' lunar sphere). The domain of theology resides within the heavens, while that of the physical sciences resides with the corruptible. For Ptolemy, mathematics belongs to both the mortal and the immortal and maintains the privileged position of a bridge between the perfect and corruptible domains. Ptolemy expounds further:

---

[7] Translation of *The Almagest* by Taliaferro, Great Books of the Western World.
[8] Translation of *The Almagest* by G. J. Toomer, Springer-Verlag, 1984.
[9] I have taken the liberty to alter Taliaferro's original translation. Taliaferro again uses "Him," which I replace with "the deity."

And therefore mediating that the other two genera of the theoretical would be expounded in terms of conjecture rather than in terms of scientific understanding; the theological because it is in no way phenomenal and attainable, but the physical because its matter is unstable and obscure, so that for this reason philosophers could never hope to agree on them; and mediating that only the mathematical, if approached enquiringly, would give its practitioners certain and trustworthy knowledge with demonstration both arithmetic and geometric resulting from indisputable procedures, we were led to cultivate most particularly as far as lay in our power this theoretical discipline.... For that special mathematical theory would most readily prepare the way to the theological.

Who is Ptolemy? He is a scientist who extends the boarders of scientific knowledge. Philosophically, he accepts the theological conventions of his age and of his culture. In this, Ptolemy is no different from the modern scientist who is brought up within a certain religious tradition. He remains faithful to the tradition and finds in his work evidence that reinforces his belief.

Was Ptolemy also a prophet? Ptolemy points to the conjectural nature of pure theological thought and warns of the shortcomings of conjecture without reason. The warning went unheeded as Europe followed Church doctrine—reasoned knowledge is not the pathway to God; rather, all knowledge comes from God with the Church as the human being's sole liaison. It would be 1300 years before Europe launched its own mathematical inquiry into the prime mover's design.

PTOLEMY AND THE *Almagest*

While culturally accepting Hellenism, Ptolemy was ethnically Egyptian. Ptolemy was among the most prodigious of the ancients with an apparent capacity for quick calculation. Perhaps the reader recalls the movie *Rain Man*, in which an idiot savant is able to perform complex mathematical computations with astounding speed. There are actually individuals with this capacity who, unlike the Rain Man, are fully functional in other capacities. The mathematician Jon Von Neumann (1903–1957) was one such individual, so we might say that he inherited the Ptolemy gene, for it is highly likely that Ptolemy was another. This capacity would be necessary for his wide-ranging, computationally intensive works.

Ptolemy was the worthy inheritor of the brilliant astronomer Hipparchus (190–120 B.C.), a predecessor at the University of Alexandria. For reasons given below, Hipparchus added features onto Apollonius' epicycles and proposed a qualitative description. The Ptolemaic system is that of Hipparchus. Then, why is it called the "Ptolemaic system" and not the "Hipparchic system?" Because there is a long distance between the qualitative description of Hipparchus and the actual calibration of the model to data. Either Hipparchus was stymied, or the computational effort was so enormous that he never completed the walk. But Ptolemy did, and so it is the Ptolemaic system.

In *The Almagest*, Ptolemy proves himself a master of geometry and computation. He begins by developing a "table of chords," which in modern parlance is a table of trigonometric functions. The table gives the trigonometric functions for angles between 0 and 90 degrees in intervals of $\frac{1}{4}$ degree, that's 360 entries, and he performs the calculation in the sextimigecimal system (base 60) out to three entries. That is to a precision of one in 60^3 or $\frac{1}{216,000}$. The entries require the use of the Pythagorean theorem, trigonometric half-angle formula, and trigonometric sums formula. These intermediate formulas, in turn, all require the calculation of square roots. In addition, Ptolemy compiles the positions of stars in the zodiac, and tabulates the motions of the sun, the moon, and five planets about Earth. Ptolemy, without paper and pencil (perhaps like Archimedes, he scratched out his mathematics in the sand), using an immature numerical system that did not contain the critical element zero, just keeps calculating. As an example of Ptolemy's fixation with computation, Ptolemy gives the Egyptian annual (365 days) movement of the sun's longitude at $359 + \frac{45}{60} + \frac{24}{60^2} + \frac{45}{60^3} + \frac{21}{60^4} + \frac{8}{60^5} + \frac{35}{60^6}$ degrees[10]. For those without the Ptolemy gene, compilation of these results would require more time than a life has to give, and the task would be insurmountable.

While it is easy to be impressed by the magnitude of Ptolemy's calculations, the quality is equally impressive. Both are on display in his calibration of model parameters to match observations of Mars. Anyone with experience in fitting parameters to models knows the pitfalls, particularly when the equations are nonlinear. Today we seek refuge in the computer and use an optimization process to minimize error. Ptolemy develops a skillful iterative method that uses a complex geometric argument as its intermediary. (Ptolemy would once again rely on the Ptolemy gene to execute the exhaustive computations of his iterative process.) The reader may consult Ptolemy's works to judge Ptolemy's skills. Below, is a presentation of Ptolemy's Mars that utilizes Cartesian coordinates so that is more accessible to the modern reader. It is the sun that illuminates Mars' motion, so we begin with Ptolemy's description of the sun.

Ptolemy's Sun

To understand Ptolemy's model of Mars' motion, it is necessary to begin with his model for the sun's motion. The simple model of Eudoxus does not quite match the standard of Ptolemy. Eudoxus' model requires that the sun's oscillatory motion be perfectly symmetric about the equator, but it is not. Every year, the sun graces the Northern Hemisphere for more days than it graces the Southern Hemisphere. Ptolemy accounts for the difference by centering the circular orbit of the sun, not on Earth, but at a point slightly removed from Earth. The circular orbit is known as the *deferent*. Figure 5.1 illustrates the configuration of the deferent with respect to Earth. In Figure 5.2, the North's summer falls on the negative side of the x axis while the North's winter falls on the positive side. The deferent's arc on the positive side is shorter than that on the negative side, and as the sun maintains a uniform speed about its deferent, it passes through winter more quickly than summer.

---

[10]The sextimigecimal system displays itself. The accuracy of the computation is to within 1 in $60^6 = 46,656,000,000$. The computation itself demonstrates that the solar year in which the sun completes one revolution about Earth is a bit longer than 365 days.

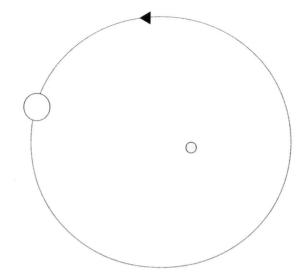

Figure 5.1. The earth is not the center of the sun's deferent.

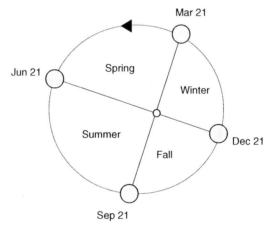

Figure 5.2. The arc associated with the northern winter is the shortest among the seasons, yielding the shortest season.

By choosing the correct center for the deferent, one can calibrate the model of the sun's motion to actual observations of the sun's motion. Hipparchus made the choice that Ptolemy accepts. We will not go through the calculations, but state that Hipparchus found that the sun passes closest to Earth in early January, while the sun is at its most distant in early July, giving the points of the sun's perigee or perihelion (closest point) and apogee or aphelion (farthest point). He also determined that the radius of the sun's deferent is about 24 times the distance between the center of the sun's deferent and Earth.

This juxtapositioning of the sun's orbit on its own upset orthodoxy; Earth was no longer at the universe's absolute center. By our modern standards, this minor displacement is irrelevant. We know that Earth is not the universe's core. But in the days of Hipparchus and Ptolemy, this must have stirred significant controversy. The minor displacement is a leap toward truth-Hipparchus and Ptolemy sacrifice orthodoxy on behalf of observation and not the other way around. Centuries after the Ptolemaic model had congealed into its own social orthodoxy, others would leap away from convention and topple the Ptolemaic universe.

Ptolemy's Mars

The Ptolemaic system relies heavily on Ptolemy's knowledge of retrograde motion, in particular the alignment between the sun, the earth, and the planet in retrograde. Ptolemy viewed retrograde through a geocentric perspective and had a clear picture of its occurrence in mind; this was his starting point. To understand his choices, an understanding of the cause of retrograde is necessary, and so we begin with a description that suffices for our purposes. Since the modern reader is much more comfortable with a heliocentric arrangement, the description is within the heliocentric framework.

The sun sits at the center of an amphitheater while, like well-behaved admirers who know their place, the planets revolve about the sun, hoping to catch their idol's attention. The closer planets, well positioned to receive attention, excitedly revolve about the sun with greater speed than do the more distant planets, while the stars outside the amphitheater do not move, but rather they, like the sun, rest. As Earth is closer to the sun than Mars, it revolves about the sun with greater speed than Mars. With the sun's absence during the night, eyes might peer toward Mars and measure its movement with respect to the stationary stars. Figure 5.3

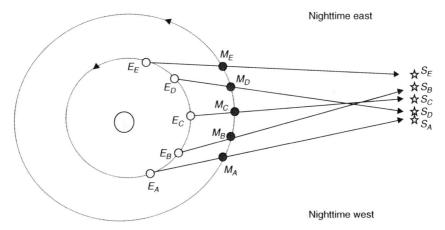

Figure 5.3. Mars' progression through the celestial sphere. The subscripts denote the time sequence: *A*—first, *B*—second, *C*—third, *D*—fourth, *E*—fifth and last.

illustrates the progression of Mars through the celestial sphere before, at the time that, and after Earth centers itself between the sun and Mars.[11]

Figure 5.3 depicts five time-sequenced positions, with mark *A* occurring first and mark *E* occurring last. Perfect alignment between the sun, Earth, and Mars occurs at the third mark, where Earth is at E_C, Mars is at M_C, and Mars' neighboring star in the celestial sphere is S_C; that is, the Earthbound observer sees Mars in the proximity of S_C. The chain of stars S_A through S_E indicates Mars' apparent motion through the celestial sphere. The chain begins moving in an easterly direction between the first and second marks. Then, at the second mark, the chain and along with it Mars' apparent motion, shifts direction, moving back toward the western sky. This western movement continues until the fourth mark (M_D), when Mars once again resumes its eastward drift. Retrograde motion occurs between the second and fourth marks when Mars moves in opposition to its dominant direction. At the third mark, during the alignment, Mars is close to the midpoint of its retrograde motion. This is critical to Ptolemy's model, and so we once again state and later apply the fact that the alignment of the sun, Earth, and Mars with Earth in the middle, so that the sun and Mars are in opposition, centers the retrograde motion.

While Ptolemy understood the critical relation between retrograde motion and the alignment of the sun, Earth, and Mars, he attributes the alignment to movement by the sun and Mars, and allows Earth to rest. It is Apollonius' epicycle that the geocentrist relies on. The epicycle is a secondary circle that revolves about Mars' deferent while Mars rotates about the epicycle. Retrograde

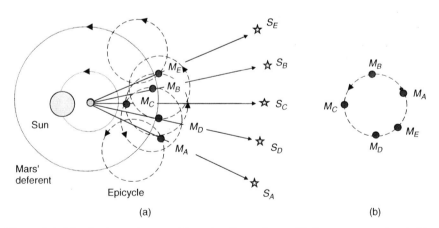

Figure 5.4. Mars' retrograde caused by epicyclic motion, with time sequence denoted by subscripts *A*—first, *B*—second, *C*—third, *D*—fourth, *E*—fifth and last, showing mars' positions (a) in celestial sphere and (b) on epicycle.

[11] Eudoxus' universal sphere motivates our interpretation of the universal sphere. It is a sphere centered around Earth of unlimited radius on which one can project the position of any star or planet from the perspective of the sphere's center. One determines the position of Mars within the universal sphere by establishing its location with respect to the fixed stars.

is the result of a universe that rests on gears, each having a specific role in orchestrating the motion of the heavens, each rotating with regularity.

Figure 5.4 a geocentrists' version of Figure 5.3, illustrates the geocentric vision of the coupling between Mars' deferent and epicyclic motions. From the perspective of the Earthbound eye, the motion of Mars as it dances through the stars is identical in both figures. Initially at mark A, there is a general eastward drift as the epicycle makes its way clockwise about the deferent. But at the second mark, mark B, Mars begins its retrograde motion. The western motion of Mars caused by its rotation about the epicycle dominates the eastward drift of the epicycle's center, causing the Earthbound observer to view a shift in Mars' motion. When the sun and Mars are in opposition with Earth in between, the third mark (C), Mars is in the middle of its retrograde behavior, and at the fourth mark (D), Mars gets back on track and resumes its eastward drift. As noted above, Mars' pathway through the celestial sphere from star A through star E is identical to the pathway displayed in Figure 5.3. As with the heliocentric version, perfect alignment between the sun, Earth, and Mars centers the retrograde motion. Note that Figure 5.4 gives the sun's position only at the third mark.

To demonstrate the correctness of their model, the geocentrist must specify the nature of both the motions, that of the epicycle's center, henceforth called *Mars' mean*, about its deferent, and that of Mars' motion about its epicycle. We begin with the motion of Mars about its epicycle. For this it is necessary to introduce the anomaly, the angle of Mars with respect to its deferent (see Figure 5.5). Because of our familiarity with Cartesian coordinates, it is natural to measure this angle with respect to the 3:00 o'clock position on the epicycle. Ptolemy, who lived

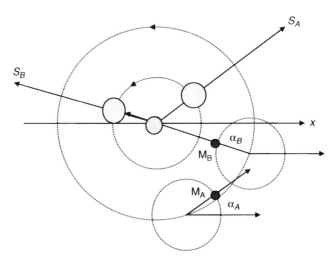

Figure 5.5. Coordination of Mars' anomaly with the solar angle. Positions of bodies at two different times, A and B. Opposition occurs at time B.

1500 years before Descartes[12], this was not a natural selection. Indeed, Ptolemy measures the anomaly with respect to Mars' aphelion (the line between Earth and the farthest point on the epicycle). Ptolemy's choice yields insight, which we highlight below. Nevertheless, we continue to measure the anomaly using the Cartesian standard.

Ptolemy chooses Mars' anomaly to be the same as the sun's longitude. There are several consequences of this choice. The first consequence is that the angular speed of Mars about its epicycle is the same as the angular speed of the sun's mean about its deferent, 360 degrees per solar year. The second consequence is that the line between Mars' mean and Mars' actual position is always parallel to the line between Earth and the sun. At opposition, this choice causes Mars, Mars' mean, Earth, and the sun to be collinear with the sun and Mars in opposition from Earth. Figure 5.5 illustrates the coordination between the solar angle and Mars' anomaly; the two are always equal so that Mars' position with respect to the 3:00 o'clock position on its epicycle is the same as the sun's position with respect to the 3:00 o'clock position on the sun's deferent. Note the alignment of the sun, Earth, Mars, and Mars' mean at opposition, mark B (M_B).

Let us at first explore a possibility with two simplifying assumptions. We first assume that the sun's deferent is centered on the earth and that the sun traverses the heavens at uniform speed. This assumption is nearly in agreement with observation and does not affect the argument concerning Mars' motion. The assumption facilitates other calculations. Accordingly, for the remainder of this chapter, we maintain this assumption.

The next simplifying assumption is that the center of Mars' deferent lies at the center of Earth and that Mars' mean speed about its deferent is slower than the sun's speed about its deferent. The sun then travels in excess of one cycle between central retrogrades. Let's try to calculate the time lapse between two occurrences of alignment in opposition. As Figure 5.6 illustrates, by symmetry the time lapse between two such alignments is always the same. This fact does not agree with the observations that Ptolemy had available, so Ptolemy knew that the assumptions of an Earth-centered deferent and constant speed about the deferent were wrong. Before continuing to the devices that Ptolemy employed to address this discrepancy, we nevertheless carry out the calculation and introduce notation as it is instructive. Concerning notation, we follow Ptolemy in measuring angles by degrees as opposed to radians.

We assume that opposition occurs at time t_0; that is, at time t_0 Mars' mean, Mars, Earth, and the sun are collinear with Mars and the sun in opposition to Earth. We seek future times at which opposition occurs. Let $M = (M_1, M_2)$ be the coordinates of Mars, $m = (m_1, m_2)$ be the coordinates of Mars' mean (the center of Mars' epicycle), and $S = (S_1, S_2)$ be the coordinates of the sun. Furthermore, let α be the angle of Mars' anomaly, θ be the longitude of Mars' mean, and λ be the solar longitude. Finally, we assume that the radius of Mars' deferent

---

[12]Descartes developed the Cartesian coordinate system to unite algebra and geometry into a single framework.

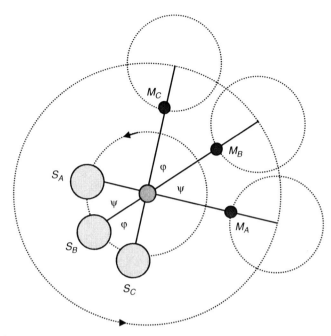

Figure 5.6. Three consecutive oppositions with Earth at the center of both the solar deferent and Mars' deferent. By symmetry the time between oppositions is the same, and so the angles φ and ψ are also the same.

is 1, the radius of Mars' epicycle is r_e, and the radius of the sun's deferent is r_S (see Figure 5.7). Then the position of Mars' mean, Mars, and the sun satisfy the following equations:

$$(m_1, m_2) = (\cos \theta, \sin \theta) \tag{5.1}$$

$$(M_1, M_2) = (m_1 + r_e \cos \alpha, m_2 + r_e \sin \alpha) \tag{5.2}$$

$$= (\cos \theta_m + r_e \cos \alpha, \sin \theta + r_e \sin \alpha) \tag{5.3}$$

$$(S_1, S_2) = (r_S \cos \lambda, r_S \sin \lambda) \tag{5.4}$$

$$= (r_S \cos \alpha, r_S \sin \alpha) \tag{5.5}$$

Note that the final equality uses Ptolemy's relation $\alpha = \lambda$. The opposition of the sun and Mars with respect to Earth occurs whenever $\theta = \alpha - (2n + 1)180$ in which n is any integer, that is, whenever the reduced angles differ by 180 degrees as in Figure 5.6.[13]

---

[13]We choose the longitude for the mean of Mars to be less than the anomaly because the longitude's angular speed is lower.

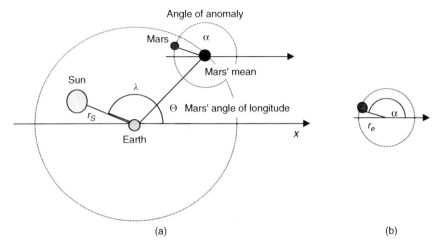

Figure 5.7. Earth-centered model parameters: (a) parameters on deferent (note that solar angle of longitude λ is identical to Mars' anomaly α; (b) Mars' position on its epicycle.

The rate of change in α is known at 360 degrees per year, and we set the rate of change of θ at ω degrees per year, in which $\omega < 360$. With this notation the following relations hold for the angles α and θ:

$$\alpha = 360t + \alpha_0$$
$$\theta = \omega t + \theta_0$$
$$= \omega t + (\alpha_0 - 180)$$

where the following conventions have been used. The value t is the time measured in solar years, α_0 is the initial anomaly at time t_0, and θ_0 is the initial longitude of Mars' mean. The final equality follows from the assumption that Mars is initially at its central retrograde position, $\theta_0 = \alpha_0 - 180$. Using the relations between the angles and the time allows one to solve for the occurrences of opposition:

$$\theta = \alpha - (2n + 1)180$$
$$\omega t + (\alpha_0 - 180) = 360t + \alpha_0 - (2n + 1)180$$
$$\omega t = 360t - 360n$$
$$(360 - \omega)t = 360n$$
$$t = \frac{360}{360 - \omega}n$$

Opposition occurs at all integer values of n and the time difference between intervals is the constant value $\frac{360}{(360-\omega)}$ solar years.

TABLE 5.1. Occurrences of Opposition

Observation	Year	Month	Day	Hour	Years Elapsed
A	15 of Hadrian	Tybi	27	1:00 A.M.	0
B	19 of Hadrian	Pharmouthi	6	9:00 P.M.	4.191324
C	2 of Antonine	Epiphi	12	10:00 P.M.	8.454452

Table 5.1 presents Ptolemy's recordings of consecutive occurrences of opposition from his own observations. The final column gives the number of years[14] from the first occurrence. Note that the time difference between occurrences is not constant; whereas the time difference between the first and second occurrence is about 4.19 years, the time difference between the second and third occurrences is about 4.26 years. While one could choose ω to match the elapsed time between the first and second observations, it is not possible to then match the next elapsed time. One cannot calibrate the simple Earth-centered model to the given data. It was Ptolemy's task to enhance the model and match it with his observations. Anyone familiar with modeling recognizes what Ptolemy must do. He must increase the degrees of freedom in order to match the model to the data. Below, we follow Ptolemy's method to its conclusion.

The Eccentric and the Equant. To explain the asymmetry between seasons, recall that Hipparchus set the center of the sun's deferent at a distance from Earth. Similarly, to explain asymmetries between observations of retrograde, Ptolemy offsets the center of Mars' deferent so that it lies at a yet undetermined distance from Earth. The off-centered circle is referred to as an *eccentric*, but as the term applies to any off-centered circle, we will continue to call the orbit of the mean the *deferent*. As an arbitrary choice, we set the center of the deferent on the x axis.

The off-centered deferent alone would not yield a model in agreement with observations; another device was necessary. Hipparchus then suggested that while Mars' mean trots along the deferent, the vertex that one uses to measure Mars' mean longitude θ sits at neither Earth's center nor the center of Mars' deferent. Instead, the vertex, known as the *equant*, sits at another point on the x axis and Ptolemy sets the center of the deferent precisely halfway between Earth and the equant. Ptolemy is less than succinct in his explanation of this alignment. He assures the reader that the choice is consistent with observations. However, maintaining the configuration, not only for Mars but also for Venus, Jupiter, and Saturn, does seem to be a stretch. Perhaps the midway compromise was a compromise between the need to match the data and the complexity of the calculations required to more precisely place the deferent's center. But perhaps, Ptolemy had insight that he could not explain. As we shall see, Kepler was initially skeptical, but then the choice is central to Kepler's discovery of the

---

[14]Egyptian years are fixed at 365 days.

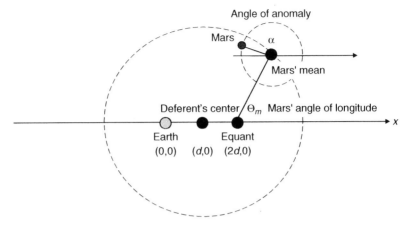

Figure 5.8. Model parameters.

ellipse. Figure 5.8 illustrates the center of the deferent, the equant, the equant-centered longitude, and the anomaly.

It is incumbent on us to specify the coordinates of Mars, Mars' mean, and the sun with these new points of reference. While the previous coordinates for the sun [eq. (5.4)] still hold, those for the mean [eq. (5.1)] no longer apply, and the equation for Mars' coordinates [eq. (5.2)] must be modified. The mean lies at the intersection of the deferent and the ray emanating from the equant at an angle given by θ. Using the convention that the radius of the deferent is 1, the equation for the deferent is that of a circle of radius 1 centered at the position $(d, 0)$:

$$(x - d)^2 + y^2 = 1$$

The equation of the line containing the ray that passes through the equant coordinates is as follows:

$$y = \tan \theta (x - 2d)$$

Note that the coordinates of the equant are $(2d, 0)$.

The mean with coordinates (m_1, m_2) satisfies both the equations above. Solving the equations simultaneously yields the following coordinates for the mean:

$$m_1 = \left(-d \cos \theta + \sqrt{1 - d^2 \sin^2 \theta}\right) \cos \theta + 2d \qquad (5.6)$$

$$m_2 = \left(-d \cos \theta + \sqrt{1 - d^2 \sin^2 \theta}\right) \sin \theta \qquad (5.7)$$

Note that the equations yield real values as long as $d < 1$, meaning that the distance of the deferent's center from Earth is smaller than the deferent's radius.[15]

The equation for Mars' actual coordinates [eq. (5.2)] is correct if one uses the values m_1 and m_2 of equations (5.6) and (5.7).

Before we continue, one feature of the model is noteworthy. As the longitudinal angle θ is not centered at the center of Mars' deferent, the speed of Mars' mean about the deferent is not uniform. Figure 5.9 illustrates the manner in which the speed varies. The equant-centered angles A and B are equal, so Mars' mean travels along each segment embraced by the angles' arms in an equal time. Because the segment embraced by B is longer than that embraced by A, the speed of Mars' mean is faster along the segment embraced by B. In general, the speed of Mars' mean decreases as Mars' mean approaches the equant and increases as Mars' mean recedes from the equant. As we will see in the next chapter, Copernicus found this feature most distasteful, desiring instead uniform circular motion. Kepler, on the other hand, viewed this feature favorably.

The Martian Year. Ptolemy's Martian year is the time that Mars' mean requires to trot once about the deferent. This is the same as the requirement for the longitude θ to advance by 360 degrees. Ptolemy uses an ingenious relationship between his choice of angle to represent the anomaly (referencing the apogee) and the occurrence of opposition. Figure 5.10 illustrates this relationship. The

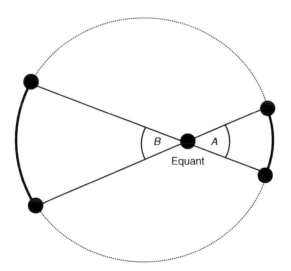

Figure 5.9. Equal angles about the equant indicate equal travel times for Mars' mean about the deferent. The size of the arc increases with distance from the equant corresponding to a higher speed.

---

[15]A quadratic formula yielding two solutions arises. We choose the solution lying on the relevant ray.

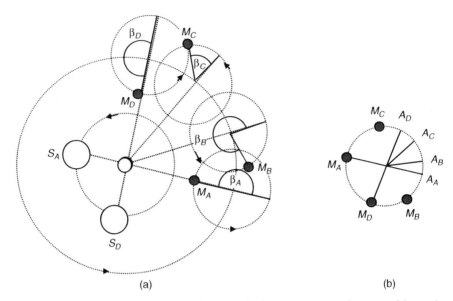

Figure 5.10. The change in Ptolemy's anomaly between consecutive oppositions; the anomaly completes one full circuit as both the initial and final anomalies are 180 degrees: (a) four observations; (b) corresponding positions of apogee and Mars on epicycle.

angle β is Ptolemy's measurement of the anomaly. As the figure illustrates, the angle executes one cycle between occurrences of opposition.

Because the speed of Mars' mean is not uniform about the deferent, the angle β does not vary with uniform speed. Indeed, if it did so, then the time between occurrences of central retrograde would be constant. However, Ptolemy reasoned that the rate of change in β has a long-run average. He also correctly reasoned that the long-run average of the rate of change in the angle β would be the same as the constant rate of change given by the angle β_Q of Figure 5.11:

$$\beta_Q = \alpha - \theta \tag{5.8}$$

By using a long-term historical record that provides the long-run changes in both α and β, the long-run change in θ is available. Dividing this change by the timeframe of the historical record yields the rate of change in θ. Consistent with previous notation, we indicate the rate of change in θ by ω. Once ω is known, one can calculate the Martian year. The following equations express the relevant relations:

$$\theta_0 = \alpha_0 - \beta_{Q0} \tag{5.9}$$

$$\theta_1 = \alpha_1 - \beta_{Q1} \tag{5.10}$$

$$\Delta\theta = \theta_1 - \theta_0 \tag{5.11}$$

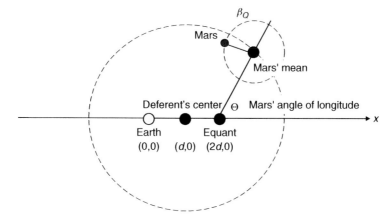

Figure 5.11. The angle β_Q.

$$= (\alpha_1 - \alpha_0) - (\beta_{Q1} - \beta_{Q0}) \tag{5.12}$$

$$\omega = \frac{\Delta\theta}{T} \tag{5.13}$$

in which the subscript 0 denotes observations at the beginning of the historical record, the subscript 1 denotes observations at the end of the historical record, and T denotes the time difference between the initial and final observations.

Ptolemy had a historical record available. Hipparchus had proposed this calculation and initiated the necessary observations. The record indicates 38 occurrences of opposition in 79.0095615[16] solar years or equivalently 37 cycles of the angle β during the timeframe $T = 79.0095615$. The angle β then changed by $37 \times 360 = 13{,}320$; that is, $\beta_{Q1} - \beta_{Q0} = 13{,}320$. We know that the angle α changes by 360 degrees in one solar year's time, so over the time range of observations, α changed by $360 \times 79.0095615 = 28{,}443.44214$ degrees; specifically, $\alpha_1 - \alpha_0$. Using equation (5.11), we find that the angle θ changed by 15,123.44214 degrees, from which one obtains the rate $\omega = \frac{15123.44214}{79.0095615} = 191.4128094$ degrees per solar year [eq. (5.12)]. With this rate of change in the equant-centered longitude, one obtains the Martian year at $\frac{360}{191.4128094} = 1.880752$ solar years.[17]

Ptolemy's Juggernaut. Ptolemy set the objective of determining all of the parameters of his model in a manner that is consistent with the observations of

---

[16]Ptolemy presents the historical record in Chapter IX of *The Almagest*. I backed into the number of years in order to match the rate of change in the Martian longitude that my translated version of *The Almagest* presents. The number of solar years differs slightly from that in the numbers I read where it sets the time frame at 79 solar years, 3 and $\frac{13}{60}$ days or equivalently, 79.008807 solar years. Confusing matters further, the translation of *The Almagest* states that the equivalent number of days is $28857\frac{53}{60}$, which is inconsistent with the number of solar years.

[17]The preceding calculation is made with a solar year equaling 365.25 days. Below, calculations are made by the year in accordance with the ancient Egyptian calendar that fixes each year at 365 days. The rate of rotation with respect to the Egyptian year is $\omega = 191.281794599$.

Table 5.1. For convenience, we once again write the full model and emphasize the variables that are time-dependent:

$$M_1(t) = m_1(t) + r_e \cos \alpha(t) \tag{5.14}$$

$$M_2(t) = m_2(t) + r_e \sin \alpha(t) \tag{5.15}$$

$$m_1(t) = \left(-d \cos \theta(t) + \sqrt{1 - d^2 \sin^2 \theta(t)}\,\right) \cos \theta(t) + 2d \tag{5.16}$$

$$m_2(t) = \left(-d \cos \theta(t) + \sqrt{1 - d^2 \sin^2 \theta(t)}\,\right) \sin \theta(t) \tag{5.17}$$

$$S_1(t) = r_S \cos \lambda(t) \tag{5.18}$$

$$S_2(t) = r_S \sin \lambda(t) \tag{5.19}$$

$$\lambda(t) = \lambda_A + \omega_\lambda t \tag{5.20}$$

$$\alpha(t) = \lambda(t) \tag{5.21}$$

$$\theta(t) = \theta_A + \omega_\theta t \tag{5.22}$$

The model variables are

t	Time measured in solar years
$(M_1(t), M_2(t))$	Coordinates of Mars
$(m_1(t), m_2(t))$	Coordinates of Mars' mean (center of Mars' epicycle)
$(S_1(t), S_2(t))$	Coordinates of the sun
$\lambda(t)$	Solar longitude
$\alpha(t)$	Mars' anomaly
$\theta(t)$	Mars' equant-centered longitude

Note that the model centers the sun's deferent on Earth.
There are eight model parameters:

r_e	Radius of Mars' epicycle
d	Distance from center of Mars' deferent to Earth, which is also half the distance from Mars' equant to Earth
r_S	Radius of the sun's deferent
λ_A	Initial solar longitude, which is also Mars' initial anomaly
θ_A	Initial longitude of Mars' mean
ω_λ	Angular speed of the sun's longitude
ω_θ	Angular speed of Mars' equant-centered longitude

An additional parameter has already been set; the radius of Mars' deferent is 1.
Ptolemy must find values for all of the parameters that ensure coincidence between the model and the observations. Aside from the radius of the deferent of Mars', two others have been set: the angular speed of the sun's longitude,

$\omega_\lambda = 360$ degrees per solar year and the angular speed of Mars' equant-centered longitude, which was found above to be 191.281794599 degrees per year. The values of the solar radius and the radius of Mars' epicycle do not impact the observations of Table 5.1. Table 5.1 gives observations of opposition, and only the sun's longitude along with Mars' mean and Mars' anomaly matter; the observations would occur with the correct solar longitude, location of Mars' mean, and anomaly regardless of the solar and epicyclic radii[18]. Accordingly, we will not consider the solar radius or the radius of Mars' epicycle. There are only three remaining parameters to specify: d, the distance from the center of Mars' deferent to Earth; λ_A, the initial solar longitude, which is also Mars' initial anomaly; and θ_A, the initial longitude of Mars' mean. Following Ptolemy, we enforce consistency between the observations and the model by selecting appropriate values for the undetermined parameters.

Figure 5.12 illustrates the model parameters and notation. The figure places Mars' mean at several observations of central retrograde; the mean lies on the line that passes through the sun, Earth, and Mars with the sun and Mars in opposition from Earth. The figure illustrates the notation that is used below. A variable with a letter as a subscript indicates the observation associated with that variable. For example, the figure illustrates Mars' mean longitude at the second observation θ_B and the sun's longitude at the third observation λ_C. With this notation, the parameters that we must determine are θ_A, Mars' initial longitude λ_A, and d.

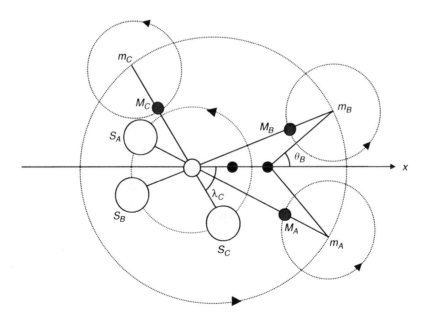

Figure 5.12. Three consecutive oppositions. Angles θ_B and λ_C are displayed.

---

[18]The radii could not be too big. The sun always remains the closest body to Earth.

Coincidence of the model and observations requires that the model's time lapse between occurrences of opposition is in agreement with the observed elapsed times of Table 5.1. Ptolemy requires a solution method to determine the parameters that ensure coincidence. Before continuing we remark on the number of free parameters. While there appear to be three free parameters, there are in fact only two, either d and λ_A or d and θ_A. To see this, let us assume that the two parameters d and λ_A are determined. Then the remaining angle θ_A is not free, but must be set by the condition that the observations are in opposition and only one value for θ_A modulo 360 degrees meets the requirement. Using a similar argument, one notes that a determination of the parameters d and θ_A sets the parameter λ_A.

As noted below, only two data points are used to determine two free parameters, the time interval between the first and second observations, and then the time interval between the first and third observations. Three observations are necessary to gather these two data points. Note that once the position of the sun and Mars' mean are correctly determined, one reconciles Mars' actual position by setting the initial value of Mars' anomaly equal to the initial value of the solar longitude, $\alpha_A = \lambda_A$. Accordingly, in the calculations below we need not further consider Mars' actual position, only its mean position.[19]

Table 5.2 gives the difference in the equant-centered Martian longitude $\Delta\theta$, where the difference is with respect to the first observation. This value is determined by multiplying the rate of change in θ (i.e., ω_θ) by the elapsed time between the observations. An equivalent calculation is made for the difference in solar longitude: $\Delta\lambda = \Delta\alpha$. Table 5.2 also displays the reduced angles, whereby the corresponding angle is reset to zero whenever it surpasses a multiple of 360 degrees. As noted above, the table gives two data points: the time interval between the first and second observations, and then the time interval between the first and third observations.

We begin by establishing a relation between the distance from Earth to the equant $2d$ and the remaining parameters. At central retrograde, when Mars' mean is in opposition to the sun, the coordinates of Mars' mean with reference to the

TABLE 5.2. Differences in Critical Angles

Observation	Year	Years Elapsed	$\Delta\lambda = \Delta\alpha$	$\Delta\theta$
A	15 of Hadrian	—	—	—
B	19 of Hadrian	4.191324	1507.83 → 67.83	801.73 → 81.73
C	2 of Antonine	8.454452	3041.57 → 161.57	1617.20 → 177.20

[19]The reader should note that none of the calculations that follow utilize the radius of the epicycle r_e. As described above, this means that coincidence between the observations of the table and the model is independent of the radius. Ptolemy then has one extra parameter that he may freely choose to match additional observations, separate from Table 5.1. The reader may consult Ptolemy's *Almagest* for details.

solar longitude are

$$m_1 = -h \cos \lambda$$
$$m_2 = -h \sin \lambda$$

where h is the distance between Earth and Mars' mean, $h = \sqrt{m_1^2 + m_2^2}$.

Equations (5.15) and (5.16) are used to determine the following expression for h:

$$h = \left\{ \left[\left(-d \cos \theta + \sqrt{1 - d^2 \sin^2 \theta} \right) \cos \theta + 2d \right]^2 \right. $$
$$\left. + \left(-d \cos \theta + \sqrt{1 - d^2 \sin^2 \theta} \right)^2 \sin^2 \theta \right\}^{1/2} \qquad (5.23)$$

Mars' mean resides on the line $m_1 = m_2 \cot \theta + 2d$, where $2d$ is the distance from Earth to the equant.

Substituting the expressions for Mars' mean coordinates into the line and simplifying yields the relation for $2d$ that we are seeking:

$$-h \cos \lambda = -h \sin \lambda \cot \theta + 2d$$
$$2d = h(\sin \lambda \cot \theta - \cos \lambda)$$
$$2d \sin \theta = h(\sin \lambda \cos \theta - \cos \lambda \sin \theta)$$
$$2d \sin \theta = h \sin(\lambda - \theta)$$
$$2d = \frac{h \sin(\lambda - \theta)}{\sin \theta} \qquad (5.24)$$

Equation (5.24) is valid at all three observation points. Equating the result at the first observation A with the result at either of the other observations and simplifying yields an expression relating the angle $\lambda_A - \theta_A$ with remaining parameters:

$$\frac{h_A \sin(\lambda_A - \theta_A)}{\sin \theta_A} = \frac{h \sin(\lambda - \theta)}{\sin \theta}$$
$$\frac{h_A \sin(\lambda_A - \theta_A)}{\sin \theta_A} = \frac{h \sin(\lambda_A - \theta_A + \Delta\lambda - \Delta\theta)}{\sin(\theta_A + \Delta\theta)}$$
$$\frac{h_A \sin(\lambda_A - \theta_A) \sin(\theta_A + \Delta\theta)}{h \sin(\theta_A)} = \sin(\lambda_A - \theta_A + \Delta\lambda - \Delta\theta)$$
$$\frac{h_A \sin(\lambda_A - \theta_A) \sin(\theta_A + \Delta\theta)}{h \sin(\theta_A)} = \sin(\lambda_A - \theta_A) \cos(\Delta\lambda - \Delta\theta)$$
$$+ \cos(\lambda_A - \theta_A) \sin(\Delta\lambda - \Delta\theta)$$

$$\frac{h_A \sin(\theta_A + \Delta\theta)}{h \sin\theta_A} = \cos(\Delta\lambda - \Delta\theta) + \cot(\lambda_A - \theta_A)\sin(\Delta\lambda - \Delta\theta)$$

$$\cot(\lambda_A - \theta_A)\sin(\Delta\lambda - \Delta\theta) = \frac{h_A \sin(\theta_A + \Delta\theta)}{h \sin\theta_A} - \cos(\Delta\lambda - \Delta\theta)$$

$$\cot(\lambda_A - \theta_A) = \frac{h_A \sin(\theta_A + \Delta\theta)}{h \sin\theta_A \sin(\Delta\lambda - \Delta\theta)} - \cot(\Delta\lambda - \Delta\theta)$$

$$(5.25)$$

As noted above, the final expression holds with the right-hand side (RHS) evaluated at either the second or third observation. Equating the RHS at the second and third observations and simplifying results in an expression for θ_A:

$$\frac{h_A \sin(\theta_A + \Delta\theta_B)}{h_B \sin\theta_A \sin(\Delta\lambda_B - \Delta\theta_B)} - \cot(\Delta\lambda_B - \Delta\theta_B) = \frac{h_A \sin(\theta_A + \Delta\theta_C)}{h_C \sin\theta_A \sin(\Delta\lambda_C - \Delta\theta_C)}$$
$$- \cot(\Delta\lambda_C - \Delta\theta_C)$$

$$\frac{h_A[\sin\theta_A \cos\Delta\theta_B + \cos\theta_A \sin\Delta\theta_B)]}{h_B \sin\theta_A \sin(\Delta\lambda_B - \Delta\theta_B)} - \cot(\Delta\lambda_B - \Delta\theta_B)$$
$$= \frac{h_A[\sin\theta_A \cos\Delta\theta_C + \cos\theta_A \sin\Delta\theta_C)]}{h_C \sin\theta_A \sin(\Delta\lambda_C - \Delta\theta_C)} - \cot(\Delta\lambda_C - \Delta\theta_C)$$

$$\frac{h_A[\cos\Delta\theta_B + \cot\theta_A \sin\Delta\theta_B)]}{h_B \sin(\Delta\lambda_B - \Delta\theta_B)} - \cot(\Delta\lambda_B - \Delta\theta_B)$$
$$= \frac{h_A[\cos\Delta\theta_C + \cot\theta_A \sin\Delta\theta_C)]}{h_C \sin(\Delta\lambda_C - \Delta\theta_C)} - \cot(\Delta\lambda_C - \Delta\theta_C)$$

$$\left[\frac{h_A \sin\Delta\theta_B}{h_B \sin(\Delta\lambda_B - \Delta\theta_B)} - \frac{h_A \sin\Delta\theta_C}{h_C \sin(\Delta\lambda_C - \Delta\theta_C)}\right]\cot\theta_A$$
$$= \frac{h_A \cos\Delta\theta_C}{h_C \sin(\Delta\lambda_C - \Delta\theta_C)} - \frac{h_A \cos\Delta\theta_B}{h_B \sin(\Delta\lambda_B - \Delta\theta_B)}$$
$$+ \cot(\Delta\lambda_B - \Delta\theta_B) - \cot(\Delta\lambda_C - \Delta\theta_C)$$

The result for $\cot\theta_A$ is the following.

$$\cot\theta_A = \left[\frac{h_A \cos\Delta\theta_C}{h_C \sin(\Delta\lambda_C - \Delta\theta_C)} - \frac{h_A \cos\Delta\theta_B}{h_B \sin(\Delta\lambda_B - \Delta\theta_B)}\right.$$
$$\left. + \cot(\Delta\lambda_B - \Delta\theta_B) - \cot(\Delta\lambda_C - \Delta\theta_C)\right]$$
$$\div \left[\frac{h_A \sin\Delta\theta_B}{h_B \sin(\Delta\lambda_B - \Delta\theta_B)} - \frac{h_A \sin\Delta\theta_C}{h_C \sin(\Delta\lambda_C - \Delta\theta_C)}\right] \qquad (5.26)$$

If only we knew the factors h, we would have a solution. Placing the values from Table 5.2 into the RHS of equation (5.26) yields the value of θ_A. With θ_A

available, equation (5.25) yields the value of λ_A. Finally using both θ_A and λ_A in equation (5.24), one obtains the value of $2d$. However, we don't know a priori the values of h.

In *The Almagest*, Ptolemy, using far more geometric methods, confronts a similar problem. He then reveals a solution. Using an initial guess for the unknown factors that are akin to our values of h, Ptolemy establishes an initial estimate of the undetermined parameters. Then, using the estimate of the undetermined parameters, he updates the unknown factors. With the updated factors, Ptolemy once again refines his estimate of the undetermined parameters and continues in iterative fashion. We formalize and execute this iterative process.

Step 0. Choose initial values for h_A, h_B, and h_C.

Step 1. Place the latest value of h_A, h_B, and h_C in equations (5.26), (5.25), and (5.24), respectively, to determine estimates for θ_A, λ_A, and d.

Step 2. Using the latest values of θ_A, λ_A, and d, determine the longitude of Mars' mean and solar longitude at each observed position. Then use these values in equation (5.23), to update the current estimates of h_A, h_B, and h_C. If the updated estimates change materially from the previous estimates, return to step 1. Otherwise accept the estimates as the model parameters.

We perform one iterate of the algorithm and then present results in a table.

Step 0. Choose an initial values for h_A, h_B, and h_C.

If the value of d is small, then the values for h are all nearly one. Hoping that this is the case, we take $h = 1$ at each observation as the initial choice.

Step 1. Use the latest value of h_A, h_B, and h_C in equations (5.26), (5.25), and (5.24), respectively, to determine estimates for θ_A, λ_A, and d.
Equation (5.26) yields the estimate for θ_A:

$$
\cot \theta_A = \left[\frac{h_A \cos \Delta\theta_C}{h_C \sin(\Delta\lambda_C - \Delta\theta_C)} - \frac{h_A \cos \Delta\theta_B}{h_B \sin(\Delta\lambda_B - \Delta\theta_B)} \right.
$$
$$
\left. + \cot(\Delta\lambda_B - \Delta\theta_B) - \cot(\Delta\lambda_C - \Delta\theta_C) \right]
$$
$$
\div \left[\frac{h_A \sin \Delta\theta_B}{h_B \sin(\Delta\lambda_B - \Delta\theta_B)} - \frac{h_A \sin \Delta\theta_C}{h_C \sin(\Delta\lambda_C - \Delta\theta_C)} \right]
$$
$$
\cot \theta_A = \left[-\frac{\cos 177.2}{\sin 15.63} + \frac{\cos 81.73}{\sin 13.9} - \cot 13.9 + \cot 15.63 \right]
$$
$$
\div \left[-\frac{\sin 81.73}{\sin 13.9} + \frac{\sin 177.2}{\sin 15.63} \right]
$$
$$
= -0.97448
$$
$$
\theta_A = -45.740429
$$

Equation (5.25) yields the estimate for λ_A:

$$\cot(\lambda_A - \theta_A) = \frac{h_A \sin(\theta_A + \Delta\theta_B)}{h_B \sin\theta_A \sin(\Delta\lambda_B - \Delta\theta_B)} - \cot(\Delta\lambda_B - \Delta\theta_B)$$

$$= \frac{\sin 36.0}{\sin 45.7 \sin 13.9} + \cot 13.9$$

$$= 0.133313$$

$$\lambda_A - \theta_A = 7.638289 + 180$$

$$\lambda_A = 7.638289 - 45.740429 + 180$$

$$= 141.897860$$

Equation (5.24) yields the estimate for d:

$$d = \frac{h_A \sin(\lambda_A - \theta_A)}{2 \sin\theta_A}$$

$$= -\frac{\sin 187.6}{2 \sin 45.7}$$

$$= 0.092796$$

Step 2. Using the latest values of θ_A, λ_A, and d, determine the longitude of Mars' mean and solar longitude at each observed position. Then use these values in equations (5.23) to update the current estimates of h_A, h_B, and h_C. If the updated estimates change materially from the previous estimates, return to step 1. Otherwise accept the estimates as the model parameters.

We provide the calculation for h_A only. The other factors h_B, and h_C, follow in an identical manner:

$$h_A = \sqrt{m_1^2 + m_2^2}$$

$$m_1 = \left(-d\cos\theta_A + \sqrt{1 - d^2 \sin^2\theta_A}\right)\cos\theta_A + 2d$$

$$= \left(-0.09\cos 45.7 + \sqrt{1 - 0.09^2 \sin^2 45.7}\right)\cos 45.7 + .18$$

$$m_2 = \left(-d\cos\theta_A + \sqrt{1 - d^2 \sin^2\theta_A}\right)\sin\theta_A$$

$$= \left(-0.09\cos 45.7 + \sqrt{1 - 0.09^2 \sin^2 45.7}\right)\sin 45.7$$

$$h_A = 1.070834$$

$$h_B = 1.079119$$

$$h_C = 0.946415$$

TABLE 5.3. Results

Iteration	θ_A	λ_A	d
1	−45.74042905	141.8978603	0.092796342
2	−41.94622205	145.1827914	0.099407705
3	−41.57671571	145.4795767	0.099954443
4	−41.54442592	145.5051929	0.099998729
5	−41.54174508	145.5073096	0.10000231
6	−41.54152639	145.507482	0.100002599
7	−41.54150865	145.507496	0.100002623
8	−41.54150722	145.5074971	0.100002625
9	−41.5415071	145.5074972	0.100002625
10	−41.54150709	145.5074972	0.100002625
11	−41.54150709	145.5074972	0.100002625

Using a computer, the iteration method has been programmed to generate the results of Table 5.3.

INSPIRATION

Using his more geometric method, Ptolemy passes through three iterations and finds the values $\theta_A = -41.55$, $\lambda_A = 145.5$, and $d = 0.1$. With these values, the model and the observations coincide to a remarkable degree of accuracy. The discrepancies between Ptolemy's values and the final result of Table 5.3 are immaterial and well within the tolerance of rounding error. For all intents and purposes Ptolemy, hits the mark.

The difficulties that Ptolemy had to overcome stymied Ptolemy's predecessors. Ptolemy derived his method prior to the advent of elementary algebra and Cartesian coordinates. He carried out exceedingly difficult calculations without the use of a computer, calculator, or even pencil and paper. After my own battles using a computer, I am dumbfounded and cannot imagine performing this feat under the conditions of Ptolemy's times. Not only did Ptolemy carry out his method for Mars; he also carried out similar calculations for Jupiter and Saturn. In addition, he used different but equally, if not more, complex methods for Mercury and Venus.

Let us jump briefly ahead to Kepler, whose remark concerning a similar calculation of his own is a testimony to Ptolemy. Initially, Kepler didn't buy into Ptolemy's midway compromise that fixes Mars' deferent's center at half the distance to the equant. He believed that data would show that the Ptolemaic system requires seating the equant at a different location. Kepler considered the position of the equant as another parameter and set about finding this parameter in addition to the other undetermined parameters. As he adds an additional free parameter, he requires an additional data point from an additional observation at opposition. After devising an iteration scheme inspired by, but more complicated than, Ptolemy's, Kepler calculates. More than 70 iterations later, Kepler arrives at the answer. In his book *New Astronomy*, Kepler describes the toll it has taken on him:

> If this wearisome method has filled you with loathing, it should more properly fill you with compassion for me, as I have gone through it at least seventy times at the expense of a great deal of time, and you will cease to wonder that the fifth year has gone by since I took up Mars....

In performing his own calculation, Kepler found sympathy for Ptolemy, who, as noted above, carried out his calculations for all the planets. What did Ptolemy receive in return for his efforts? *The Almagest* is foundational; Ptolemy leaves a legacy that both Copernicus and Kepler require for their works. One imagines that Ptolemy believed he had crossed the divide between mathematics and theology and found perfection.

CHAPTER 6

REVOLUTIONARY

POSSIBILITIES

Charles Freeman opens his book *The Closing of the Western Mind* with two quotes, one from a fifth-century-B.C. playwright and one from a fifth-century-A.D. bishop:

> Blessed is he who learns how to engage in inquiry, with no impulse to harm his countrymen or to pursue wrongful actions, but perceives the order of immortal and ageless nature, how it is structured.
>
> —Euripides, fifth century B.C.

> There is another form of temptation, even more fraught with danger. This is the disease of curiosity .... It is this which drives us to try and discover the secrets of nature, those secrets which are beyond our understanding, which can avail us nothing and which man should not wish to learn.
>
> —Augustine, fifth century A.D.

The achievements of Chapters 2–5 flowed on the currents of Euripides. Then Augustine displaced Euripides and European science stagnated. Eleven centuries after Augustine, Euripides' spirit returned in the form of a church canon:

> So if the worth of the arts were measured by the matter with which they deal, this art—which some call astronomy, others astrology, and many of the ancients

Shifting the Earth: The Mathematical Quest to Understand the Motion of the Universe,
First Edition. Arthur Mazer.
© 2011 John Wiley & Sons, Inc. Published 2011 by John Wiley & Sons, Inc.

the consummation of mathematics- would be by far the most outstanding . . . . For who in applying himself to things which he sees established in the best order and directed by divine ruling, would not through diligent contemplation of them and through a certain habituation be awakened to that which is best and would not wonder at the Artificer of all things, in Whom is all happiness and every good?
—Nicolaus Copernicus, sixteenth century A.D.

That the Roman Catholic Church shepherded Europe into an intellectually barren era cannot be denied. However, there is some controversy concerning the time when Europe awoke from its stupor. Some point to Thomas Aquinas' (1225–1274) Aristotelian tryst as a turning point. Aquinas, a Dominican friar, lived when Europeans rediscovered the works of Aristotle. This stimulated some bloodflow into Europe's atrophied brain. One outcome was a pro-Greek movement, Averroism, a movement based on Islamic efforts to reconcile Aristotle and Islam. This open challenge to the Church's control of the intellect spawned a contra-Aristotelian–contra-Averroist movement. Aquinas sought middle ground; in his writings Aquinas brings Aristotle closer to Christianity and tugs Christianity toward Aristotle. The middle of two passionate movements is a dangerous place to be. The contra-Averroists within the Church associated Aquinas with Averroism, sullying his reputation for the remainder of his life and beyond. Still, the Church could not staunch the advance of Aristotle. Unable to contain interest in Aristotle, the Church decided to baptize him and coopt his works. Aquinas' writings became critical in this effort. Years after his death, the Church restored Aquinas' reputation, and Pope John XXII conferred sainthood on Aquinas.

By this means, Aristotle, the great seeker of knowledge, became an instrument of continued ignorance. Aristotle had seeded this role for himself in his own writings. By any measure, his output is prolific, covering all disciplines. Unfortunately, at least in the physical sciences, Aristotle presents his ideas as conclusive, requiring no further inquiry.[1] Concerning their objective to prevent the curious mind from evolving into an inquisitive mind, all the Church need do was to put forth the Bible as the repository of all theological knowledge and Aristotle as the repository of scientific and other knowledge—some editing of Aristotle was necessary to ensure that the public version did not contradict scriptures. Whereas before the fourteenth century, the Church's theology was perfect, now, with the canonization of Aristotle, all knowledge was perfect and complete. In the fourteenth century, as in the centuries before, knowledge came from God and there was little independent inquiry. Instead, what passed for scholarship was no more than commentary on Aristotelian and theological dogma that wallowed in a sea of minutiae.[2]

---

[1] Aristotle seems not to be concerned with verification of his ideas. For example, the simple experiment of dropping objects of different weights and shapes from a bridge could have tested Aristotle's hypothesis that heavier objects fall faster than light objects. Aristotle never performs the experiment and poses the hypothesis as fact.

[2] There are admirable exceptions of excellence. Perhaps the most prominent among the exceptions is Nicolas Oresme (1323–1382). Among other works, he first enunciated the law of motion for a falling object that Galileo later rediscovered.

In the summer of 1492 white smoke wafting from the site of the College of Cardinals signaled that the college had come to a decision. The newly elected pope, the Spaniard Borgia, would assume the position of a somewhat diminished office. King Ferdinand of Aragon favored the Spaniard in the same manner that an ambitious corporate politician wishes to place his own loyal men throughout the company. Toward this effort, King Ferdinand provided Cardinal Borgia with sufficient bullion to bribe his way to the Papacy.[3] The plan succeeded brilliantly, under the assumed name of Pope Alexander VI, Borgia placed the influence of the Church at Ferdinand's disposal. The election highlighted the shift of power from the Church to European monarchs.

Although the Catholic Church would rather we forget him, we will remember Borgia. As Ferdinand's agent in Rome, Borgia negated a previous papal Bull that bequeathed all the world's heathen lands to Portugal and in its stead issued his own Bull, or that requested by Ferdinand, that divvied up the world's heathens between the Spaniards and the Portuguese. When Ferdinand contended with Charles VIII of France to assume sovereignty of Sicily and Neapolitan Italy, Borgia's support assisted Ferdinand in his victory. When Ferdinand demanded that Portugal follow Spain's example and exile its Jewish population, Borgia was powerless to intercede, but to his credit, Borgia did open Church lands to Jews seeking refuge from persecution. But the reason why we will best remember Borgia and why the Catholic Church most wishes that we forget him is his sexual appetite, which made a mockery of the vows of celibacy.

Prior to Borgia, clergymen in high office had active sex lives, but they were for the most part discrete. Borgia, to the contrary, was most public. Such was his reputation that the public knew his lovers, and his offspring, and suspected him of fathering his own grandchild. His papal entertainment would have left a previous inhabitant of Rome, Caligula, in a mad state of jealousy. A diarist notes an event attended by clergymen, leading citizens, and an entourage of prostitutes. Borgia encouraged public fornication and offered a prize to the man who could boast the most ejaculations. When, at the close of the event, Borgia found himself not satiated with debauchery, he asked that a band of criminals be released into a courtyard, whereupon he and his son shot them.

Perhaps the late fifteenth century was singular in that it was the only time that a man of Borgia's well-known lust could have achieved the high office of the Pope. During the fourteenth century disasters visited Europe in the form of the plague followed by the Hundred Years' War.[4] The fifteenth century then opened with a spiritual disaster as two rival claimants to the office of the Papacy

---

[3] King Ferdinand was equally successful in placing his offspring in prominent positions. Through marriages to royal houses in England, the Netherlands, Portugal, and the Holy Roman Empire, Ferdinand created a network of political influence that supported Spain's domination of Europe in the sixteenth century.

[4] The plague left a depleted population, perhaps the European population successful greeting the midfourteenth century was half of what all those alive only 5 years earlier.

split the Church and Europe into opposing camps of allegiance.[5] The impact of misfortunes caused changes in attitudes, paving the way for the Renaissance.

Concerning attitudes toward sex, views had loosened. Having weakened its moral authority, the Church's admonitions toward a moral life did not carry the weight of its past. Europe was not afraid of going to hell, it had just been there. Sex in the fifteenth century was not an indication of human weakness to temptation, not a sinful inheritance of Adam and Eve, and it was not bound up in guilt. By the end of the fifteenth century the ancient Greek hetairas had been revived in the form of Venetian courtesans. Clergymen along with their congregations participated in both church and bedroom activities. Indeed, as a young man, Borgia had shown a special capacity to combine the passions of the bedroom with those of the church by throwing orgies within the cathedral at Sienna.

Accompanying the disrobing of Europe's clergymen was the unshackling of the European mind. The other inheritance of Adam and Eve, the sin of knowledge, was, like its physical counterpart, no longer sinful. Europeans were free to bite the apple and to enjoy it. The God of Europe was not a God who imposed limits, but one who showed possibilities. It was under this God that Borgia became pope and the Renaissance flourished.

The list of Renaissance achievements is admirable and begs the question which of all the achievements is the most remarkable. In answering this question, let us weigh the achievement's positive impact on the human condition. We are looking for the game changer, and I will take this down to a personal level when considering several candidates.

The works of the Renaissance artists are certainly noteworthy. But taking it down to a personal level, I'm not too terribly impacted by them. I don't see my lifestyle in much different terms with or without Renaissance artwork.

The Renaissance explorers are of a different nature altogether. Their discoveries altered world politics. The discoveries were first steps toward European colonization across all continents and mass migration of Europeans to the Americas. While the political impact is undeniable, it is arguable whether the world order that emerged actually benefited the human condition. The enslavement of Africans; indentured servitude of South American Indians; and dispossession of native lands in the Americas, Africa, and Asia that followed this change in the world order illustrate that the outcome was not universally favorable.

An argument in favor of the explorers' discoveries as game changers relies less on their political impact and more on their inspirational role. Aboard their nearly insignificant watercraft, the explorers challenged and conquered the oceans. To be sure, the oceans did not give way without a battle. Poseidon colluded with Lucifer, and together they confronted the explorers with ill winds, unknown dimension, a labyrinth devoid of any distinguishing features, disease, hunger,

---

[5]The event is known as the *great schism*. One claimant, Clement VII, backed by a group of French cardinals, resided in Avignon, while the other claimant, Urban VI, backed by Italian cardinals, resided in Rome. Both occupied the office of humanity's sole direct liaison with God.

ocean currents, and underwater rocks that seemed to be in the arms of spirits hell-bent on tearing through a ship's hull. Yet the explorers outwitted them, demonstrating that, indeed, God did not impose limits but showed possibilities. Did the Renaissance ride on the breeze that filled the explorer's sails? If so, the explorers' achievements paved the way for the ultimate game changer in the human condition. But their discoveries were not themselves the game changer.

There is another achievement remarkable for its political and philosophical impact. Without this achievement, Martin Luther would have been an unknown, and university bookshelves would be empty. This was the invention of neither a politician nor a philosopher, but a technologist, Johan Gutenberg (1398–1468). His invention, the printing press, does illustrate the reliance of politicians and philosophers on technology. I do see my life in different terms without technology in general and the printing press in particular. Although it was of great significance, I do not see Gutenberg's invention as the game changer. Historically, China's technology had been ahead of Europe's in many areas, and the Chinese had printed books for centuries.[6] Yet the game changer eluded the Chinese.

Imagine world leadership in the hands of the most enlightened among men. Imagine the most enlightened philosophies underpinning civilizations. Imagine peaceful coexistence among nations. Imagine beautiful artistic expression in literature, music, painting, sculpture, and architecture. Now imagine all of this without modern science. Well over 90% of humankind would live in a self-sufficient, survivalist economy by scratching the earth with a stick, planting their crops, and praying for good weather. Few would have contact with artistic expression, and equally few would find relevance in abstract philosophical teachings. On a personal level, without modern science, I see my lifestyle in significantly different terms. Science sets us apart from our preindustrial predecessors, who, given the chance to see our world, would view us as superior alien beings. As such, the most remarkable and revolutionary achievement of the Renaissance is the one that set humankind on its path toward modern science. The remainder of this book follows the path from its Renaissance originator, Nicolaus Copernicus, and his heliocentric proposal. The heliocentric proposal was the game changer.

That the earthly game changer came in an obscure fashion by way of the heavens is somewhat surprising. Somewhat more surprising is the man who initiated the effort. As presented by Arthur Koestler in *The Sleepwalkers*, Nicolaus Copernicus does not come across as someone who would share a seat at the scientific table with the other notables of this book. He is pictured as very capable and an admirable technician, but one who does not share the brilliance, creativity, scope of works, or prodigious output of the men in his company. He expresses his timid sentiments at the outset of his most famous scientific work. Yet he sits at the table with very outspoken men, Eudoxus, Archimedes, Apollonius, Ptolemy, Kepler, and Newton, right alongside Aristarchus in a well-deserved position. He deserves the position because he delivered what others did not, a coherent and convincing

---

[6]Johannes Gutenberg's invention improved on Chinese printing techniques. Gutenberg had adopted pressing instruments used for extracting vegetable oils to imprint letters and also developed a system for aligning letters.

analysis in support of the heliocentric theory. Perhaps Koestler's perception that Copernicus was less than a genius is wrong.

Although orphaned at an early age, Copernicus had a charmed youth. The Renaissance breeze was an exciting time for an intelligent young man to grow up, and Copernicus was fortunate enough to have the wherewithal to be at its vortex. After their father's death, the Copernicus orphans grew up under the care of their maternal uncle, Waltzenrode. Waltzenrode held an influential position in the Church, eventually becoming the Bishop of Warmia. Waltzenrode had special affection for Copernicus. Waltzenrode financed Copernicus' education and then secured Copernicus a canonship that provided a comfortable lifetime stipend. Waltzenrode further used his influence to secure Copernicus a leave of absence that allowed Copernicus to continue graduate studies in Borgia's Italy over a 10-year period.[7]

In Renaissance fashion, Copernicus' studies were wide-ranging. He studied medicine, law, mathematics, and astronomy. He wandered somewhat, leaving the University of Bologna prior to completing his law degree and later entered Padua University for further studies in medicine while also doing a stint at Ferrara. While Borgia upheld no moral standard and undressed Italy's women for his pleasure, he did not enforce a false dogmatic standard but allowed Italy's intellectuals to uncover their own truths. It was in Italy where Copernicus studied Ptolemy's universe. It was in Italy where there were open discussions on other possibilities. It was in Italy where Copernicus learned that ancient Greeks had considered a moving earth and a heliocentric universe. While Borgia seeded children in many women's wombs, his Italy seeded the heliocentric system in Copernicus' mind.

Copernicus left Italy in 1506. He was personal secretary to his uncle Waltzenrode in Ermland until 1512. Afterward, he assumed his duty as church canon in the town of Frauberg, which Copernicus refers to by its Greek translation, Gynopolis. In Ermland, Italy's heliocentric seeds came to life, and in Gynopolis they went into hiding. Around 1511, Copernicus circulated a pamphlet, entitled *Commentariolus* (Little Commentary). *Commentariolus* is somewhat of a marketing experiment for a forthcoming work. The pamphlet explains the difficulties that Copernicus has with Ptolemy's universe and then provides a preview of Copernicus' heliocentric world. In the introduction, Copernicus makes a promise to provide details in a larger work that he hints is already underway:

> I shall endeavor to show how uniformity of the motions can be saved in a systematic way. However, I have thought it well, for the sake of brevity, to omit from this sketch mathematical demonstrations, reserving these for my larger work.

Perhaps the marketing experiment went awry, for Copernicus just barely kept his promise. In 1543 he permitted a friend to oversee the publication of *On Revolutions* and then died on receiving a preliminary copy. In the introduction

---

[7]It is my conjecture that while in Italy, Copernicus collected his stipend. Like the parents of many professional students, Uncle Waltzenrode may have grown tired of footing the bill.

to *On Revolutions*, Copernicus explains that he had held his views privately for 36 years before publishing them. What had happened in the ensuing years?

What would be more disorienting, being forced to move midlife to a foreign country without knowledge of its language and customs, or witnessing a transformation within one's homeland that is so momentous, that the experiences guiding one since one's youth no longer seem relevant? At least in the first circumstance one would be keenly aware of one's shortcomings and could adapt. In the second, it may take a few jarring lessons before one loses one's own confidence and does not know how to proceed. An insidious energy source heated the refreshing breeze of the Renaissance and encompassed Europe in a chaotic whirlwind that ensnared Copernicus along with an entire generation.

In 1517 Johann Tetzel traveled between German towns accompanied by acrobats, musicians, and accountants. One has to imagine life in Germany as it stumbled through the Middle Ages into the Renaissance. Entertainment was sparse, and one day was indistinguishable from another. Each day, a German ate the same food, put on the same set of clothes, performed the same tasks, saw the same people, and complained about the same bad rainy weather. As Tetzel entered a town with much fanfare—with his acrobats juggling and his musicians performing—he radiated energy and excitement. Aside from his acrobats, musicians, and accountants, the expected retinue of a circus troupe, Tetzel brought with him a giant cross that he planted in the middle of a town, and a sack containing documents with the seal of Pope Leo impressed on them. Tetzel was not a circus man; he was a property salesman, selling real estate in heaven.

The documents in Tetzel's sack were indulgences that guaranteed the holder access through Saint Peter's gate. Tetzel, a Dominican friar, sold indulgences of all sorts. One could purchase an indulgence that would absolve past sins, or a more expensive indulgence that would apply to future sins. One could also purchase an indulgence on behalf of another individual, including the already deceased. All of these indulgences had the official stamp of approval from the highest authority, God's single point of contact between heaven and Earth, the occupant of the Office of the Pope, Pope Leo X. The indulgence was a very successful product, and the Church, along with the Church's banker, an influential family named Fugger, split the profits from their sales.

While enriching the Church so that it could support its Renaissance construction, the sale of indulgences indicated an underlying moral bankruptcy that did not go unnoticed by a Catholic priest, Martin Luther. When in 1617, Martin Luther challenged the Church's authority by posting his antiindulgence thesis on the wall of the Church of All Saints, like the jingling of Tetzel's ill-begotten cash, his words resonated. The words accused the Church of susceptibility to corruption. If the Church had in the past been the reflection of a perfect God, like the circle unquestionably true, it would no longer be so. Luther exposed the dents for all to see. With the assistance of the printing press, Luther's thesis circulated across Europe, setting in motion events that Luther himself could never have imagined.

The Roman Catholic Church could not ignore this challenge to papal authority, nor could the pope ignore the popularity of the challenger. In 1518, Pope Leo X summoned Luther to Rome for disciplining. Luther showed himself to be an exceptionally adept politician. He managed to shift the location of a trial from Rome, where Leo could stage-manage the event and ensure complete humiliation of Luther, to Augsburg, where local sentiment was dominant. The pope sent a legate and Luther, much to the delight of his many accompanying followers, turned the council over like an hourglass. Whereas the council was supposed to extract a confession of guilt from Luther, Luther used the occasion to place the Roman Catholic Church on trial. Luther then evaded arrest by slipping away at night. Leo had shown himself to be no Borgia. Undoubtedly, Borgia would have ensured Luther's execution, if not through legal process, then by an assassin. But Leo allowed the sheep thief to escape, and the thief would steal half of Leo's flock.

For nearly 700 years, Catholic bishops had unquestioned privileges across Europe. They ruled on Church lands, collecting taxes and holding court. Priests under their authority heard confessions and granted absolution. They were political appointees on whom their secular counterparts relied to maintain order. But in the mid-1520s, throughout Germany, Bohemia, parts of Poland, and Scandinavia, they were the foxes that angry mobs pursued and hunted down like packs of foxhounds. Once in their grip, the mobs lynched bishops, priests, monks, and nuns, and absconded with church property. Across northern Europe, the Catholic Church collapsed.

In 1519, the responsibilities of the church canon of Frauberg increased in scope from administrator, tax collector, adjudicator, and physician to administrator, tax collector, adjudicator, physician, and military commander. Nicolaus Copernicus was ordered to set up the defenses of Olzstyn. The aggressor was Prince Albert, who challenged his uncle, the Polish king, Sigismond, by rallying the Order of the Teutonic Knights to his cause. Copernicus successfully defended Olzstyn against the Teutonic Knights. A stalemate induced a truce in 1521, but tensions lingered and the carefree atmosphere of Copernicus' younger years never returned. The tensions turned religious as Albert, seeking autonomy from Sigismond, and the Holy Roman Empire converted his territory of Prussia to Lutheranism.

We might conclude that Copernicus withheld publishing *On Revolutions* because the revolutionary tumult of the times, one that explosively mixed politics and religious fervor with economics, required caution. This is a plausible explanation, but the timeline just doesn't support it. Copernicus writes that he had withheld publishing his views for 36 years, which means that Copernicus joined Aristarchus on a moving planet in 1507. By 1512 there was limited circulation of *Commentariolus*, and Copernicus indicates that by that time he already had an advanced theory with further results. Furthermore, as noted below, by 1512 Copernicus was taking observations with the specific intention of completing his work. While later observations stretch beyond 1520, it is certain that before completing the observations, Copernicus had sufficient material for a publication. Copernicus could have expanded *Commentariolus* and published it. He could have used Ptolemy's data, gone through his calculations, and published an initial work with a follow-up publication once his observations were complete. The Teutonic–Polish War did not start until 1519, and the turmoil unleashed by Luther did not affect Europe's established order until 1520. Furthermore, Pope Leo, like Copernicus, was a man molded in Borgia's Italy. He continued Borgia's policy of allowing independent intellectual development, so long as Church authority was not challenged. Nor did Leo instigate investigations into the loyalty of the clergy after Luther's challenge. Over a 7-year stretch, Copernicus sat on his results, during which time there was little hint of the subsequent chaos.

There is a perception that Copernicus feared being accused of heresy by his own church. Let us briefly address this issue. As noted, Leo was a man molded in Borgia's Italy, and he continued Borgia's policy of allowing independent intellectual development. In fact, by 1530, Copernicus' *Commentariolus* had circulated widely enough that the intelligentsia within the Roman Catholic Church was aware of Copernicus' heliocentric hypothesis. Far from decrying the theory as heresy, the Church intelligentsia, who were also molded in Borgia's Italy, seemed fascinated. In Rome, there were lectures on *Commentariolus*, and the theory had reached the top, Pope Clement VII, Leo's successor. Showing that the pope was not much different from any other man, Clement VII responded like those around him, with fascination. He arranged to be personally tutored in the topic and also asked the highly respected Cardinal Schonberg to look into the matter.

Cardinal Schonberg also looked favorably on Copernicus' theory, so much so that in 1536 Cardinal Schonberg penned a letter of admiration to Copernicus requesting further information:

> Some years ago word reached me concerning your proficiency, of which everybody constantly spoke. At that time I began to have a very high regard for you, and also to congratulate our contemporaries among whom you enjoyed such great prestige. For I had learned that you had not merely mastered the discoveries of the ancient astronomers uncommonly well but had also formulated a new cosmology. In it you maintain that the earth moves; that the sun occupies the lowest, and thus the central, place in the universe; that the eighth heaven[8] remains perpetually motionless and fixed on the outermost sphere, and that, together with the elements included in its sphere, the moon, situated between the heavens of Mars and Venus, revolves around the sun in the period of a year. I have also learned that you have written an exposition of this whole system of astronomy, and have computed the planetary motions and set them down in tables, to the greatest admiration of all. Therefore with the utmost earnestness I entreat you, most learned sir, unless I inconvenience you, to communicate this discovery of yours to scholars, and at the earliest possible moment to send me your writings on the sphere of the universe together with the tables and whatever else you have that is relevant to this subject. Moreover, I have instructed Theodoric of Reden to have everything copied in your quarters at my

[8]Reference to the eighth heaven is a Eudoxian holdover in which all planetary bodies are ranked by their distance from Earth. The stars, which are the most distant, reside in the last heaven, which is the eighth heaven.

expense and dispatched to me. If you gratify my desire in this matter, you will see that you are dealing with a man who is zealous for your reputation and eager to do justice to so fine a talent. Farewell.

—Rome, 1 November 1536

It is possible that some individuals with whom Copernicus shared his heliocentric belief did view the idea as heretical. It is possible that some individuals around Copernicus let him know of their disdain. But Cardinal Schonberg indicates a clear understanding of Copernicus' revolutionary theory and shows admiration. Copernicus had support from the highest authority within the Church and received assurances that not only would there be no charges of heresy, there would more likely be assistance with publication of the theory. The Church would treat Copernicus with respect and offer him fame.

Copernicus did not respond to Cardinal Schonberg's request. He never forwarded his works, never communicated with Theodoric of Reden, and never sought any recognition. At times, the absence of response sends a loud message. The message appears 7 years later in the introduction of his work *On Revolutions*. In the introduction, Copernicus addresses Pope Clement's successor, Pope Paul III:

> I can readily imagine, Holy Father, that as soon as some people hear that in this volume, which I have written about the revolutions of the spheres of the universe, I ascribe certain motions to the terrestrial globe, they will shout that I must be immediately repudiated together with this belief. For I am not so enamored of my own opinions that I disregard what others may think of them. I am aware that a philosopher's ideas are not subject to the judgment of ordinary persons, because it is his endeavor to seek the truth in all things, to the extent permitted to human reason by God. Yet I hold that completely erroneous views should be shunned. Those who know that the consensus of many centuries has sanctioned the conception that the earth remains at rest in the middle of the heaven as its center would, I reflected, regard it as an insane pronouncement if I made the opposite assertion that the earth moves. Therefore I debated with myself for a long time whether to publish the volume which I wrote to prove the earth's motion or rather to follow the example of the Pythagoreans and certain others, who used to transmit philosophy's secrets only to kinsmen and friends, not in writing but by word of mouth, as is shown by Lysis' letter to Hipparchus. And they did so, it seems to me, not, as some suppose, because they were in some way jealous about their teachings, which would be spread around; on the contrary, they wanted the very beautiful thoughts attained by great men of deep devotion not to be ridiculed by those who are reluctant to exert themselves vigorously in any literary pursuit unless it is lucrative; or if they are stimulated to the nonacquisitive study of philosophy by the exhortation and example of others, yet because of their dullness of mind they play the same part among philosophers as drones among bees. When I weighed these considerations, the scorn which I had reason to fear on account of the novelty and unconventionality of my opinion almost induced me to abandon completely the work which I had undertaken.

In this introduction to his work, Copernicus introduces himself and shows that his personality is consistent throughout his life—throughout the formative years when peace prevailed and possibilities seemed endless as well as the latter years, when chaos overtook Europe. He is the student who, rather than carving out his own career, spends 10 years in graduate studies. He is the young church canon who, rather than politically positioning himself for advancement, assumes an unambitious posting in a remote location. He is a scholarly man who does not wish to become the target of controversy. He finally relents and publishes his work because he knows that his impending death will spare him the scorn that he both foresees and fears. The man who instigated the game changer was an unassuming, modest man. This is one of the gems in history's mineshaft that I find most pleasing.

COPERNICUS' *On Revolutions*, THE PURSUIT OF ELEGANCE

What had caused Copernicus to dispense with Ptolemy and reformulate astronomy along heliocentric coordinates? On technical grounds, science demands the vigilant defense of truth. On technical grounds, science demands exactitude. On technical grounds, science demands consistency. Was there evidence that Ptolemy's universe did not meet these technical demands of science that caused Copernicus to drift toward Aristarchus? Concerning the technical demands, in his introduction to *On Revolutions*, Copernicus concedes that Ptolemaic methods work: "But even if those who have thought up eccentric circles seem to have been able for the most part to compute the apparent movements numerically by those means ...."

It is not technical merit that Copernicus finds lacking. There is a non-technical essence that may be considered unscientific that nevertheless influences scientists and scientific thought—call it elegance. Let us complete Copernicus' thought and find out what troubles him:

> But even if those who have thought up eccentric circles seem to have been able for the most part to compute the apparent movements numerically by those means, they have admitted a great deal which seems to contradict the first principles of regularity of movement. Moreover, they have not been able to discover or to infer the chief point of all, i.e., the form of the world and the commensurability of its parts. But they are in exactly the same fix as someone taking from different places hands, feet, head, and the other limbs—shaped beautifully but not with reference to one body and without correspondence to another—so that such parts made up a monster rather than a man.

Here, Copernicus states his problem. The Ptolemaic theory has no underlying elegance. Instead, it is unsightly, like a monster with the most unsightly feature its contradiction of the regularity of movement. Copernicus seeks a unified system of regular motions, not the polyglot of unrelated irregular motions.

When Copernicus penned these words, he was true to the original spirit with which he began his investigations. In his publication *Commentariolus* (three decades before writing the passage above presented) Copernicus informs the reader that Ptolemy's unsightly universe disturbs him:

Yet the planetary theories of Ptolemy and most other astronomers, although consistent with the numerical data, seemed likewise to present no small difficulty. For these theories were not adequate unless certain equants were also conceived; it then appeared that a planet moved with uniform velocity neither on its deferent nor about the center of its epicycle. Hence a system of this sort seemed neither sufficiently absolute nor sufficiently pleasing to the mind.

Copernicus has no qualms with Ptolemy's technical ability to match observed phenomena. In fact, Copernicus relies heavily on Ptolemy's methods. It is Ptolemy's equants that upset Copernicus' sense of aesthetics, and so Ptolemy's universe does not please Copernicus' mind. Recall from Chapter 5 that as the longitude moves uniformly about the equant, the planet's motion about the deferent, which is not centered at the equant, is not uniform. So the introduction of equants contradicts the regularity of movement about a planet's deferent, and this is precisely Copernicus' complaint three decades later in his introduction to *On Revolutions*.

Let us complete the preceding passage of *Commentariolus*:

Having become aware of these defects, I often considered whether there could perhaps be found a more reasonable arrangement of circles, from which every apparent inequality would be derived and in which everything would move uniformly about its proper center. After I had addressed myself to this insoluble problem, the suggestion at length came to me how it could be solved with fewer and much simpler constructions than formerly were used, if some assumptions were granted me. The follow in this order .... Assumption 3. All spheres revolve about the sun as their midpoint, and therefore the sun is the center of the universe.

Here, Copernicus directly informs us that his desire to eliminate the equant caused him to move Earth and fix the sun at the universe's center.

Before proceeding to our mathematical exposition of Copernicus' universe, let us dwell further on several philosophical points of relevance. While Ptolemy's equants offend Copernicus, the two do agree that exploration of the heavens is a bridge to theological discovery. Copernicus refers to God in several passages, often in terms that combine human and heavenly qualities. In addition to being the "Best and Most Orderly Workmen," God is also the "Artificer of all things," the "Universal Artisan," and the "Best and Greatest Artist."[9] Ptolemy believes that one seeks God by expanding the frontiers of knowledge. During the intervening centuries between Ptolemy and Copernicus, one did not seek to expand the frontiers of knowledge but learned what God had already made clear to leaders of the Church. In circular fashion, the intellectual landscape familiar to Copernicus had returned to the landscape familiar to Ptolemy—a landscape where science and religion were at peace with one another.

Concerning circles, it is noteworthy that Copernicus was able to shed the geocentric myth, but following the tradition of the Greeks, Copernicus fully

---

[9] I am convinced that Copernicus' Artisan is himself an astronomer since Copernicus refers to astronomy as the "noblest of all arts."

embraced the circle as the perfect shape, as the only shape worthy of God's grand design. Copernicus begins Book 1 of *On Revolutions* by revealing his reverence of circular motion:

> Among the varied literary and artistic studies upon which the natural talents of man are nourished, I think that those above all should be embraced and pursued with most loving care which have to do with things that are very beautiful and very worthy of knowledge. Such studies are those that deal with the godlike circular movements of the world.[10]

Copernicus' reverence for the circle causes him to seek the circle as the cause of irregular motions. He justifies this choice with the following false argument:

> We must confess that these movements are circular or are composed of many circular movements, in that they maintain these irregularities in accordance with a constant law and with fixed periodic returns: and that could not take place if they were not circular. For it is only the circle which can bring back what is past and over with.

Only a prejudicial preference for the circle could have clouded Copernicus' views to the extent that he would claim that no other shape could repeat its own pattern.

Copernicus' efforts were motivated by his desire to unify the motions and his distaste for the equant that upsets the purity of uniform circular motion. The mild-mannered scholar is revolutionary in his approach; he fixes the sun and moves the earth along with all the planets on revolutionary pathways about the sun. A common misunderstanding of Copernicus is that his planets travel about perfectly circular pathways. Certainly Copernicus would have preferred such an arrangement, but the "best and most orderly workmen of all" did not construct the universe in this fashion—irrefutable astronomical observations informed Copernicus so. Yet Copernicus believes that the "master workmen" crafted the universe using regular motion of the planets about perfect circles. Upholding these two principles, heliocentricity and perfectly circular motions, Copernicus unites Aristarchus with Apollonius. The planets move about a network of epicyclic gears that revolve around the sun. Like Ptolemy's "Prime Mover," Copernicus' God is also the "Perpetual Organ Grinder."

Refutation

One has to be sympathetic with Copernicus' caution. Unlike Ptolemy, who adopted the commonly accepted model of his ancestors and only had the task of taking the model to maturity, Copernicus proposes a revolutionary concept. His first task is to convince an audience that the fundamental way in which they view the universe is all wrong. While this is daunting, Copernicus is skillful.

---

[10]In Copernicus' day, *world* meant universe.

Copernicus refutes the Ptolemaic model in Section 8 of Book 1. One geocentric argument that he must address is the Ptolemaicists' fear that a moving Earth would not be capable of sustaining motion without breaking apart. Like a jujitsu master, Copernicus redirects the weight of this argument against the geocentric arrangement. Copernicus answers that the geocentric model demands that the heavens move with far greater fury than Earth's required motion in the heliocentric model. For example, given their distance from Earth the 24-hour circular motion of the geocentric model requires them to move at frenzied speeds. It is more likely that a geocentric universe itself would break apart, than that Earth would break apart from a far less rapid motion demanded by a heliocentric model. Copernicus finds the geocentric response that the universe is finite, so the stars are bound within a finite envelope beyond which there is nothing somewhat lacking. He points out that "in that case it is rather surprising that something can be held together by nothing." Copernicus concludes his refutation with the following paragraph:

> In addition there is the fact that the state of immobility is regarded as more noble and godlike than that of change and instability, which for that reason should belong to Earth rather than to the world (universe). I add that it seems rather absurd to ascribe movement to the container or that which provides the place and not rather to that which is contained and has a place, i.e., Earth. And lastly since it is clear that the wandering stars (planets) are sometimes nearer and sometimes further from the Earth, then the movement of the one and the same body around the center—and they mean the center of Earth—will be both away from the center and toward the center. Therefore it is necessary that movement around the center should be taken more generally; and it shall be enough if each movement is in accord with its own center. You see therefore that for all these reasons it is more probable that the Earth moves than that it is at rest—especially in the case of the daily revolution, as it is the Earth's very own.[11]

THE MODEL

Copernicus' Mars shares common features with Ptolemy's Mars, those that elegantly imbibe the circle into Mar's motion. As with Ptolemy's Mars, the mean of Copernicus' Mars revolves about a deferent whose center is somewhat offset from the center of the universe while Mars itself circles about an epicycle that is centered at Mars' mean. In contrast with Ptolemy's Mars, regularity governs the mean of Copernicus' Mars about its deferent. Mars' mean revolves about its center with uniform speed, and Ptolemy's most distasteful equants have no role.

Copernicus' Mars rotates about its epicycle somewhat differently from Ptolemy's. While Ptolemy coordinates epicyclic motion with the occurrence of retrograde, requiring that Mars' mean, Mars' actual position, the sun, and Earth are collinear in the middle of Mars' retrograde, by endowing Earth with

[11]The words in parentheses are not Copernicus' but my efforts to clarify Copernicus' verbiage.

freedom to move, Copernicus frees Mars' mean of the collinear requirement. Instead, as Mars' mean revolves once about the deferent, Mars performs two uniform rotations about its epicycle. Figure 6.1 illustrates the composite motion. The mean revolves counterclockwise about the deferent as Mars likewise rotates counterclockwise about its epicycle. Starting with the mean at 3:00 o'clock, Mars is at the 9:00 o'clock position on its epicycle. During the time that the mean had jogged 3 hours to 12:00 o'clock, Mars has run 6 hours to 3:00 o'clock. When at its leisurely pace, Mars' mean completes half its revolution to 9:00 o'clock, Mars itself has completed one full rotation and returned to the 9:00 o'clock position. The return to the initial position mirrors the first half of Mars' periodic journey, the mean completes the remaining half of its cycle, while Mars completes one full rotation about the epicycle.[12]

Figure 6.1 also illustrates the deferent-centered mean longitude of Mars θ and Mars' anomaly about its epicycle α . Copernicus imposes the following relation between these angles:

$$\alpha = 2\theta + 180 \qquad (6.1)$$

We wish to specify Mars' position and mean position given its mean longitude θ, the distance of the center of Mars' deferent from the sun d, and the radius of

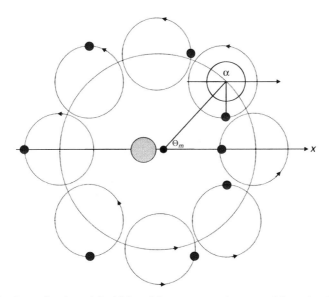

Figure 6.1. Copernicus' model of Mars. Mars rotates twice around its epicycle for every single revolution of the mean about the sun. Note that the deferent's center is not at the sun, but there is no equant.

---

[12]As before, we place the initial position at 3:00 o'clock so that it is aligned with the x axis of a Cartesian coordinate plane. This allows us to formulate positions using Cartesian coordinates, simplifying calculations. Copernicus lived before Descartes and did not utilize the modern coordinate system. He places the initial position at 12:00 o'clock.

Mars' epicycle r_e. Let $m = (m_1, m_2)$ denote the coordinates of Mars' mean and $M = (M_1, M_2)$ denote the coordinates of Mars. We have the following relationships for the mean that correspond to the Ptolemaic equations (5.6) and (5.7) of Chapter 5:

$$m_1 = \cos \theta + d \tag{6.2}$$

$$m_2 = \sin \theta \tag{6.3}$$

The coordinates of Mars follow by adding Mars' epicycle-centered coordinates to the mean coordinates:

$$M_1 = m_1 + r_e \cos \alpha$$

$$M_2 = m_2 + r_e \sin \alpha$$

Noting the relation between α and θ yields the following:

$$M_1 = m_1 + r_e \cos(2\theta + 180)$$

$$= m_1 - r_e \cos 2\theta \tag{6.4}$$

$$= \cos \theta + d - r_e \cos 2\theta$$

$$M_2 = \sin \theta - r_e \sin 2\theta \tag{6.5}$$

The Martian Year

By ascribing sun-centered motion to all the planets, Copernicus correctly endows the planets with their own heliocentric year. The time of the planet's year may be expressed in terms of Earth's solar year. We present the calculation for the planet Mars.

Copernicus uses the data that Hipparchus bestows to Ptolemy; however, Copernicus' interpretation of the data is different. For Copernicus, on its inner track, Earth laps Mars 37 times in 79 solar years and 2.450833 days, which Copernicus equates to 79.008158 solar years. The average number of years required for Earth to lap Mars is $79.008158 \div 37 = 2.135356$ solar years. Between laps, the mean position of the earth then moves $360 \times 2.135356 = 768.728026$ degrees around its circuit while on its outer, slower lane, the mean position of Mars moves $768.728026 - 360 = 408.728026$ degrees. Mars' speed of 408.728026 degrees per 2.135356 solar years equates to 191.409815 degrees per solar year, which, in turn, yields the length of a Martian year at 1.880781 solar years.

The reader should note that the geocentric Martian year of Ptolemy and the heliocentric Martian year of Copernicus are the same.[13] How is this so? A comparison of the Ptolemaic and Copernican calculations provides the answer. Each

---

[13]There is a slight difference in the observed times, indicating that Copernicus read a different version of Ptolemy than my translation.

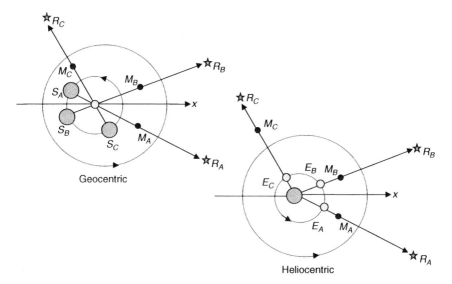

Figure 6.2. Geocentric and heliocentric positioning of Mars within the celestial sphere are the same at opposition.

calculation uses the same data: time differences between occurrences of opposition. Figure 6.2 compares the geocentric and heliocentric interpretations of the data. At the moment of opposition, both the geocentric interpretation and the heliocentric interpretation place Mars in the same position with respect to the universal sphere. One can consider the Martian year as the time required for Mars to complete one cycle around the universal sphere. Since the data from which one makes this calculation are the same in both the heliocentric and geocentric models, both models yield the same same timespan for the Martian year.

The simplicity and elegance of Copernicus' heliocentric model expresses itself in the ease of Copernicus' calculation of the Martian year. By contrast, Ptolemy's calculation is less intuitive.

Calibrating with Ptolemy

Below is Copernicus' model of Mars and Earth:

$$M_1(t) = m_1(t) + r_e \cos \alpha(t) \tag{6.6}$$

$$M_2(t) = m_2(t) + r_e \sin \alpha(t) \tag{6.7}$$

$$m_1(t) = \cos \theta(t) + d \tag{6.8}$$

$$m_2(t) = \sin \theta(t) \tag{6.9}$$

$$E_1(t) = r_E \cos \xi(t) \tag{6.10}$$

$$E_2(t) = r_E \sin \xi(t) \tag{6.11}$$

$$\xi(t) = \xi_A + \omega_\xi t \tag{6.12}$$

$$\theta(t) = \theta_A + \omega_\theta t \tag{6.13}$$

$$\alpha(t) = 2\theta(t) + 180 \tag{6.14}$$

A comparison with equations (5.14)–(5.22) highlights the differences with Ptolemy. Expressions for the position of Mars ($M_1(t)$, $M_2(t)$) are identical. As there is no equant, the mean of Mars ($m_1(t), m_2(t)$) is simpler than the corresponding Ptolemaic expression. Rather than calibrating with the position of a moving sun, Copernicus calibrates observations against a moving Earth with coordinates ($E_1(t)$, $E_2(t)$). Consistent with the Ptolemaic simplification of an Earth-centered deferent for the sun, we apply a sun-centered (heliocentric) deferent for Earth. The parameter r_E is the radius of Earth's deferent, while the variable $\xi(t)$ is the sun-centered longitude of earth.

A description of Mars' orbit requires specification of several parameters, including

r_e Radius of Mars' epicycle
d Distance of center of Mars' deferent from the sun
ω_ξ Angular speed of Earth's longitude
ω_θ Angular speed of Mars' mean motion;
ξ_A Earth's initial longitude
θ_A Initial longitude of Mars' mean.

With the calculation of the Martian year, we have already determined the angular speed of Mars' mean motion: $\omega_\theta = 191.4098$ degrees per solar year. By definition, the angular speed of Earth's longitude ω_ξ is 360 degrees per solar year. What remains is to determine the initial longitude of Earth with respect to the x axis ξ_A, the initial longitude of Mars' mean θ_A, the distance of the deferent's center from the sun d, and the radius of Mars' epicycle r_e (see Figure 6.3). In this section we are concerned with Copernicus' calibration to Ptolemy's observations at opposition. Just as the sun's radius did not enter into Ptolemy's calibration, the earth's radius does not enter into Copernicus', and so we do not consider the parameter r_E until a later section.

As a first step, Copernicus synchronizes his model with the Ptolemaic system. One way to visualize the synchronization is to imagine that Copernicus does not eliminate the equant, but rather forces the equant and center of Mars' deferent to merge into a single point. For Copernicus, the natural point where the two merge is midway between Ptolemy's equant, which sits at 0.2 along the x axis, and the deferent, which sits at 0.1 along the x axis. Copernicus accordingly selects $d = 0.15$. Copernicus then accepts Ptolemy's initial position of the sun and transforms Ptolemy's geocentric longitude of the sun into a heliocentric longitude for Earth. It is an easy transformation; add 180 degrees to Ptolemy's solar longitude to arrive at Copernicus longitude ξ (see Figure 6.3). Copernicus

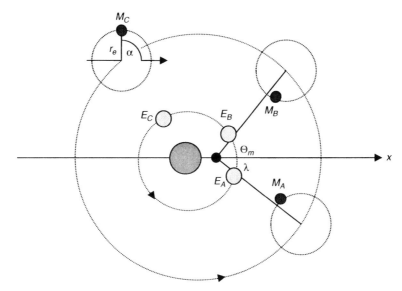

Figure 6.3. The parameters λ, Θ_m, α, and r_e shown: λ at position A, Θ_m at position B, α at position C, and r_e at position C.

also accepts Ptolemy's initial value for the longitude of Mars' mean with respect to the deferent's center (now also the equant) θ_A. As the anomaly and its speed are set by its relation to Mars' mean [eq. (6.14)], there remains only one parameter left to determine, the radius of the epicycle r_e. Copernicus explains that the choice of $r_e = 0.05$ most closely maintains Ptolemy's geometry, although this is not so obvious.

Copernicus next demonstrates that using these values, his heliocentric model and Ptolemy's observations coincide. We go through our own calculation and verify Copernicus' claim. Because the mean longitude of Mars is initially equal for both Ptolemy and Copernicus and in both systems the longitude also changes at the same rate in both systems, it follows that in both systems Mars' mean longitude must be the same at all times. A similar statement relating the sun's geocentric longitude and Earth's heliocentric longitude holds; in both systems they are always commensurate because they are initially synchronized and revolve at the same speeds. All that remains to demonstrate is that the position of Mars is in opposition to the sun in accordance with Ptolemy's observations.

Table 6.1 gives a heliocentric perspective of Table 5.1, Ptolemy's observations at opposition. The value ξ, Earth's longitude, is obtained by adding 180 degrees to Ptolemy's solar longitude. The last two columns give values of the Martian longitude in accord with Copernicus' model. Copernicus determines the Martian longitude using the relationship longitude $= \arctan \frac{M_2}{M1}$, where, as above, (M_1, M_2) are the coordinates of Mars.

When Mars and the sun are in opposition, the heliocentric longitude of Earth and Mars must coincide so that $\xi = \arctan(\frac{M_2}{M_1})$ (see Figure 6.2). Using

TABLE 6.1. A Comparison of Copernicus' Model with Ptolemy's Observations

Year	θ: Ptolemy	ξ: Ptolemy	arctan $\dfrac{M_2}{M_1}$: Computer	arctan $\dfrac{M_2}{M_1}$: Copernicus
15 of Hadrian	−41.55	−34.5	−34.514230	−34.5
19 of Hadrian	40.18	33.333333	33.346894	33.333333
2 of Antonine	135.65	127.06667	127.087000	127.083333

equations (6.6) and (6.7) to evaluate Mars' coordinates at the first, second, and third observations allows one to calculate Mars' longitude at each observation to arrive at the third column of Table 6.1. We present a calculation of coordinates at the first observation, as the others are similar. At the first observation. $\theta_A = -41.55$ and $\xi_A = -34.5$:

$$M_1 = \cos\theta + d - r_e \cos 2\theta$$
$$= \cos 41.55 + 0.15 - 0.05 \cos 83.1$$
$$= 0.892370$$
$$M_2 = \sin\theta - r_e \sin 2\theta$$
$$= -\sin 41.55 + 0.05 \sin 83.1$$
$$= -0.613636$$
$$\arctan \frac{M_2}{M_1} = -34.514230$$

The final column of Table 6.1 is Copernicus' calculation from *On Revolutions*. While the result given above and Copernicus' result perform the same computations, there is a difference due to inaccuracies inherent in representing irrational numbers with rational numbers specifically, roundoff differences. As we see below, Copernicus did not give heed to this issue.

Calibrating with Copernicus' Own Observations

Copernicus' calibration of his model with Ptolemy's observations serves two purposes: (1) given the general respect for Ptolemy's work, Copernicus felt obligated to demonstrate that his model could achieve the same results; and (2) calibrating with Ptolemy is a warmup exercise for Copernicus' calibration with his own observations. In this manner Copernicus not only replaces Ptolemy's model but also updates Ptolemy's results.

Copernicus prepares another table similar to Table 6.1 using three of his own observations at opposition. Copernicus takes his first observation in June 1512 and the final observation in March 1523. Now it is Copernicus' turn to accept his punishment in the form of an iterative calculation, and he does.

As his first encounter with Ptolemy was successful, Copernicus fixes the epicycle's radius at $r_e = 0.15$. Then he follows Ptolemy's iterative procedure to arrive at values for the center of the deferent d, Earth's initial longitude ξ_A, and the longitude of Mars at its initial observation θ_A. Copernicus gives an idea of the tedium of this task:

> Accordingly, by the same method which we employed in the case of Saturn and Jupiter—let us pass over in silence the multitude, complication, and boredom of the calculations.

After an adjustment, the iterative method that we use in Chapter 5 is applicable to Copernicus' model as well. Following Copernicus, we defer from presenting the details and skip to Copernicus' results.

Copernicus reports that $\theta_A = 125.583333$ degrees and $d = 0.146$. Noting a discrepancy of 0.004 between Copernicus' own calibration of d and the Ptolemaically tuned calibration of d, Copernicus writes the following:

> Moreover, we have found a lesser distance between the centre, i.e. 40, whereof the radius of the eccentric circle is given as 10,000—not because either Ptolemy or ourselves made a slip, but manifestly because the center of the orbital circle of the Earth has approached the center of the orbital circle of Mars, while the Sun has remained immobile.

Copernicus must have felt a need to explain this difference. He could have attributed the difference to other sources such as differences in the Ptolemaic and Copernican models, the imprecision of the data, or roundoff errors in the calculations. Instead, Copernicus is steadfast in the accuracy of both the data and the calculations, preferring to attribute the difference to a physical cause. Copernicus' conclusion must have been most disturbing for it presents a movement that his model does not explain.

Relative Radii of the Deferents of Earth and Mars

In January 1512, Copernicus makes an observation of Mars and later uses the observation to determine the ratio between the radii of Mars' and Earth's deferents. Before following Copernican geometry, let us remark on the date of the observation. The date is consistent with Copernicus' introductory statement, where he claims that he held his work for more than three decades prior to publication in 1543. Copernicus did not log data as a matter of course. If he took an observation, it must have been with a specified intention. One can surmise that Copernicus took this observation with the specific intention of calculating the radius of Mars' deferent from which he could deduce Mars' minimum and maximum distances from the sun. By 1512, Copernicus had, indeed, worked out the geometry of his universe. From 1512 onward Copernicus took observations to calibrate his model.

With the intention of determining Mars' relative distance from the sun, relative to Earth's distance from the sun, Copernicus' observation is not at central retrograde, so Mars and the sun are not in opposition. Prior to performing this calculation, Copernicus must parameterize his model as specified in the preceding section. Copernicus' third and final observations at opposition occurred in 1523. It was possible to calibrate his model and determine the ratio between the radii of Earth's and Mar's deferents only after the final observation. One can conjecture that before completing his own set of observations, Copernicus calculated the relative radii using his Ptolemaically tuned model.

Figure 6.4 illustrates that one may determine the coordinates of Mars using the coordinates of Earth $E = (E_1, E_2)$, the distance between Earth and Mars h, and the geocentric angle of Mars φ:

$$M_1 = E_1 + h \cos \varphi$$
$$M_2 = E_2 + h \sin \varphi$$

The coordinates of Earth are respectively $E_1 = r_E \cos \xi$ and $E_2 = r_E \sin \xi$, where r_E is the distance from Earth to the sun. It is the value r_E that we wish to determine.

Using equations (6.10) and (6.11) to substitute for E results in a pair of equations that are linear in r_E and h:

$$r_E \cos \xi + h \cos \varphi = M_1$$
$$r_E \sin \xi + h \sin \varphi = M_2$$

One eliminates the terms containing h by multiplying the top equation by $\sin \varphi$, multiplying the bottom equation by $\cos \varphi$, and subtracting the resulting equations. Further simplification gives the value r_E in terms of the remaining

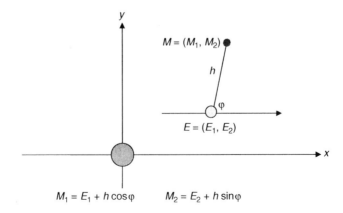

Figure 6.4. Calculation of the coordinates of Mars using Earth's coordinates h and φ.

values:

$$r_E(\cos \xi \sin \varphi - \sin \xi \cos \varphi) = M_1 \sin \varphi - M_2 \cos \varphi$$

$$r_E \sin(\varphi - \xi) = M_1 \sin \varphi - M_2 \cos \varphi \qquad (6.15)$$

$$r_E = \frac{M_1 \sin \varphi - M_2 \cos \varphi}{\sin(\varphi - \xi)}$$

In the remainder of this section we endeavor to determine the right-sided values M_1, M_2, ξ, and φ from which we determine the left-sided value r_E. The convention that the radius of Mars' deferent is 1 still applies. As a first step it is necessary to determine the longitude of Mars' mean at the time of the uncentered observation θ. From the preceding section, we know the longitude of Mars' mean several months later at opposition: $\theta_A = 125.583333$. With $\omega_\theta = 191.281794599$ degrees per year, the speed of rotation for the longitude and $\Delta t = 0.426920$, the timespan between observations the following relation must hold:

$$\theta_A = \theta + \omega_\theta \Delta t$$

$$\theta = \theta_A - \omega_\theta \Delta t = 43.866667$$

We next determine the position of Mars using equations (6.6–6.9). Copernicus uses his calibrated value of $d = 0.146$. The result is $M_1 = 0.864977$ and $M_2 = 0.643022$.

Copernicus measures the position of Mars within the celestial sphere at $\frac{1}{8}$th of a degree east of a star that he refers to as the "bright star of Chelae." The geocentric longitude of Mars is then $\varphi = 71.7758267$ degrees with respect to the x axis.[14] Additionally, Copernicus measured the sun's position in the celestial sphere at 262 degrees, 142.309160 degrees with respect to the x axis.[15] We note that Earth's heliocentric longitude is in opposition to the sun's geocentric longitude by 180 degrees: $\xi = 142.309160 + 180 = 322.309160$ degrees.

All the right-sided values of equation (6.15) are known, and Table 6.2 summarizes the values. Placing the table values into equation (6.15) results in $r_E = 0.658115$. Since the radius of Mars' deferent is 1, the value 0.658115 also represents the ratio of Earth's deferent radius to Mars' deferent radius.[16]

---

[14]Copernicus reports the measurement $191°28'$. Copernicus' measurements are with respect to the position of the sun at spring equinox as the reference point; thus, within the celestial sphere, at the moment of the spring equinox, the sun is at 0 degree. This gives a standard reference point from which to measure the position of stars. The angles that we provide are with respect to the x axis, which hosts the center of Mars' deferent. The difference between our angles and Copernicus' is 119.690840 degrees.

[15]As the sun passes through zero degree at the spring equinox, its position at any time may be determined by the time that has passed since the spring equinox. Alternatively, one can look for the position of a star in opposition to the sun, for example, at sunset, or at midnight. Knowing the celestial longitude of the star yields the celestial longitude of the sun, which is in opposition by 180 degrees.

[16]Once again imprecision inherent in the calculations displays itself. Copernicus' result is $r_E = 0.6580$

TABLE 6.2. Required Values

M_1	M_2	φ	ξ
0.864977	0.643021	71.775826	322.309160

AN ELEGANT RESULT

Calibration of the Ptolemaic and Copernican models both apply similar geometric methods to the exact same data; data taken at central retrograde when Mars and the sun are in opposition to Earth. At the moment of opposition the alignment of the three bodies in relation to the celestial sphere is identical in both the heliocentric and geocentric systems, and so the geocentric longitude of Mars coincides with the heliocentric longitude of Mars. The bottom line is that at the moment of opposition, the data from either perspective are identical.

What if Copernicus had not found the equant so displeasing? What if he accepted this feature within a heliocentric system? If that were the case, his book would have been much shorter. Copernicus would only have to exchange positions of the sun and Earth, so that as Copernicus puts it himself, the sun sits on its "throne" and Earth circles about the sun along its deferent. The position of Earth on the deferent could be ascertained by adding 180 degrees to Ptolemy's value of the sun's position. That is all Copernicus need do to create a heliocentric system that is consistent with Ptolemy's observations.

But Copernicus did find the equant very displeasing, forcing him to rework everything to ensure that his new configuration was in agreement with Ptolemy's observations. It was most definitely an arduous task. Given that Copernicus' system proved incorrect, was it worth the effort? One wonders whether a heliocentric Ptolemaic system would have attracted attention. Would anyone pay heed to work that replicates an icon with only an alteration that affects the mathematics not one iota? Would this not be akin to an artist unveiling a painting that is the mirror image of the Mona Lisa, an unveiling that would not receive any acclaim? It is not possible to know how Europe would have received a heliocentric version of Ptolemy. We do know that Copernicus' rework attracted sufficient attention to further its cause and become the game changer.

By the time of Copernicus' death, Europe was no longer the same continent that it was at Copernicus' birth. The Roman Catholic Church was not infallible, and it could not be stated with certitude that the universe was geocentric. There was still one refuge of certainty—the "universal governor" regulated the heavens about perfect circles.

CHAPTER 7

RENEGADES

COUNTERING THE REFORMATION

On September 16, 1621, a cardinal and future saint confronted his immortality. The cardinal was preeminent among the Church intelligentsia; he once directed the Collegio Romano, the Jesuit university considered the intellectual center of the Church. The cardinal had left for posterity a number of well-respected treatises, works that to this day continue to serve as a guiding inspiration within the Catholic Church. Aside from his intellectual capacities, the cardinal was known for his pious and modest character, which was centered on devotion to God. Supporting this reputation are accounts of an austere lifestyle and daily prayer sessions of up to 3 hours. When Pope Urban VIII became aware of the cardinal's deteriorating health, he granted permission for the cardinal to take a leave from his posting as the pope's councilor. The pope then placed one of his palaces at the cardinal's disposal and directed the cardinal to rest and seek comfort as the last day of the cardinal's residence on Earth drew nearer.

In his bed on the final day, one must imagine a man so fervent in his belief and dedication would be at peace with himself. It would be only a short moment before he would join his maker for eternity. Yet, it is possible that the cardinal trembled with fear and was tormented by a dark chapter that blemished the

Shifting the Earth: The Mathematical Quest to Understand the Motion of the Universe,
First Edition. Arthur Mazer.
© 2011 John Wiley & Sons, Inc. Published 2011 by John Wiley & Sons, Inc.

cardinal's resume. The chapter is not one that a pious man would like to take to the final authority on judgment day. Perhaps the torment had taken hold of the cardinal on February 9, 1600 and he lived with the torment until his end. On that date a man of equal devotion to God and equal intellectual capacity looked into his inquisitors' eyes and taunted the inquisitors with the words "You pass this sentence with greater fear than I receive it." Were the sentenced man's words reverberating off the canyon walls as the cardinal stepped through the pass between life and death? Was the image of the man being subject to the sentence that the cardinal had ensured in his capacity as inquisitor for the Holy Office of the Inquisition haunting the cardinal on the cardinal's deathbed? Did the cardinal see the prisoner muzzled with a spike piercing his jaw, chained naked to a stake, burning on the pyre? How could the cardinal reconcile this punishment on a man who had caused no harm to anyone, with the teachings of Christ, and how could the cardinal account for his leading role in the sentencing? Would the maker hold the cardinal accountable?

Had Giordano Bruno—ordained priest, theological scholar, university lecturer, author, philosopher, and convicted heretic—and Robert Bellarmine—ordained priest, theological scholar, university lecturer, author, philosopher, inquisitor, cardinal, and saint—both have come of age aside Borgia, they might have formed a longlasting friendship, one based on mutual respect for the other's intellectual capacity and deeply felt convictions. But both came of age in the aftermath of the Lutheran revolt. Under Pope Paul IV the Roman Catholic Church did its best to return to the pre-Renaissance era, the age of Augustine. The Renaissance had sufficient momentum to excite the inquiring mind of Giordano Bruno, and the Church had sufficient authority to ensnare Robert Bellarmine within its conservatism. Rather than developing a friendship, these two men found themselves on opposite sides of a divide.

The Counter-Reformation is an era defined by divisions. Religious divisions abounded as Catholics, Lutherans, Calvinists, Hugenots, and Anglicans, all claiming to know the true path to heaven, competed for souls. Nationalist divisions occurred among factions contending for power, and European divisions surfaced as nations braced themselves in opposition to Spanish hegemony.[1] These overarching divisions caused subdivisions as well as divisions that cut across both religious and national sentiment. There were philosophical divisions among Catholics with Erasmus, leading the way among reformers, opposed by Pope Sixtus. Civil strife arose in France as the French Catholics, lead by the Guise faction and French Hugenots, following the Navarres, pitted their mercenary armies against one another. England separated itself from the Roman Catholic Church and all continental protestants by establishing its own state-sponsored religion. Italian sentiment toward their Spanish overlords quelled any efforts to unite Catholics against the British upstarts. While acrimony describes the relations between Elizabeth of England and Pope Sixtus, the pope did not lend his moral support to

---

[1] The Spanish financed their European power strike on the backs of native Americans whom the Spaniards impressed into service as unpaid silver miners.

Philip of Spain in his efforts to conquer England.[2] The Habsburg Empire was another focal point of conflict. Lutherans, both German and Bohemian, resented oversight by the Catholic Habsburgs with their connections to the Spanish crown.[3]

Strife-ridden times require discipline. Among contending factions, each established its own disciplining process. The Roman Catholic Church strengthened the Holy Office of the Inquisition. Spanish Catholics continued to control their own Inquisition on behalf of the Spanish crown.[4] Calvinists established neighborhood watchdogs who would report any suspicious behavior to an investigative office with disciplinary authority.[5] The Lutherans were far less organized than either the Catholics or Calvinists, with the result that perhaps their disciplining was the most unsettling. At least under the Catholic Inquisition, Spanish or Roman Catholic, and Calvinist religious police, the populace knew the boundary that delineated the safety zone from one that would attract the attention of the authorities. It was among German Lutherans where the dispensation of justice was most unpredictable. It was among German Lutherans where accusations of witchcraft and sorcery were most likely to initiate a victim's path toward the pyre.

THE TRIO: TYCHO, KEPLER, AND RUDOLF

Lining the channel between the North Sea and the Baltic is a craggy, discontinuous coastline. The city of Copenhagen lies on the island of Zealand that itself appears as a fragmented piece from a jigsaw puzzle. Zealand divides the North Sea into straits. Across the Oresund Strait is Sweden, and within the Oresund Strait lie several scattered islands fixed to the Strait's floor like anchors that had been randomly dropped from the hidden sky above the clouds. These islands were beyond the reach of the turmoil that engulfed much of Europe during the Counter-Reformation. On one of these islands, using generous financing from a family connection, King Frederick, Tycho Brahe constructed his own sanctuary, an astronomical observatory. It is within this sanctuary that Tycho Brahe and his retinue charted the heavens and gained renown throughout Europe. Encrypted within Tycho's observations was the true pathway of the heavens. But the observations' message would not be decoded on the island Hven. In order for the contents of the observations to find a worthy decipherer, the observations would have to pass through the heartland of Counter-Reformation divisiveness.

There is nothing normal about the life or death of Tycho Brahe. Bizarre circumstances may have been a prerequisite to acquiring the wherewithal to devote

---

[2]That effort failed with the defeat of the Spanish armada in 1588.

[3]For King Ferdinand of Aragon, his offspring were political capital. Ferdinand betrothed his children to royalty across Europe. His progeny became kings, queens, and emperors in France, the Netherlands, the Holy Roman Empire, and England. A strong link between the Spanish Court and the Habsburgs endured throughout the Counter-Reformation.

[4]Efforts by Spanish authorities to assume control of the Inquisition within their Italian territories fueled resentment among both the populace and the Italian clergy. The efforts failed.

[5]Bruno was well acquainted with the Calvinist courts. He happily skidaddled out of Geneva to avoid further punishment after his excommunication and banishment.

one's life to charting the heavens; the occupation was expensive and demanded dedication. In the midsixteenth century, Europe had no institutions that would support an astronomical observatory. Only the independently wealthy nobility had the means to finance such an endeavor. But the nobility were by and large a political class engaged in the affairs of state. Who among the nobility would devote their lives to astronomy?

Tycho Brahe's upbringing differed from that of other noblemen. At age 2, Tycho was kidnapped by his paternal uncle Joergen Brahe, an event that framed Tycho's future. Tycho shuttled between the household of his biological parents and his otherwise childless adopted parents. On a cold day in 1565, while crossing a bridge, a man in Joergen Brahe's company fell to his certain death in the icy waters below. Joergen Brahe dove after the victim and in doing so exchanged his life for that of the king. King Frederick felt indebted to the Brahe family and to Joergen's favorite, Tycho, in particular. When Tycho Brahe came of age, King Frederick offered Tycho a dream. He would bequeath to Tycho a fiefdom of significant size, further enriching Tycho beyond his already considerable wealth. Tycho would have none of it.

During his university years, Tycho developed a passion for astronomy, and his sole desire was to dedicate himself to that occupation. Tycho wore a highly visible badge of distinction that displayed his passion for science. The badge was an artificial nose crafted from gold and silver. Tycho required the badge because during his student years, his passion got the best of him. An argument over an equation between himself and a fellow student escalated into a sword fight that ended when Tycho's real nose hit the ground.[6]

Perplexed by Tycho's rejection of his generous offer, King Frederick became Tycho's genie and requested a wish. When King Frederick learned that Tycho wished for an astronomical observatory, the king personally set about finding an ideal location. Tycho enthusiastically accepted King Frederick's gift, the island Hven. The king also provisioned Tycho with the necessary resources for Tycho to build and maintain his observatory. The observatory became operational 1577 and on Hven, Tycho lived a charmed and eccentric[7] life for many years.[8]

In 1577, while supervising the construction of his observatory on Hven, Tycho Brahe noted a comet of unusual intensity. Halfway across the European continent in the imperial city of Vienna, the inexperienced emperor, Rudolf, also viewed the comet. A comet upset the perfection and predictability of the heavens. On Earth,

---

[6]It's too bad that nobody walked around with cell phones and digital cameras. The sight of Tycho standing agape watching his nose as it struck the ground would be one for posterity.

[7]Tycho had a pet moose who enjoyed beer. Unfortunately, like many human counterparts, the moose did not always know when to quit. After imbibing more than he could handle, the moose took a fatal tumble down a flight of stairs.

[8]Tycho may have been happy, but the farmers who inhabited the island prior to Tycho's arrival were not. Before Tycho, the kingdom ignored the island and the peasants lived undisturbed. Once Hven was Tycho's property, the inhabitants were under his charge. Tycho obliged them to perform construction and maintenance services pro bono. Tensions escalated to the point that the farmers registered a formal complaint with the king's court. The king's irrevocable support of Tycho left the farmers worse off.

this uninvited visitor portended chaotic and confusing times. That year, in 1577, Rudolf felt the full weight of uncertainty and became the victim of psychological depression. Throughout his life, Rudolf, a Catholic who under his Spanish mother's influence was educated in the court of Spain's King Philip, sought solace from a cadre of astrologers, mystics, magicians, and Jewish kabbalists in his service.

From a knoll in the town of Leonberg at the edge of the Black Forest, a poor commoner, the 6-year old Johannes Kepler, held his mother's hand and peered with wonder at the comet. It was a rare moment of pleasure in an otherwise abused life. As an enthusiastic mercenary, Kepler's father frequently abandoned the family, enlisting in the war efforts of the highest bidder. As the father beat his wife and children, the absence of Kepler's father was an improvement on the father's presence. Kepler could not look to his mother to provide emotional support, as she generally neglected the children. Once, while Johannes suffered a near-death childhood illness, Kepler's mother left Johannes in the care of his grandparents, who neglected the boy even more than the mother had.

In 1577, Tycho Brahe, Emperor Rudolf, and Johannes Kepler lived in separate worlds, sharing only a common view of the heavens and the comet that alighted the nighttime sky.[9] Neither the astrologers, mystics, nor kabbalists could divine that the paths of these three men would intersect, let alone divine the change in human history that would result.

Among the three men, Johannes Kepler is preeminent. Tycho was a facilitator who bequeathed Kepler a record of astronomical observations. Rudolf was also a facilitator, facilitating the introduction between Tycho and Kepler. But Kepler, whose spiritual sensibility cohabited the same mind as a rare analytic ability and equally rare integrity, was the man who directed his capacities toward determining the correct configuration of the heavens.

Perhaps it was Johannes Kepler's contentious family life and the absence of a father figure that caused him to devote his life to God in all his endeavors. Kepler's family was Lutheran, and Lutheran philosophy appealed to and influenced Kepler. While Luther was the charismatic spokesman and politically adept leader of the Reformation, Melancthon was the Reformation movement's philosophical guide. It was Melancthon who diverged from the Catholic philosophy that required a command and control approach to interpretation of scriptures. Melancthon encouraged Christians to seek their own path toward God. For Melancthon, God revealed himself[10] not only in scriptures but also in nature. Seeking God through both avenues was the highest service that a man could perform. Throughout Kepler's life, Melancthon's philosophy was Kepler's guide. While this was a critical element of Kepler's success, it also handicapped Kepler's social relations. During the divisive years of the Counter-Reformation, the Lutheran's own command and control tactics displaced Melancthon's

---

[9] James Conner in his wonderful book, *Kepler's Witch*, introduces the reader to Tycho and first connects Tycho with Kepler through the comet of 1577. I am following this precedent.

[10] The world of Melancthon did not have any pretensions concerning equality of the sexes. Of course, men were superior, and God was a man.

enlightened philosophy. Kepler, the independent thinker guided by Melancthon, would pay for his insolence.

Concerning Kepler's integrity, he was honest to the point of self-brutality. The following punishment that he served on himself in a written self-assessment (his own horoscope) demonstrates his commitment to honesty, even if the trail of honesty goes through a dark forest of pain:

> Since being stingy with money deters one from play, he often plays by himself. One has to note the following here, holding on to money does not have the goal of wealth, but rather, the alleviation of the fear of poverty. Of course most greediness grows out of unfound worry. Or perhaps not; rather, the love of money possesses many. Perhaps it is the fear of poverty that can be blamed for much. Because he is presumptuous and contemptuous of mass opinion, he tends to be hard.

> By nature he is very well suited toward pretense of all kinds. There is also a tendency toward disguise, deceit, and lies. It has its root in common with jokes and jest. Mercury does this, instigated by Mars. But one thing prevented these disguises: the fear for his reputation. Because foremost he yearns for true recognition, and every type of defamation is unbearable to him. He would pay good money to buy himself free of even harmless, but wicked gossip, and poverty frightens him only because of the shame.

> This person harbors dark thoughts about his enemies. And why would he do this? Could it be because his enemies compete with him for industry, success, distinction, and fortune? Or could it be because the sun and Mercury are in the seventh house?

> This person has the nature of a dog. He is just like a spoiled little pet.... He likes to gnaw on bones and chew on hard crusts of bread. He is voracious without discipline. When something is put before him, he snatches it up.

Kepler's integrity forced him to confront his character flaws. Later, it would force him to confront the flaws of his analysis of Tycho's data. His willingness to explore and expose his own blemish is what led Kepler to the truth.

On July 31, 1600 Kepler's facility for self-introspection would provide a measure of comfort as he required his family to follow him down a difficult path. Archduke Ferdinand, brother to the emperor Rudolf and sovereign of Austria, heeding the call of his Spanish ancestry, demanded that all citizens of Graz, a Lutheran town where Kepler resided and taught mathematics, report to the town church where all citizens would proclaim their faith. Those who chose Catholicism could continue with their lives. Those who did not would face exile.

Looking back on his life, Kepler knew himself. He was a child of an unloving family who had sought love through a relation with God. While education was normally reserved for the wealthy, he, through his God-given abilities, was one of few to receive a scholarship that secured him a Lutheran education from elementary school through his university years at Tubingen, a preeminent Lutheran university. He was the author of *Mysterium Cosmographicum*, a work that in Melancthonian fashion sought to find an understanding of God by deciphering God's arrangement of the heavens. He was the loving stepfather of his Lutheran wife's daughter by a former marriage. He was the steadfast believer of a loving, compassionate God even as he watched his own two children die in their infancy. The man who feared poverty because of the shame that accompanies poverty would reaffirm his Lutheran faith, choose exile, and, if need be, poverty.

As years have gone past and the dead do not speak, we are left to conjecture as to why King Frederick's heir, Christian IV, did not share Frederick's admiration for Tycho and in fact disdained the man. A conjecture discussed in the literature[11] is that in his youth, during a visit to Hven, Christian saw his mother, the queen, and Tycho in a compromising position. The impression left by this stark vision lead Christian to question the identity of his own father; was Christian the legitimate heir to the throne or merely a bastard son to sinners? Following the logic of this conjecture, Christian worried that others might also doubt his legitimacy. Christian felt compelled to rid Denmark of Tycho.

After years of haranguing Tycho, in 1597 Christian struck decisively. Christian dispossessed Tycho of Hven and forced Tycho into exile. On Tycho's evacuation from Hven, as Scipio treated Carthage, Christian incinerated Tycho's observatory, leaving no trace of its existence, save fragments of the foundation. Indeed, Christian harbored intense enmity of Tycho.

Wishing to relocate his equipment and open a new observatory, Tycho sought the sponsorship of courts throughout Europe. Tycho's résumé was impressive; he was Europe's foremost astronomer. The same inexplicable forces that compelled Emperor Rudolf to seek the guidance of astrologers, mystics, magicians, and kabbalists, and for whom astronomy and astrology were identical occupations, also compelled Emperor Rudolf to bring Europe's foremost astronomer to his court. In 1598, Tycho and the emperor concluded their negotiations. From Tycho's perspective, the terms and conditions of the agreement were impressively favorable—he would receive even more support than he had from his deceased patron and admirer, King Frederick. Tycho had to view this as a remarkable coup, especially when one considers that Tycho was negotiating from an untenably weak position; he was no more than a refugee. But then Tycho did not know with whom he was negotiating. Emperor Rudolf was a master of bait and switch. Once Tycho resettled to the outskirts of Prague, a very difficult undertaking that committed Tycho to the emperor, Tycho discovered that the emperor effortlessly and unilaterally altered the terms and conditions of Tycho's employment.[12]

In 1599, while en route to Prague, Tycho received by mail a book authored by a newcomer to the field of astronomy.[13] The book held a most unconventional theory claiming that God configured the solar system around the five Platonic

---

[11] In her book *Tycho and Kepler*, Kitty Ferguson delves into this conjecture.

[12] On one point Rudolf kept his word. Rudolf endowed Tycho's children with titles and privileges of nobility. For Tycho's progeny, who, because of their commoner mother, were themselves commoners in Denmark, this was an irrevocable upgrade in social status.

[13] Following Copernican astronomy, Kepler noted that the sun's force engulfed six planets: Mercury, Venus, Earth, Mars, Saturn, and Jupiter. This leaves five empty regions between the planets. Kepler maintains that the spatial distances between the planets are determined by the five Platonic solids: the tetrahedron, cube, octahedron, dodecahedron, and icosahedron. These are the only convex polyhedra with geometrically similar faces.

solids. For the author, God was a master geometer. What did Tycho see in this work? The work was bold and exposed the author's creative and original mind. The technical aspects of the geometry were exceedingly difficult, demonstrating the author's brilliant analytic skills. Perhaps, too, Tycho may have noted that the author of *Mysterium Cosmographicum* lived in the town of Gratz, not too far from Tycho's destination. This coincidence, brought about through the inadvertent collusion of King Christian and Emperor Rudolf, may be the real cosmic mystery.

Tycho had been struggling with his data for years. Like Kepler, he wished to discover the universal order. Tycho was predisposed to his own theory, a compromise between Ptolemy and Copernicus. Earth was the center of Tycho's universe, while the sun was a supreme planet. All of the planets recognized by Ptolemy are mere moons of the sun, wheeling around the sun on their Apollonian epicycles, as the sun revolves around Earth. While a brilliant empiricist, Tycho did not have the originality or the analytic skills displayed by the author of *Mysterium Cosmographicum*. Lacking these skills, Tycho was unable to demonstrate his theory. Tycho recognized that he needed Kepler's mind.

For Kepler, Tycho's arrival had the potential to address two concerns. The first concern was scientific. Kepler had been thinking deeply about the configuration of the cosmos. He wanted access to Tycho's observations so that he could confirm his own vision. Kepler's other concern was the much more nitty-gritty financial concern for his family. Frederick's order demanding universal conversion to Catholicism was not an unforeseen event. Tensions had been mounting, and Kepler knew that it was only a matter of time before his employment in Gratz would come to an end. He had been lobbying his mentor at Tubingen to assist in arranging a position at that university, which, on publication of *Mysterium Cosmographicum*, Kepler did merit. Tubingen's faculty overlooked Kepler's formidable intellect and his devotion to Lutheranism. Instead, they recalled the troublesome, independent-thinking student who did not conform to the call of obedience that intensified along with the tensions of the Counter-Reformation. When Tycho arrived in Prague, Kepler had no job prospects, which troubled him deeply.

It appears that destiny set its path, that the empiricist and the theoretician would join and together discover the architecture of the heavens. But destiny's path is never clear. It is tangled, twisted, and obscure, and bifurcates in many directions, leaving the traveler with endless possibilities, all ill-defined. Trouble had seeded itself into Kepler and Tycho's relation before Kepler and Tycho had a chance to meet. On the very day that Tycho received a copy of *Mysterium Cosmographicum*, Tycho received another manuscript from a fellow named Ursus who was Tycho's predecessor as emperor Rudolf's imperial mathematician. Ursus had visited Tycho on Hven, and Tycho suspected that Ursus had stolen Tycho's proprietary data. Ursus' manuscript confirmed these suspicions; there Ursus presented Tycho's Ptolemaic–Copernican compromise theory as his own. Enclosed with Ursus' manuscript was a letter that with exaggerated admiration commended the talents of Ursus—the signature at the bottom of this letter of praise read "Johannes Kepler."

During his job search, Kepler had networked broadly. In a subsequent letter to Tycho, Kepler explained his former letter that praised Ursus. The subsequent letter to Tycho displays Kepler's self-deprecating humor, a hallmark of his later work:

> That nobody that I was then searched for a famous man who would praise my discovery. I begged him for a gift and, behold, and it was he who extorted a gift from the beggar.[14]

Satisfied by Kepler's explanation, in January 1600 Tycho invited Kepler to visit his location outside of Prague, the site of Tycho's new observatory. Conditions for the visit were unfavorable for both men. Tycho fought a losing battle against the forces that conspired against the construction of his new observatory. At the time of Kepler's arrival he was still full of fight, but a sense of despair may well have affected his disposition. Tycho learned about Emperor Rudolf's empty promises the hard way, and many of his Danish entourage became homesick and returned to Denmark. Wisened by his experience with Ursus, Tycho was overtaken by suspicion and kept Kepler on a leash. During Kepler's visit, Tycho treated Kepler like a student, giving Kepler responsibilities that were not commensurate with Kepler's stature. Tycho also safeguarded his treasured observations from Kepler, frustrating Kepler's deep desire to confirm his cosmological views. Financially, Tycho provided room and board, but offered no salary. The nitty-gritty concern for financial survival overwhelmed Kepler. Destiny placed the nobleman and the commoner on a collision course.

The collision occurred on April 5, 1600. During discussions aimed at securing a position for Kepler, Kepler lost his temper. It is not wise for a perspective employee to offend the potential employer. It is even less wise for a commoner to not heed protocol and openly demonstrate disdain for nobility. Despite his dire circumstances, Kepler, in "I am my own worst enemy" fashion, diverged far from accepted social protocol and said many unwise things. Like recoiling billiard balls, Kepler and Tycho distanced themselves from one another. The breach seemed irreparable, and destiny's path was at a dead end.

With time, the overriding needs of each party reasserted themselves. Tycho proved that his scientific passion, the passion that caused him to lose his nose, was stronger than his sensibilities as a nobleman. Recognizing Kepler's gifted abilities, Tycho worked to secure terms of employment that were favorable to Kepler. For his part, Kepler recognized the efforts of Tycho on Kepler's behalf and came to appreciate Tycho's patronage. The Archduke Frederick may have kindled Kepler's appreciation for Tycho with his summertime anti-Protestant proclamation. In September, after publicly affirming his Lutheran faith, Kepler removed his wife and daughter-in-law from their native town of Graz and set out for his new position alongside Tycho Brahe in Prague.

Working under the patronage of Tycho Brahe had both advantages and disadvantages. The disadvantages were evident from the outset. Tycho had not yet

---

[14]The quote is taken from page 123 of James Connor's *Kepler's Witch* (Connor 2005).

come to fully trust Kepler and would not give Kepler access to his observations. Additionally, Tycho tasked Kepler with senseless work; Kepler was to prepare a damning critique of Ursus' book. Certainly, in Kepler's mind this was a complete waste of effort. Why focus on obvious nonsense while substantial issues go unaddressed?

The passing of summer into fall in 1601 saw a change of circumstances for both Tycho and Kepler. Emperor Rudolf insisted that Tycho leave his observatory (still under construction) and take up permanent residence in Prague. Despite concerted efforts to avoid court life, Tycho found himself imprisoned within Emperor Rudolf's court as a highly trusted adviser. With Rudolf obliging Tycho to serve his time in court, the time that Tycho had to devote to science dwindled. As Tycho became less engaged with scientific concerns, he entrusted Kepler with his data and gave Kepler a greater degree of independence. Tycho also used his influence in the court to secure from Emperor Rudolf a commission to develop astronomical tables. Showing a capacity to do what one must, Tycho suggested that the tables be named the *Rudolphine Tables*. More relevant to Kepler, Tycho had personally introduced Kepler to the emperor as a senior member of the commission, and the emperor granted his approval.

Of all the courses destiny could have taken, it had seemingly settled on the Leibnizian best of all possible worlds. Tycho and Kepler had finally bonded into a formidable team. Rudolf's patronage of both Tycho and Kepler enabled the two Protestants to have a sense of security in Rudolf's predominantly Catholic court. One imagines the two men at long last finding deep satisfaction in each other's company. In Kepler, Tycho found a collaborator unlike any with whom he had engaged throughout his career. Kepler provided deep insights that could excite Tycho's imagination. Kepler, who, throughout his 6-year stay in Graz, had been scientifically isolated, at long last was in the company of someone who fully comprehended his ideas—and Kepler was a man who loved sharing his ideas.

Destiny would not allow Tycho a normal life, or death. For Tycho there was always a rendezvous with the bizarre. How would you like to depart this world? Would you prefer a departure like Bruno, knowing your fate and succumbing quickly, or would you prefer an uncertainty about the event while enduring days of physical and emotional torment? On October 13, 1601, after a banquet in which Tycho drank liberally, Tycho lay on his bed in pain. Despite pressure on his bladder with the accompanying sensation of utmost urgency, he was unable to pass urine. Following several days of sleeplessness, fever, and delirium, Tycho seemed to have recovered. But then there was another bout and on October 24, Tycho finally succumbed.

For centuries, it was believed that Tycho died of natural causes, a burst bladder, a urinary infection, or an obstruction of the urinary tract. But a recent examination (1991) of samples from Tycho's beard revealed that Tycho ingested sufficient dosages of mercury to have killed him. Furthermore, the symptoms of Tycho's illness are consistent with mercury poisoning. Murder is the charge. There are many investigators on the trail of the perpetrator, but teasing secrets from the dead is a difficult task. Tycho died as he lived, enigmatically.[15]

As Tycho passed from this world, what he left behind was up for grabs. On his deathbed, Tycho made a specific request of Kepler: "Let me not have lived in vain." The request reflected Tycho's recognition that Kepler was most qualified to transform Tycho's lifetime efforts from a collection of data points into an everlasting model of the universe. With this request, Tycho bequeathed his data to Kepler. The problem for Kepler is that while the request invested Kepler with moral authority over the data, legally, Tycho's heirs owned the data. For Kepler, the moral authority of Tycho and probably God himself, trumped any legal process. At a precipitous moment, Kepler secretly grabbed Tycho's Mars data. This rash action, that of an ambitious man who yearned to create his own Melancthonian destiny, did in fact accomplish Tycho's request. From Tycho's data, Kepler did in fact decode God's planetary design, ensuring an enduring legacy for both Tycho and Kepler.

There was one other gift of Tycho's that benefited Kepler. Tycho apparently gave a very positive assessment of Kepler to the emperor. On Tycho's death, the emperor bestowed the title of imperial mathematician on Kepler. The refugee evicted from Graz for steadfastly maintaining his Lutheran identity now had an impressive position in the staunchly Catholic court of Emperor Rudolf.

Uncovering God's design proved to be much more difficult than Kepler imagined. Kepler fell into the trap that had ensnared his predecessors. Kepler followed Plato with his axiomatic belief that the design of the heavens was perfect. Then, like Eudoxus, Apollonius, Ptolemy, and Copernicus before him, rather than asking God what God deemed to be perfect, Kepler imposed his own views of perfection on God. Kepler, as did those before him, approached the data with the hope of confirming his prejudices. This continued for 5 years until, at a dead end, Kepler recognized his folly and confronted the weakness in his analysis just as he had earlier confronted the weaknesses of his character. Kepler shed his prejudices and in earnest sought God's perfection. This unencumbered search for truth was the transition point that allowed Kepler to succeed where others had failed. We will revisit this point later in the chapter, where we give a mathematical presentation of Kepler's *New Astronomy*.

Besides the psychological barriers and the technical difficulties that vexed Kepler, Kepler had to cope with a legal battle as well. Two years after Tycho's death, Tycho's son-in-law, Tengnagel, took note of the missing Mars data. Although the emperor granted Kepler Tycho's title as court mathematician, the emperor could not dispossess Tycho's heirs of their inheritance. Legalities forced Kepler to forfeit the data for a 2-year period, during which Kepler wrote

---

[15]Suspicions fall on Tycho's distant cousin, Eric Brahe. Eric, a stranger to Tycho, arrived in Prague from Denmark just prior to Tycho's death. Peter Andersen, a professor at the University of Strasbourg, believes that Eric was a paid assassin in the service of King Christian IV. Other suspects include Catholic members of Rudolf's court who were jealous of the influence of the foreign Protestant. Another possibility is that Tycho concocted his own medications that contained mercury and inadvertently poisoned himself. A ridiculous charge against Kepler has also been made.

his work *The Optical Part of Astronomy*. The ghost of Tycho must have been watching over Kepler as this hiatus proved to be critical to Kepler's success. While Tegnagel seemingly distracted Kepler from his objective, he inadvertently led Kepler to the solution.

In writing *The Optical Part of Astronomy*, Kepler became the world's leading authority on the ellipse. He read Apollonius' work, *Conics*, and distilled all its contents, including Apollonius' review of Archimedes' work. Kepler became so knowledgeable that he discovered a property of the ellipse that both Archimedes and Apollonius had overlooked.[16] Kepler's familiarity with the geometric properties of the ellipse would allow him to decode Tycho's data when he subsequently recovered the observations from Tegnael.[17] Two years after recovering those data, Kepler found God's truth; the planetary pathway that he had been seeking over 7 years was the ellipse.

Both *The Optical Part of Astronomy* and *New Astronomy* are pivotal works of historical significance. Both not only present discoveries of significance but also are gateways to further discoveries. *New Astronomy* gives two laws of planetary motion. Newton generalizes Kepler's results into more fundamental laws of motion, and then demonstrates that the elliptic orbits of Kepler satisfy Newton's fundamental laws. *The Optical Part of Astronomy* marks the beginning of modern optics. In 1612, using principles established in that treatise, Kepler was the first to describe the optics of a telescope and in doing so explained to a confused Europe how the instrument works. Kepler would later contribute more to science and mathematics. In his book *Stoichiometry*, Kepler presents a method for determining volumes of objects constructed by rotating a planar shape around a central axis, This method is a precursor to integral calculus.[18] In 1619, Kepler presents another law of planetary motion, a formula relating the period of a planet's orbit with the major axis of the elliptic orbit.

By any measure of justice, these achievements deserve reward. Neither nature nor the principles of the Counter-Reformation rest on justice. After publication of *New Astronomy*, tragedy visited Kepler in both guises. In 1611 Kepler's 6-year-old son died; 2 months later, Kepler's wife Barbara followed her son.

Prague was the epicenter of the Thirty Years' War, a senseless war in which the political elite exploited all the divisions of the day for personal gain. While Europe was poised for war, the impetus that set the cannons firing was a personal rivalry between Emperor Rudolf and his brother Mathias. In the course of this rivalry, a Protestant army wrought its savagery against the populace of Prague. Fortunately, Kepler lived in a Protestant neighborhood, and the target of the

---

[16]Kepler discovered that the ellipse is the lattice of points with the property that the sum of the distances to both foci is a fixed value. This property allows one to sketch an ellipse. On a piece of paper, peg a loose string to two fixed points, the foci. Then, with a pencil, stretch the string taut and mark the point where the string bends. Do this in all directions to sketch out an ellipse.

[17]During the 2 years in which Tengnagel possessed the data, the data were ignored as an unwanted orphan. On promises that Kepler would write favorably of Tycho's contributions, Tengnagel finally relented and returned the data to Kepler, where they were once again in capable hands.

[18]Archimedes had also discovered the method and used it in his proof of the volume of a sphere.

army's wrath was Catholics. The rivalry came to an end when Mathias soundly defeated an army hastily assembled by Mathias' incompetent brother Rudolf. In 1612, Rudolf died a powerless figurehead.

With the demise of Rudolf and the ascension of Mathias, a ferociously anti-Protestant clique came to power. The clique continued to rule beyond Mathias' death as Ferdinand assumed the title of emperor. This is the same Ferdinand who caused Kepler's flight from Graz with his 1600 anti-Protestant decree, and in 1619 Ferdinand's ascendancy would confront Kepler with a déjà vu experience.

Seeing the centers of power shift, in 1611, Kepler once again initiated contact with his Lutheran community in Tubingen. As the acknowledged preeminent scientist in all of Europe, Kepler, a devoted Lutheran who placed his family at risk on behalf of the Lutheran cause, by any sense of justice, deserved a position at Tubingen. The leadership of the Lutheran Church did not consider Kepler's steadfast affirmation of the Lutheran faith in the face of eviction as sufficient proof of Kepler's commitment to the Lutheran community. Instead, they burrowed deep into Kepler's files[19] and noted his refusal several years earlier to unconditionally accept the the Formula of Concord.[20]

Kepler's signature would have been his passport to Tubingen, the place foremost in his heart. But just as Kepler could not, out of principle, convert to Catholicism, he could not, out of principle, sign a document that he did not fully believe in. The Counter-Reformation's sinister demand for discipline displaced Melancthon's philosophy as the guiding spirit of the Lutheran Church. With Kepler's refusal to sign the document, he unknowingly sealed the space between himself and the leadership of the Lutheran Church. All of Kepler's future efforts to reconcile himself with the Lutheran Church ended with bitter disappointment. The upstart who resided in the Catholic court of Emperor Rudolf could not be trusted.

Once again, unable to procure a position at a Lutheran university, Kepler looked for alternative employment outside of the virulently anti-Protestant Prague. In 1612, he was able to arrange a position at a school in the nearby town of Linz. The arrangement suited both Mathias, who retained Kepler as imperial mathematician on account of Mathias' need for a superb astrologer, and Kepler. Both parties saw an advantage in Kepler's absence from Prague while at the same time being within calling range when the emperor needed his services. In Linz, Kepler rather cluelessly continued to attempt to resolve his difficulties with the Lutheran church. Kepler contacted the local pastor, a man named Hitzler, with the best of intentions. Forewarned of Kepler's insolence by higher authorities of the Lutheran church, Hitzler skimmed the more controversial aspects of his conversations with Kepler and placed their contents into Kepler's file until the case for excommunication was complete. On July 31, 1619 the theological leaders of Tubingen had sufficient evidence

---

[19] Keeping files on subjects is a long tradition that the philosophically irreconcilable Calvinists, Catholics, and Lutherans all agreed on.

[20] The Formula of Concord was the Lutheran Church's answer to Catholicism's Council of Trent. It was a dogmatic set of principles that Church leadership established to define the Lutheran faith.

and cause to finalize Kepler's excommunication, and Kepler was officially not a member of any religious community.

Excommunication was not the only punishment that the Lutheran community meted out to Kepler. In September 1620, after a 5-year investigation, the judge in the case accusing Kepler's mother of witchcraft pronounced his verdict: guilty. That the investigation occurred during the same years as the Lutheran leadership was preparing the Kepler files cannot be a coincidence. Wishing to place Kepler within the clamp of religious authority, the Lutherans waged an assault against Kepler.

Although Kepler's efforts on his mother's behalf did not allow her to escape the verdict, the sentencing, while cruel, was far less severe than that of others found guilty of the same charge. Unlike six others who during 1615–1616 the very same court sent to the pyre, thanks to her son's dedicated effort, Katherina Kepler's punishment would be torture incognito. The executioner displayed an arsenal of instruments of torture, all the while giving a graphic description of their application. Throughout the ordeal, the executioner solicited a confession from the victim. At the end of the executioner's performance, Katherina Kepler, showing the steely disposition that she bequeathed to her son Johannes, looked the executioner in the eye and said, "Do what you want to me. Even if you were to pull one vein after another from my body, I would have nothing to confess."[21] A week after the punishment, the authorities released Katherina Kepler. She died shortly after at the age of 76.

Kepler's residence in Linz was at the mercy of a temperamental Europe. The divisions of the Counter-Reformation pulled on the Holy Roman Empire like an overly taut guitar string that is being stretched to an even higher pitch, an eventual break was inevitable. War raged between Protestant armies, raised among the Holy Roman Empire's Protestant states and the Catholic army of the emperor. In 1620 the battlefront was at the edge of Prague. In 1626, the battlefront moved to Linz, and Kepler wanted out. When order was restored, Kepler petitioned the emperor to leave his posting; the reason offered by Kepler was that he was unable to complete the printing of the *Rudolphine Tables* that Rudolf had commissioned before Tycho's death. The printer's house, along with his equipment, had been destroyed during the battle of Linz. In fact, Kepler's school, along with all of Linz, had become dysfunctional.

Kepler finished the printing of the *Rudolphine Tables* later in 1927, and it may have been sensible for him to deliver a copy to the emperor, except the Counter-Reformation compelled one to follow a different sensibility. Around the time of the *Tables'* completion, emperor Ferdinand resumed his anti-Protestant decrees, this time targeting government officials who were not Catholic. Kepler, squarely within Ferdinand's crosshairs, fit both these categories. Not having any other options, Kepler with some trepidation traveled to Prague. While in Prague, Kepler found a patron. General Wallenstein sought the astrological services of Kepler, and as potentate of Sagan in Selesia, the general offered Kepler employment away

[21] This quote comes from James Connor's *Kepler's Witch*, page 18.

from Prague. In Selisia, somewhat relieved of the emperor's decrees, Kepler took residence until his death.

Death came with the arrival of winter in 1630. Hoping to secure payment on a financial note, Kepler journeyed to Linz through inclement weather. He became ill with a raging fever. When a Lutheran minister present at Kepler's bedside asked Kepler how he would gain salvation, after all protocols of the Lutheran Church did not permit the minister to administer last rites to the excommunicated, Kepler replied that his salvation was in Christ. With that tidy summary of his lifetime conviction, Kepler transited to death.

MEANWHILE IN ITALY

A peculiar triangle defined Kepler's social relations in the latter part of his life. The three elements of the triangle are the Lutheran Church, the Jesuits, and Kepler himself. The relations are very simple to state. Kepler loved the Lutheran Church, which answered Kepler's love with scorn. The Lutherans hated the Jesuits, who returned their enmity. The Jesuits loved Kepler, who rejected them with kindness. As a group, the Jesuits were among the best educated men in the world. Many admired Kepler's accomplishments. Kepler had made the most convincing case for Copernican astronomy, an argument that many Jesuits found convincing. Furthermore, Kepler reconciled his findings with biblical interpretation in an elegant and discrete manner; he published a scholarly book. Until his death, the Jesuits attempted to convert Kepler.

This wooing of Kepler is somewhat at odds with the Jesuits' treatment of another famous individual. In 1616, Richard Bellarmine, Catholicism's highest-ranking Jesuit, sat in the palace that housed his office, awaiting a man that he would admonish on behalf of Pope Paul V. There would be differences between this admonishment and the punishment that Bellarmine had dished out to Bruno only 16 years earlier. The cardinal was no longer a member of the Holy Office of the Inquisition, and this was not a formal investigation, but a warning. Nevertheless, to convey the gravity of the offense, members of the Holy Office of the Inquisition escorted the offender to the office of Richard Bellarmine, and another high-ranking official from the inquisition, Father Seghizzi, was present. Richard Bellarmine issued a cease-and-desist order requiring the offender to abandon his defense of Copernican astronomy. Father Seghizzi punctuated the cardinal's order with his own "you better follow this injunction or else" warning. The offender, Galileo Galilee, previously somewhat out of touch with church sentiment, had tuned in and clearly received the signal.

How does one reconcile the injunction against Galileo with efforts to convert Kepler, the man who had given the most convincing defense of Copernican astronomy? The answer is multifaceted, incorporating political, personal, spatial, and temporal dimensions; we emphasize personality. We all remember two very different personality types from the time when we were growing up. One was charismatic, a boy who could break the rules and get away with it. This character could toss pie at his targets while everyone laughed along, but not a single pie ever came his way. The other hapless character type sat in opposition to his charismatic counterpoint. He was the boy that everyone loved to throw pie at. Galileo, a man who possessed a scientific mind alongside artistic sensibilities, was the charismatic boy who had grown into a proud man. Kepler, the poor, unloved boy targeted by social circles above his own, learned how to temper his sense of righteous indignation with humility, but would never compromise his integrity.

When, from the pulpits of Florence's lesser-known clergy, insults against the Copernican scientist Galileo became public, the proud man was not disposed toward turning his cheek. The man with an impressive résumé: mathematics professor at Pisa; mathematics professor at Padua; court mathematician to Cosimo Medici, the duke of Florence; honored guest of Pope Paul V; celebrated author throughout Europe, was going to throw pie in his detractors' faces. Galileo's plan was to use his fame and connections to gain another audience with the pope. Once in the pope's company, Galileo would apply his formidable skills in persuasion to convert the pope to Copernicanism. With this accomplished, the unworthy upstarts whose insults toward Galileo only displayed their own ignorance would be silenced.

Galileo's plan was spectacularly untenable, and his supporters around him to no avail did their best to impress Galileo of its folly. If I were there I would have slapped his face, reminded him of the touchiness of the Counter-Reformation as well as of Bruno's death just 16 years before, and...Galileo would have done what Galileo was wont to do anyway.

Controversies of the Counter-Reformation inundated the Holy See. Protestant discontent in the Holy Roman Empire percolated throughout Europe, challenging Catholic authority. French jealousy of Spain threatened to split the Catholic nations. Spanish arrogance only exacerbated both of these problems. With these issues preoccupying the pope, the pope on his own would not have given any thought whatsoever to Copernican astronomy. But unwillingly pressed into the controversy, the pope had no room for maneuver. He would most certainly take a conservative, biblical interpretation consistent with the Council of Trent.[22] The charismatic Galileo, who had always had his way, was in for a rude shock. After Bellarmine's very direct anti-Copernican message, perhaps Galileo had a "phew" moment. He had instigated a losing battle, but emerged without wound. There would be no inquisition, and while the battle resulted in the censorship of several books on Copernican astronomy, including Copernicus' *On Revolutions*, Galileo's works remained uncensored. All Galileo needed to do was to abide by the terms of the injunction and remain publicly quiet on the issue of Copernican astronomy.

---

[22]Pope Paul III commissioned The Council of Trent in 1545 for the purpose of clarifying Catholic doctrine in the face of the Protestant challenge. The council held 25 sessions between 1545 and 1563, culminating in a series of edicts that served the dual purposes of differentiating Catholicism from Protestant beliefs and strengthening the Church hierarchy.

For years, Galileo abided by the injunction. Oh, but how this insult festered within. In 1623 Galileo saw changed circumstances that would allow him to redress the indignation. Following the death of Pope Gregory XV, an admirer of Galileo's Martin Barbari ascended to the papacy and assumed the name Pope Urban VIII. Soon Galileo gained an audience with the new pope, during which time Galileo proposed a project that he had been mulling over for years. As presented to the pope, the project was to give an unbiased presentation of the scientific arguments both in favor of and against Copernican astronomy. The new pope agreed that Galileo should pursue this undertaking.

Eight years after striking this agreement, 11 years after the death of Robert Bellarmine, 2 years after the death of Johannes Kepler, Galileo, himself 68 years of age, completed this project. The pope received a copy of the work, *Dialog of Two World Systems*. What he read was not the unbiased presentation that Galileo had described to him years earlier. The presentation was in the form of a debate. The chief debater on the side of the geocentric divide had the name Simplicius. This simpleton was always bested by the superior logic of his Copernican counterpart, Salviati—at long last the pie that Galileo had been baking for so long landed in the faces of his detractors. The pope took offense at being misled, and he was not the only one offended. The Jesuit community, whose stomachs had not yet recovered from their share of the Galilean pie,[23] was also smarting for revenge. The Jesuits pulled out Galileo's file, uncovered the injunction of 1616, and presented it to the pope. Now Pope Urban felt fully deceived by Galileo, for this was one bit of information that Galileo neglected to bring to Pope Urban's attention at the outset of the project. The Jesuits had no problem convincing Pope Urban that this affair belonged in the capable hands of the Office of the Inquisition.

Galileo, the proud man from Tuscany who as a boy could get away with breaking the rules, discovered his limitations and the Inquisition's authority. In 1634 the Inquisition found Galileo guilty of harboring heretical sentiment and violating the 1616 injuncture. His sentence was not brutal in the same vane as Bruno's. Rather, the Inquisition subjected Galileo to humiliation. Through a public statement confessing his crimes, Galileo was forced to abjure. Afterward he returned to Tuscany, where he was placed under house arrest. At age 84 in 1642, the same year that saw the birth of Isaac Newton, Galileo passed away. His book *Dialog of Two World Systems*, while banned in Italy, was well received elsewhere throughout Europe. Arguments from that work concerning relative motion shaped Einstein's ideas on his theory of relativity.

In an ideal world justice is blind and impersonal. In a world of relative peace, this ideal is difficult to achieve. In the world of the Counter-Reformation, where Bruno, Kepler, and Galileo resided, obedience trumped justice. Each was punished, not for their beliefs or any misdeeds, but for their independent dispositions.

---

[23]In 2010, Galileo wrote less than kind words about opinions held by high-ranking Jesuits. The fact that the Jesuits had previously shown Galileo gracious hospitality only caused the insults to burn with greater heat.

While Kepler and Galileo may have suffered the consequences of their dispo-sitions, we have reaped their rewards. Kepler's unequaled technical mastery of the subject of astronomy as exhibited in *New Astronomy* presents the case for a heliocentric universe that nobody could refute. Galileo's *Dialogs of Two World Systems* is an enjoyable read, accessible to the layperson, that explains why the universe is heliocentric.[24] On the eve of the death of Galileo and Newton's birth, the Roman Catholic Church stood alone on its geocentric planet. The rest of Europe following Kepler and Galileo's lead, joined Aristarchus and Copernicus.

New Astronomy, IN KEPLER'S OWN LIKENESS

New Astronomy is in some sense the book that Galileo promised to Pope Urban. It presents the geocentric model of Ptolemy, Copernican astronomy, and Tycho's intermediate compromise, pinpointing the weaknesses in each. While at the end, Kepler corrects the deficiencies of the Copernican model and demonstrates its superiority, he does so not in the "geometry in your face, Simplicius" manner of Galileo, but with a subtler poison administered while the witnesses watch unknowingly, even as the corpses lie at their feet.

Beyond the astronomy, analytics, and physics, there is another element within *New Astronomy*. It is as much a lens on the heavens as a mirror of self-reflection. Kepler did not write an end-of-journey septic, Euclidean argument. Kepler, in somewhat the fashion of a diarist, takes the reader along the whole journey, through all the false leads, wrong assumptions, and incorrect arguments, as well as the brilliant insight—and he allows the reader to peek at his image in the mirror. Concerning his own character, Kepler reveals a complex personality that is somewhat of a riddle. Kepler is deeply pious and spiritual:

> I too, implore my reader, when he departs from the temple and enters astronomical studies, not to forget the divine goodness conferred upon men, to the consideration of which the psalmodist chiefly invites. I hope that, with me, he will praise and celebrate the Creator's wisdom and greatness, which I unfold for him in the more perspicacious explanation of the world's form, the investigation of causes and the detection of errors of vision. Let him not only extol the Creator's divine beneficence in His concern for the well being of all living things, expressed in his firmness and stability of the earth, but also acknowledge His wisdom expressed in its motion, at once so well hidden and so admirable.[25]

Kepler demonstrates that he knows the scriptures, possibly by heart:

> Joshua makes mention of the valleys against which the sun and moon moved, because when he was at the Jordan it appeared so to him...David was describing

---

[24]While not in technical style, *Dialog of Two World Systems* gives a penetrating analysis of relative motion and inertia. There is also a demonstration of what is now known as the *Coriolis force*, a force that effects motion as viewed in a rotating frame. Stephen Hawking identifies Galileo's work as the beginning of modern physics. My opinion is that Kepler's works predate Galileo's and initiate modern physics.

[25]Kepler, *New Astronomy*, p. 65.

the magnificence of God made manifest (and Syracides with him), which expressed so as to exhibit them to the eyes...Joshua meant that the sun should be held back in its place in the middle of the sky for an entire day.[26]

Yet Kepler is of an independent mind, as evidenced by his own interpretation of scriptures:

> Now the holy scriptures, too, when treating common things (concerning which it is not their purpose to instruct humanity), speak with humans in a human manner, in order to be understood by them. They make use of what is generally acknowledged, in order to weave into other things more lofty and divine.... Now God easily understood from Joshua's words what he meant, and responded by stopping the motion of the earth, so that the sun might appear to him to stop.[27]

Kepler also shows a Socratic capacity to review the sacrosanct with fresh logic and cause consternation among the dogmatists.

> Joshua meant that the sun should be held back in its place in the middle of the sky for an entire day with respect to the sense of his eyes, since for other people during the same interval of time it would remain beneath the earth.[28]

While demonstrating his clarity of thought as above, Kepler could be maddeningly incomprehensible:

> It is true with respect to the shape that this was a fiction, since the path of the planet is not a circle. But with respect to the position, and the center B, it is not a fiction, but true: thus this fiction described about B is the opposite of the prior fiction described about C.[29]

While independent minded, Kepler is self-deprecating. If Kepler throws pie in anyone's face, it's his own:

> Galatea seeks me mischievously, the lusty wench: She flees to the willows, but hopes I'll see her first.[30] It is perfectly fitting that I borrow Virgil's voice to sing about Nature. For the closer the approach to her, the more petulant her games become, and the more she again and again sneaks out of the seeker's grasp just when he is about to seize her through some circuitous route. Nevertheless, she never ceases to invite me to seize her, delighting in my mistakes.[31]

Although self-deprecating, Kepler has a Galilean moment where he demonstrates a level of contempt for common opinion:

---

[26] Ibid., p. 61.
[27] Ibid., pp. 60, 61.
[28] Ibid., p. 61.
[29] Ibid., p. 465.
[30] Kepler quotes Virgil, *Eclogues*, 3, 64.
[31] Kepler, *New Astronomy*, p. 573.

Advice for idiots. But whoever is too stupid to understand astronomical science, or too weak to believe Copernicus without affecting his faith, I would advise him that, having dismissed astronomical studies and having damned whatever philosophical opinions he pleases, he mind his own business and scratch in his own dirt patch, abandoning this wandering about the world. He should raise his eyes (his only means of vision) to this visible heaven and with his whole heart burst forth in giving thanks and praising God the Creator. He can be sure that he worships God no less than the astronomer, to whom God has granted the more penetrating vision of the mind's eye, and an ability and desire to celebrate his God above those things he has discovered.[32]

Of note in this quote are both the consistency with Kepler's earlier self-appraisal, his disdain for common opinion, and the Melancthonian mission of the astronomer, a celebration of God.

There is one quality of Kepler's that he displays with consistency, it is his integrity. Throughout the book, we find confessions of error, often accompanied by his pie-in-my-face self-deprecating humor:

My first error was to suppose that the path of the planet is a perfect circle, a supposition that was all the more noxious a thief of time the more it was endowed with the authority of all philosophers[33] . . . it must appear strange to the astronomer that there remains yet another impediment in the way of astronomy's triumph. And me, Christ!—I had triumphed for two full years. Nevertheless. . .it will easily be apparent that we are still lacking.[34]

But then what they say in the proverb, "A hasty dog bears blind pups", happened to me. . . . Now having seen from the observations that the planet's orbit is not perfectly circular, I immediately succumbed to this great persuasive impetus, believing that those things that were called absurd in fabricating the circle of chapter 39, now transmuted into a more probable form. . . . If I had embarked upon this path a little more thoughtfully, I might have immediately arrived at the truth of the matter. But since I was blind with desire, I did not pay attention to each and every part of chapter 39, staying instead with the first thought to offer itself. . .and thus entered into new labyrinths from which we will have to extract ourselves.[35]

A PATH STREWN WITH CASUALTIES

Kepler launched into his investigation of the heavens as others did before him. Beginning with *Mysterium Cosmographicum*, following precedent, Kepler sought beauty and perfection while encumbered by preconceived notions. On noting that the five spaces between the six planets matches the number of Platonic solids, Kepler found beauty in an arrangement whereby the Platonic solids fill the gaps

[32] Ibid., p. 65.
[33] Ibid., p. 417.
[34] Ibid., p. 451.
[35] Ibid., p. 455.

between the planets. Kepler was so enamored with the beauty of this arrangement that he felt God could have arranged the planets in no other way.

Moving onto his investigation into Mars' motion, Kepler, again following precedent, accepted the circle of his forebearers as the perfect shape on which God configured the motion of the heavens. For 5 years Kepler maintained this assertion and after 5 years' effort, Kepler believed that he had found God's circular design. Let us take some time to investigate Kepler's method.

While burdened by a preconceived idea, Kepler also initiated his studies with an insightful concept, the simpler explanation is the more likely candidate. Attributing this concept to common wisdom, Kepler states "it is the most widely accepted axiom in the natural sciences that Nature makes use of the fewest possible means."[36,37] Kepler places three proposals before his reader, Ptolemy's geocentric proposal, the Tychonic compromise, and Copernicus' heliocentric model[38] and leaves no doubt in the mind of the reader as to which is the simpler model. In chapter 1, Kepler presents a diagram of Mars' trajectory (see Figure 7.1) as prescribed by Ptolemy's and Tycho's universe. Just above the figure is the following explanation articulating why it is that the geocentric and compromise models are implausible:

> If one were to put all this together, and were at the same time believe that the sun really moves through the zodiac in the space of a year, as Ptolemy and Tycho Brahe believed, he would have to grant that the circuits of the three superior planets through the ethereal space, composed as they are of several motions, are real spirals, not (as before) in the manner of balled up yarn, with spirals set side by side, but more like the shape of pretzels, as in the following diagram [Figure 7.1].

Kepler lays plain that the trajectory as a circle containing the sun is a far simpler design. Underlying this simplification is the unifying theme that the sun supplies the force of motion to all the planets. This is Kepler's starting point. While favoring the sun as a source of locomotion proved to be invaluable, acceptance of his forebearers' circular prejudice would cause a significant detour in Kepler's journey of discovery.

Although a devout Copernican, Kepler did not share Copernicus' disdain for the equant. In fact, Kepler found the equant to be a most ingenious device that explains the varying speed of Mars as it circles about the sun. In opposition to Copernicus, Kepler disdained the epicycle, because it offended his quest for simplicity. While Kepler could attribute the motion of a planet about a deferent to the sun, he saw no physical cause for a planet to rotate about an epicycle.

Recall that Apollonius originally proposed the epicycle as an explanation for retrograde. With Copernicus' demonstration that relative motion of Earth and

---

[36]Ibid., p. 51.
[37]This statement is a precursor to the use of optimization methods in physics. Fermat's law has light following the path of minimum time, and Hamilton's extension of Euler's principle of least action gives the pathway of an object as that that minimizes the difference between the object's kinetic and potential energies.
[38]Kepler credits Aristarchus as the original Copernican.

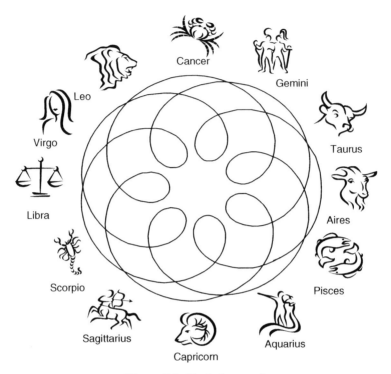

Cancer

Gemini

Leo

Virgo

Taurus

Libra

Aires

Scorpio

Pisces

Sagittarius

Aquarius

Capricorn

Figure 7.1. Kepler's pretzel.

Mars causes retrograde (Figure 5.3), epicycles are no longer necessary. The position of the planets on the heliocentric racetrack accounts for retrograde and the equant moderates a planet's speed in accordance with observation. Kepler, the astronomer, had created his vision, which he dubs the "vicarious hypothesis." It is somewhat of a heliocentric variant of Ptolemy's universe. Kepler exchanges the position of the sun and Earth, so that the sun occupies the position of the supreme entity at the center of the universe. He then eliminates Mars' epicycle, but not its equant. These changes on their own matched Ptolemy's data, and had Kepler lived at the time of Ptolemy, he could have claimed success. But Kepler had the more exacting standards of Tycho's observations, which informed Kepler that this was not good enough. Kepler, the geometer, had the task of specifying the positions of the center of Mars' deferent and the equant.

At the outset of his investigation, Kepler saw no reasoning behind Ptolemy's placement of the center of Mars' deferent halfway between the universe's center (for Ptolemy, Earth; for Kepler, the sun) and the equant. Kepler thought this to be artificial and removed the constraint. He then sought the true positions of the deferent's center and the equant. The geometric method of Kepler rests on the foundations of Ptolemy. Whereas Ptolemy required three observations of opposition, Kepler demanded four, an additional observation to yield the additional degree of freedom allotted to the distance between the equant and the sun.

Following Ptolemy, Kepler developed an iterative method to find a solution that matched his observations.

Kepler's additional observation causes his iterative method to be far more complex than that of Ptolemy. Just as foul odor accompanies sweaty feet, tedium accompanies complexity. Whereas Ptolemy was able to determine the positions of the deferent's center and equant after executing only three iterates of his method, Kepler required in excess of 70 iterations. It was an exhausting process and we emphasize the toll it took on Kepler by repeating the quote at the end of Chapter 5:

> If this wearisome method has filled you with loathing, it should more properly fill you with compassion for me, as I have gone through it at least 70 times at the expense of a great deal of time, and you will cease to wonder that the fifth year has now gone by since I took up Mars, although the year 1603[39] was nearly all taken up by optical investigations.[40]

After laboring through his difficulties and calibrating the positions of the eccentric's center and the equant, reward awaited Kepler. Kepler's comparison of his calculated trajectory with different observations from those used to calibrate his model is favorable. Kepler's 5-year investigation of Mars is near a successful completion.

> You see then, O studious reader, that the hypothesis founded by the method developed above, is able in its calculations not only to account, in turn, not only for the four observations upon which it was founded, but also able to comprehend the other observations within two minutes...I therefore proclaim that the achronycal positions displayed by this calculation are as certain as the observations made with Tychonic sextants can be.[41]

A solitary traveler determines the end of his journey. There is no company to inform the traveler that something else lies beyond, that his current position is not the endpoint, but a dead end. Kepler had gone beyond Ptolemy and Copernicus. He had modeled Mars' trajectory with greater accuracy than had any predecessor. He could claim success and give a convincing argument, one that no other individual in the world would have been able to refute, that he had found the true path of Mars. Only Kepler could refute Kepler, and he did so; "Who would have thought it possible? This hypothesis, so closely in agreement with the achronycal observations is nonetheless false."[42]

What caused Kepler to acknowledge failure rather than declare victory? Kepler went on to calibrate his model with the latitudes from Tycho's observations. The

---

[39] Kepler was still obsessed with circular motion after recovering Tycho's observations from Tegnagel. It would be another year before he would apply his lessons from Apollonius that he had discovered during his investigations into optics.
[40] Kepler, *New Astronomy*, p. 95.
[41] Ibid., p. 276.
[42] Ibid., p. 281.

calibration required an adjustment in the longitudes which caused a discrepancy, of 8 minutes, $\frac{2}{15}$ths of a degree. To illustrate the magnitude of this error, hold a butter knife vertically at arm's length with the edge of the blade facing your eye. The edge is of the same order of magnitude as Kepler's discrepancy. How tempting it must have been to attribute this nearly insignificant difference to other causes. There are so many candidates: transcription error, instrumentation error, calibration error, error caused by refraction of light. But integrity did not allow Kepler to bait himself and he instead accepted the failure:

> Since the divine benevolence has vouchsafed us Tycho Brahe, a most diligent observer, from whose observations the eight minute error in this Ptolemaic computation is shown, it is fitting that we with thankful mind both acknowledge and honor this benefit of God. For it is in this that we shall carry on, to find at length the true form of the celestial motions, supported as we are by these arguments showing our suppositions to be fallacious. In what follows, I shall myself, to the best of my ability, lead the way for others on this road. For if I had thought that I could ignore eight minutes of longitude, in bisecting the eccentricity I would already have made enough of a correction in the hypothesis found in chapter 16. Now because they could not have been ignored, these eight minutes alone will have lead the way to the reformation of all of astronomy, and have constituted the material for a great part of the present work.[43]

The preceding paragraph's characterization of Kepler's efforts as a dead end and failure ignores the relevance of those efforts. Kepler does, indeed, register a victory by taking down others with him. While Kepler prostrates at the reader's feet, confessing his errors as though the reader is the Inquisition, Kepler surreptitiously dips his dagger in the poison that he drinks and aims it at the bellies of both the Ptolemaic and Tychonic universes. For Kepler demonstrates that if his vicarious hypothesis is wrong, then both Ptolemy's and Tycho's models suffer the same error. The self-appraising horoscope writer who admits that he jests with others, jests with his reader. Kepler does not pronounce the competing models dead, but to the attuned reader,[44] there is no other possibility.

There is another larger relevance to the effort, the prophetic claim that the 8-minute error would reform astronomy. There is a perception that scientific development is inevitable. The body of knowledge grows at a steady pace as the logical extension of known ideas generates further knowledge. If one considers science as a body of knowledge, the appropriate analogy is that the body is forever adolescent and one never knows when it will hit a growth spurt. For nearly three millennia, humankind had fixated on the circle as the perfect shape on which the Gods or God configured the heavens. The perfect circle stunted scientific growth. To find the correct configuration, the circle had to be exposed for what it was, a chimera. Kepler's efforts allowed him to transit through the

---

[43]Ibid., p. 286.
[44]The most attuned reader is none other than William Donahue, who, in his introduction to his translation of *New Astronomy*, points out Kepler's victory.

circular barrier, and his honest appraisal of these efforts is how truth displaced perfection:

> Therefore, something among those things we have assumed must be false. But what was assumed was: that the orbit upon which the planet moves is a perfect circle; and that there exists some unique point on the line of apsides at a fixed and constant distance from the center of the eccentric about which Mars describes equal angles in equal times. Therefore, of these, one or the other or perhaps both are false, for the observations used are not false.[45]

The circular detour is not an exception but an omen. Mars is a prize to be grasped only by the most able but only after much sweat and unwavering pursuit. Having convicted the circle as a fiction, Kepler handicaps himself with another prejudice. In a letter to a friend, showing his deep respect for Archimedes and Apollonius, Kepler states that Mars' orbit cannot be an ellipse, for if it were Archimedes or Apollonius would have discovered it. Kepler does, however, demonstrate that the path is an oval, but the oval is a concept, not a precise geometric shape. It falls on Kepler to specify the shape.

Having dismissed the ellipse, Kepler resorts to another Apollonian device. He brings back the epicycle, even though he had previously dismissed this possibility. Copernicus had demonstrated that the trajectory of an epicycle on a circle is—well—an oval, and Kepler can think of no other way to generate such a trajectory. The vicarious hypothesis casts its shadow on this endeavor as well. After a promising start, Kepler must once again confess:

> You will say that we have come out worse, since in chapter 48 we came nearer the truth in our results. But, my good man, if I were concerned with results, I could have avoided all this work, being content with the vicarious hypothesis. Be it known, therefore, that these errors are going to be our path to the truth.[46]

More errors follow along with more confessions, before Mars capitulates:

> While I am thus celebrating a triumph over Mars, and fetter him in the prison of tables and the leg-irons of eccentric equations, considering him utterly defeated, it is announced in various places that the victory is futile, and war is breaking out again with full force. For while the enemy was in the house as a captive, and hence lightly esteemed, he burst all the chains of the equations and broke out of the prison of the tables. That is, no method administered geometrically under the direction of the opinion of chapter 45 was able to emulate in numerical accuracy the vicarious hypothesis of chapter 16 (which has true equations derived from false causes). Outdoors, meanwhile, spies positioned throughout the position of the whole circuit of the eccentric—I mean true distances—have overthrown my entire supply of physical causes called forth from chapter 45, and have shaken off their yoke, retaking their liberty. And now there is not much to prevent the fugitive enemy's joining forces with his fellow rebels and reducing me to desperation, unless I send

---

[45] Kepler, *New Astronomy*, p. 284.
[46] Ibid., p. 494.

new reinforcements of physical reasoning in a hurry to the scattered troops and old stragglers, and, informed with all diligence, stick to the trail without delay in the direction whither the captive has fled. In the following few chapters, I shall be telling of both these campaigns in the order in which they were waged.[47]

The ellipse enters Kepler's mind not as the prize, but as an assistant. The epicyclic trajectories, the candidates for the prize are complicated and not well suited to carrying out the calculations that Kepler requires. The world's leading authority on the ellipse looks to the ellipse as a tool for approximating the epicyclic trajectories. Below, the details of the required calculation are presented. Here, we note that a first approximation, the full circle yields numbers that are above the required values. Another approximation using an ellipse that is a best fit to the epicyclic trajectory of interest yields numbers below the required values. After what seems an excruciating labor, Galatea's[48] thumpings finally knock sense into Kepler. Seeing that between the circle and an ellipse is another ellipse, he discovers that the ellipse is the prize:

> My line of reasoning was like that presented in chapters 49, 50, and 56. The circle of chapter 43 errs in excess, while the ellipse of chapter 45 errs in defect. And the excess of the former and the defect of the latter are equal. But the only figure occupying the middle between a circle and an ellipse is another ellipse. Therefore, the ellipse is the path of the planet.[49]

THE PHYSICIST'S LAW

With the collapse of the vicarious hypothesis, Kepler initiated an investigation into the physical cause of Mars' motion. It is a difficult task, nature leaves the clues of its causes in its design, but without knowing the design, one cannot determine the cause. The two must be reconciled with one another. Kepler began this reconciliation and Newton would conclude it.

Kepler, the physicist, firm in his belief that the sun is the locomotive force behind the movement of the planets, begins to describe that force. There is much that he got wrong; in Kepler's day, force, like the oval, was a concept without definition. Kepler did not supply the definition, which is a cause for his shortcomings. But as first steps toward the Newtonian universe, Kepler did make significant contributions. Underlying these contributions is a penetrating intuition, the kind that distinguishes the genius from the very smart.

Kepler was the first to describe gravity as a force that could act across space:

> Every corporeal substance, to the extent that it is corporeal, has been so made as to be suited to rest in every place in which it is put by itself, outside the influence of a kindred body.[50]

[47] Ibid., p. 508.
[48] Recall Kepler's quote of Virgil; see footnotes 28 and 29 (above).
[49] Kepler, *New Astronomy*, p. 575.
[50] This comes very close to Newton's first law of motion.

Gravity is a mutual corporeal disposition among kindred bodies to unite or join together; thus, the earth attracts the stone much more than the stone seeks the earth. (The magnetic faculty is another example of this sort.)[51]

Kepler includes more passages, stating that (1) a small object seeks the center of a round body; (2) an aspherical earth would not necessarily attract objects toward its center; (3) two objects outside the influence of any other body would meet at an intermediate location, and the distance of each to that location is proportional to the mass of the other object; and (4) the tides are caused by the gravitational forces of the moon attracting Earth's waters.[52] These observations are remarkable for their time. If an apple did strike Newton's head, Kepler threw that apple.

Kepler did not attribute the force causing planetary motion to gravity. For Kepler, gravity was a force that unites objects, and he did not see how a uniting force could cause one body to hover away from the other as the planets orbit but never approach the sun. Indeed, Kepler's whole sideshow on gravity is meant to convince the reader that bodies can act on one another through distant and invisible forces. For Kepler the invisible force that causes Mars' motion is a magnetic force that emanates from the sun. Belying the uncertainty of a love–hate relationship, at times the magnetic force attracts Mars while at other times the magnetic force repels Mars and the resulting orbit is the ellipse.

Kepler attempts to quantify the strength of the magnetic force through an analogy between light and the magnetic force. A geometric argument illustrates that the intensity of light on an object is inversely proportional to the square of the object's distance form the light's source.[53] Kepler applies the same concept to the magnetic force and is the first to arrive at an inverse-square law for force. This is a point to log in memory as a controversy over priority for the inverse-square law arises in the next chapter.

There is another insight from Kepler, the physicist, that is the hallmark of genius. Kepler asserts that Mars follows a path that conserves angular momentum. Why is this the hallmark of genius? For Kepler this was an obscure result, far removed from any previous knowledge of motion. The very concept of angular momentum was unknown to Kepler, so there would be no a priori reason to look into this quantity. Conservation of angular momentum follows from Newton's laws of motion published in Newton's *Principia* 75 years after publication of Kepler's *New Astronomy*; in the following chapter we will use Newton's laws to prove conservation of angular momentum. Without Newton's laws there is no guide indicating a possibility of the conserved quantity. Conservation of

[51] Kepler, *New Astronomy*, p. 55.

[52] Kepler got the tides right, while Galileo was wrong. Galileo attributes the motion of the tides to the combined movements of Earth's revolution about the sun and its rotation about Earth's axis. While it is a clever argument that presupposes Coriolis' work (the Coriolis force), it is nevertheless incorrect.

[53] Follow a beam from its source. The area of the beam's cross section is inversely proportional to the beam's intensity. The area of the cross section increases by the square of the distance, so the intensity is inversely proportional to the square of the distance.

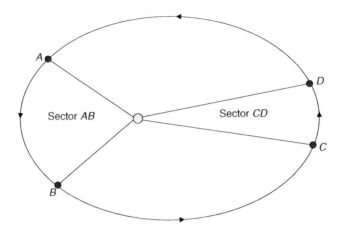

Figure 7.2. Kepler's second law relating sector areas to path time.

angular momentum does not announce itself but lies encrypted within Tycho's observations.

Kepler initially asserts the conservation of angular momentum in terms that are somewhat similar to those of modern physics. But then, foraging in the foothills of the fundamental theorem of calculus, Kepler transforms the assertion into a statement that relates speed to area, for it is this form of the conservation of linear momentum that is most useful to Kepler.[54] The statement is now known as *Kepler's second law:* The line joining the planet to the sun sweeps out equal areas in equal intervals of time.

Figure 7.2 illustrates Kepler's second law. In the figure, the planet takes the same time to travel through each of the two highlighted arcs, *AB* and *CD*. As the arc closer to the sun is longer than that more distant from the sun, the planet moves faster through the arc that is closer to the sun, an intuitive notion following from the belief that the sun is the source of the locomotive force acting on the planet. The second law states that the areas of the sectors are the same. The calculations mentioned above, one showing the circle in excess and one an ellipse that is deficient, are calculations of sector areas that Kepler then equates with times. He can then compare the times along a sector with actual observations.

This presentation of Kepler's second law begs the following question: What is his first law? The first law is that the planet's orbital pathway is an ellipse. The numbering convention of the laws throws one into a state of temporal confusion. Let us settle this confusion; the second law predates the first law. In fact, it is Kepler's genius that he uses the second law to detect the first. Once establishing

---

[54]We note that Kepler was not the first to forage in these foothills. Archimedes had been there long before. In his treatise *On Spirals*, Archimedes performs a similar operation and relates the length of a spiral path to the area enclosed within the spiral. Kepler never read any of Archimedes' works, but indicates that he had heard of them.

the second law in Chapter 40 of *New Astronomy*, Kepler invests the next 18 chapters seeking an orbit that satisfies the second law. Thus, Kepler's second law is his gateway to the first law, the ellipse.

This is so astonishing that I can't refrain from restating it. Kepler invested himself in an unproved concept, declared it sacrosanct, and used it as the vehicle by which he found God's design. Somewhere residing in Kepler's mind was a sense that conservation of angular momentum was not only correct but also the key to discovery. Was this sense physical intuition, geometric reasoning, a conviction of faith, or something else that I don't have the imagination to consider? That is the mystery of genius.

TYCHO'S GIFT

Before continuing, let's introduce some of the data that Tycho bequeaths to Kepler. Table 7.1 gives several observations at opposition.[55]

The observations give the location of Mars within the celestial sphere. The convention of the day was to divide the sphere into 12 longitudinal sectors, each spanning 30 degrees, and identify each sector by its resident constellation. Figure 7.3 illustrates the division on the ecliptic and gives the position of the first observation's longitude. Figure 7.4 illustrates the coordinitization of the entire celestial sphere and gives the position of the 1593 observation. While it is common to set the North Star at 90 degrees latitude, Kepler does not adopt this practice. Instead, he sets the latitude of the ecliptic at zero degrees. Paying homage to Copernicus longitudes and latitudes are heliocentric; a reported longitude gives the constellation of the observed as though the observer sits on the sun, and the sun is also the vertex of the reported angle of latitude.

TABLE 7.1. Observations at Opposition

Date	Time	Longitude	Constellation	Oriented Longitude	Latitude
November 18, 1580	1331	6.473889	Gemini	277.462778	1.666667 N
December 28, 1582	1558	16.925	Cancer	317.913889	4.1 N
January 31, 1585	0714	21.602778	Leo	352.591667	4.536111 N
March 6, 1587	1923	25.716667	Virgo	26.705556	3.683333 N
April 14, 1589	1823	4.383333	Scorpio	65.372222	1.2125 N
June 8, 1591	1943	26.716667	Sagittarius	117.705556	4.0 S
August 26, 1593	0527	12.266667	Pisces	193.255556	6.033333 S
October 31, 1595	1239	17.516667	Taurus	258.505556	0.133333 N
December 14, 1597	0344	2.466667	Cancer	303.455556	3.55 N
January 19, 1600	0202	8.633333	Leo	339.622222	4.513889 N
February 21, 1602	0213	12.45	Virgo	13.438889	4.166667 N
March 29, 1604	0423	18.619444	Libra	49.608333	2.433333 N

[55]The last two observations occur after Tycho's death.

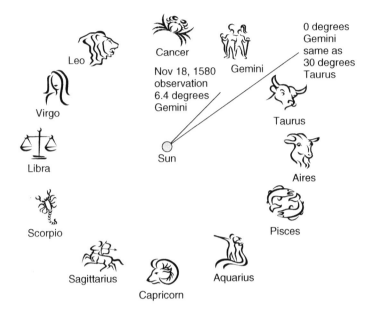

Figure 7.3. Estimation of longitude using constellations.

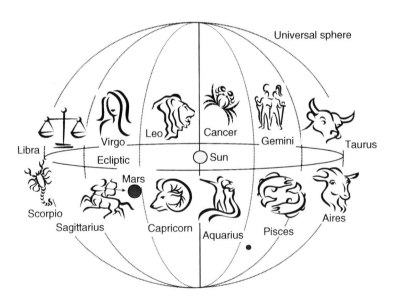

Figure 7.4. Mars' heliocentric location in Sagittarius on June 8, 1591.

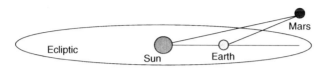

Figure 7.5. Geocentric and heliocentric Martian longitudes.

As Earthbound humans on Hven performed the observations, the observations are Earth-centric projections onto the celestial sphere. As Figure 7.5 illustrates, the Earth-centric and sun-centric latitudes differ. In describing his heliocentric model, Kepler had to account for the differences. At opposition, the Earth-centered and sun-centered longitudes are identical; however, at times other than opposition, these two longitudes differ. Kepler had to adjust an observation's longitude whenever the observation was not taken at opposition.

MARS' BED

Because of Kepler's perpetual bemoaning of ideas that proved false, *New Astronomy* does not give the reader a balanced sense of the remarkable discoveries that Kepler got right. Table 7.2 gives Kepler's estimation of Mars' orbital parameters along with the modern values in parentheses. Consider all of the resources available to the modern astronomer that were not available to Kepler, and one is left in awe by the accuracy of Kepler. Bravo Kepler.

TABLE 7.2. Mars' Estimated Orbital Parameters

1. The length of the semimajor axis is 1.52350^{56} AU (1.52368).
2. The distance from the sun to the center of the ellipse is .14116 AU (.14218).[57]
3. The length of the semiminor axis is 1.51690^{58} AU (1.51703).
4. The major axis points toward Leo at $29.011^{\circ 59}$ (28.537). This is the direction in which we orient the x axis. The oriented longitude of Table 7.1 gives the longitude with respect to the major axis; i.e., a point that lies along the positive x axis, 29.011° in Leo, has longitude zero.
5. The angle of inclination between the ecliptic and Mars' orbital plane is $1.84^{\circ 60}$ (1.85).
6. The line of intersection between the ecliptic and Mars' orbital plane points toward $16.77^{\circ 61}$ (16.562) Taurus.

---

[56] Kepler, *New Astronomy*, p. 540. AU (astronomical unit) 1 AU = mean distance between the sun and earth.
[57] Ibid. p. 540
[58] Kepler, *New Astronomy*, p. 543. Kepler does not give the length. Instead, he gives the incursion of the orbit from the circle of radius, 1.52350 AU.
[59] Kepler, *New Astronomy*, p. 535.
[60] Ibid., p. 610.
[61] Ibid., p. 606.

The results are not a simple matter of applying known geometry to the observations and carrying out the calculations. Geometry alone does not suffice to arrive at these conclusions. The raw data behind Kepler's calculations do not give true positions of the planet. As one more veil shielding her mystery, nature places the atmosphere between the Earthbound observer and Mars. The atmosphere refracts incoming light, giving the source's apparent, but not true position.[62] Using techniques that he develops in *The Optical Part of Astronomy*, Kepler adjusts the observations to account for refraction. Table 7.2 indicates how well Kepler's techniques work.

Concerning the geometry, it is complicated. Throughout this book, for simplicity, we have examined the pathways of the planets as though they orbit the sun in a common plane. In reality the Martian and Earth orbits do not share the same bed; instead, they rest on inclined surfaces and sunbathe on separate lounges. Kepler goes beyond his predecessors in establishing the relationships in three-dimensional space and encounters more difficult mathematics in doing so.

The distances given in entries 1 and 2 in Table 7.2 are presented in astronomical units. One astronomical unit (AU) represents the mean distance of Earth from the sun.

Entry 3, the orientation of the major axis, singles out the longitude of the major axis.

Using entries 4 and 5, one can specify the orbital plane of Mars. Figure 7.6 illustrates the line of intersection between the ecliptic and Mars' orbital plane. The line passes through the origin, where the sun resides, so the longitude along the ecliptic fully identifies the line. Figure 7.6 also illustrates the inclination angle between the ecliptic and Mars' orbital plane; the illustration exaggerates the angle, which Kepler measures at 1.84 degrees. This inclination angle is the same as the angle between the normal lines to each of the planes. The plane of Mars' orbit is one plane among all those containing the line of intersection, which also has the correct inclination angle. There are two such planes. Kepler knew that Mars' aphelion lies north of the ecliptic, eliminating one candidate and identifying Mars' orbital plane (see Figure 7.7). The footnote gives the equation of Mars' orbital plane.[63]

---

[62] Look at a fish in the fishbowl. Refraction of light through the water, glass, and air misleads the observer as to the fish's true location within the fishbowl. Similarly, refraction of light through the atmosphere misinforms the observer as to a stars' true location.

[63] If we set a coordinate system with the $x - y$ plane along the ecliptic and align the y axis with the line of intersection with Mars' orbital plane, all planes passing through the y axis have equations of the following form:

$$ax + bz = 0$$

We wish to specify a and b. For the ecliptic, the normal line lies along the z axis and for Mars' orbital plane, the vector with components $(a, 0, b)^t$ gives the normal line. We normalize this vector to have length 1. The scalar product of the vectors $(0, 0, 1)^t$ and $(a, 0, b)^t$ gives the cosine of ψ, the angle; hence $b = \cos \psi$. Since the vector $(a, 0, b)^t$ is a unit vector, $a = \pm \sin \psi$. Aligning the aphelion as in Figure 7.8 allows one to deduce that a is a positive value.

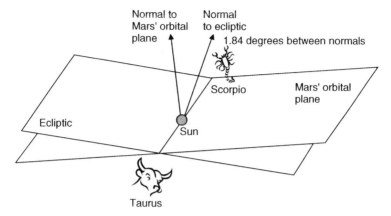

Figure 7.6. Mars' orbital plane and the ecliptic.

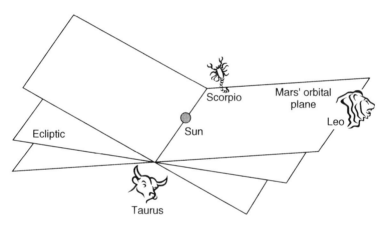

Figure 7.7. Mars' aphelion in Leo lies north of the ecliptic.

Once one determines Mars' orbital plane, one can also specify the line supporting the major axis. It points toward Leo as indicated by Table 7.1. Figure 7.8 illustrates the configuration of the major and minor axes within Mars' orbital plane. The dashed line is the line of intersection between the ecliptic and Mars' orbital plane.

CONFIGURING THE ELLIPSE: THE SECOND LAW

A physical law is to the physicist what an axiom is to a mathematician. The law by itself is not derived from any other principles; it is expounded as fundamental to the order of nature, or in Kepler's case it was God's design. Kepler extracted his laws by deciphering Tycho's observations. Although, for the most part, Kepler

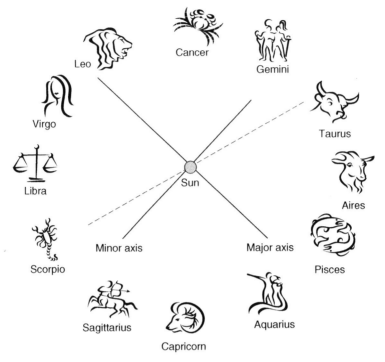

Figure 7.8. Major and minor axes of Mars' orbit.

discovered his laws while sitting at his desk in a room occupied by only himself, he was not alone. Accompanying Kepler were his predecessors, Eudoxus, Euclid, Archimedes, Apollonius, Ptolemy, and Copernicus. Kepler called on the contributions of each of these men, even the incorrect geocentric model, to discover his own laws. Below, we trace Kepler's methods.

Aside from its novelty and centrality to uncovering the ellipse, the second law is also notable because a critical step in its discovery rested on the bedrock of geocentrism, Ptolemy's *Almagest*.[64] Kepler was a devout Copernican. Any view of the heavens was from a heliocentric perspective. As Kepler asserted that the sun was the locomotive force behind Mars' movements, he intuited that there must be a relation between Mars' distance from the sun and Mars' velocity. It is this relation that Kepler sought using Ptolemy as an intermediary.

After the failure of the vicarious hypothesis, Kepler continued to use his initial Ptolemaic–heliocentric universe as a tool. Recall that this model incorporates the equant into a heliocentric system and eliminates the Apollonian epicycles. The failure of the vicarious hypothesis taught Kepler that Ptolemy's midway compromise in which the center of Mars' deferent bisects the segment from the

---

[64]As mentioned above, another instance where Ptolemy was influential was in Kepler's vicarious hypothesis.

sun to the equant was correct. In Chapter 32, Kepler states and demonstrates the following statement:

> First, the reader should know that in all the hypotheses constructed according to Ptolemaic form, however great the eccentricity, the speed at perihelion and the slowness at aphelion[65] are very closely proportional to the lines drawn from the center of the world to the planet.[66]

Using the diagram of Figure 7.9, the assertion states that the times that Mars requires to traverse two small sectors, one about the perihelion and one about the aphelion, are roughly proportional to the distance from the sun to the perihelion and aphelion. The mathematical statement is

$$t = kar_s \qquad (7.1)$$

where t is the time Mars requires to traverse a short arc of length a around the perihelion or aphelion, r_s is the distance of the perihelion or aphelion from the sun, and k is the constant of proportionality that applies to both the perihelion and the aphelion.

Let us demonstrate this claim. Figure 7.10 is a Ptolemy-inspired version of Mars in a heliocentric orbit. Mars moves about its circular deferent, where the circle's center is the midpoint between the equant and the sun. The common distance from the sun and equant to the deferent's center is q. Mars' constant angular speed about its equant is ψ. One converts the angular speed of an object

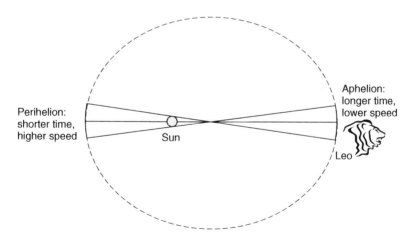

Figure 7.9. Time of travel through arcs around aphelion and perihelion is proportional to distance of arc from the sun.

---

[65] Perihelion and aphelion are, respectively, the closest and farthest points from the sun to the deferent.
[66] Kepler, *New Astronomy*, p. 373.

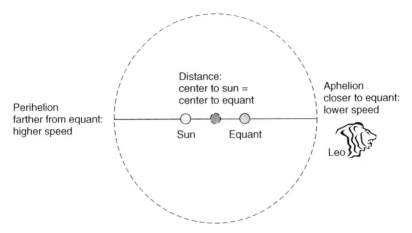

Figure 7.10. A circular orbit with speed proportional to distance from the equant.

moving about a circle to the standard speed v using the formula

$$v = r\psi \tag{7.2}$$

where r is the radial distance from Mars to the equant.[67]

Referring to Figure 7.10, the distance from the equant to the perihelion is $(1 + q)$, and the distance to the aphelion is $(1 - q)$. Setting the value of r in equation (7.2) to the distance at perihelion and aphelion yields the speeds at both points

$$v_P = (1 + q)\psi \tag{7.3}$$

$$v_A = (1 - q)\psi \tag{7.4}$$

where the subscript denotes the value at the perihelion P or aphelion A.

If a is the common length of two very small arcs about the perihelion and aphelion, the time that Mars requires to pass these arcs is approximately given by the following equations:

$$t_P \approx \frac{a}{v_P} = \frac{a}{(1 + q)\psi} \tag{7.5}$$

$$t_A \approx \frac{a}{v_A} = \frac{a}{(1 - q)\psi} \tag{7.6}$$

[67]Note the linear relationship between speed and distance. The arc length of a circle is proportional to its distance from the center, so the distance traveled in a given unit of time, the speed, is also proportional to the distance from the center of the circle.

The cause of approximation is that Mars' speed through the arc is not constant, but changes. Nevertheless, for a very small arc, the approximation is very good, and we shall continue using it.[68]

Accepting the approximation, solving for the variable a in both equations (7.5) and (7.6), and equating the results yields the following equality:

$$a = t_P(1+q)\psi = t_A(1-q)\psi$$

Dividing by $(1-q^2)\psi$ and simplifying results in the following equation:

$$a = t_P(1+q)\psi = t_A(1-q)\psi$$

$$\frac{a}{(1-q^2)\psi} = \frac{t_P}{1-q} = \frac{t_A}{1+q}$$

Noting that the distance from the sun to the perihelion is $1-q$ and the distance from the sun to the aphelion is $1+q$, we obtain the following equality:

$$\frac{a}{(1-q^2)\psi} = \frac{t_P}{r_{SP}} = \frac{t_A}{r_{SA}}$$

The resulting equations are the same at both perihelion and aphelion, so we may eliminate the subscript. Setting $k = \frac{1}{(1-q^2)\psi}$ and multiplying through by r_s yields the result, equation (7.7) [equivalent to eq. (7.1)]:

$$kar_s = t \tag{7.7}$$

For Kepler, this relationship (7.7) is an expression of the sun's locomotive force on Mars. There is nothing special about the perihelion and aphelion; equation (7.7) applies along Mars' entire orbit. Using the result about Mars' orbit, Kepler wishes to address the following problem. Suppose that at an arbitrary time t_0 Mars is at a specified position on the orbit. How does one find its position at a later time t? Kepler expresses the problem and solution in Chapter 40:

> Since, therefore, the times of a planet over equal parts of the eccentric[69] are to one another as the distances of those parts, and since the individual points are all at different distances, it was no easy task I set myself when I sought to find how the sums of the individual distances may be obtained. For unless we can find all of them (and they are infinite in number) we cannot say how much time has elapsed for any one of them. ... I consequently began by dividing the eccentric into 360

---

[68] The reader with a calculus background should note how Kepler edges toward the frontier of calculus. He makes a first-order approximation and later relates the first-order approximations to an area.

[69] The eccentric is the circular orbital path that Kepler first considers as a demonstration of his concept.

parts, as if these were least particles, and supposed that within one such part the distance does not change.[70]

To approximate the period of the entire orbit, Kepler divides the orbit into 360 segments of equal length, applies equation (7.7) to each segment using the assumption that the value r_s is constant in each segment, and then sums up all the resulting times. In equations, Kepler performs the following:

$$T = \sum_{j=1}^{360} t_j = \sum_{j=1}^{360} kar_{sj} = ka \sum_{j=1}^{360} r_{sj} \qquad (7.8)$$

where T is the period of Mars, t_j is the time required to traverse the jth arc segment, r_{sj} is the assumed constant distance of the jth arc segment to the sun, a is the common length of the arc segments and k is as given in equation (7.7).

Kepler recognizes that this is an approximation because the distance r_s is not constant in each segment, but varies. Kepler also recognizes that he would really have to go down to the level of least particles and sum along an infinite number of arc segments in order to get the true answer. This is the problem of integral calculus. It is unfortunate that Kepler did not have access to all of Archimedes' works, because Archimedes had thought deeply about such problems and did in fact develop methods that are precursors to integral calculus. Nevertheless, Kepler was aware of Archimedes' works through other sources, one of which is Apollonius' *Conics* and may have seen some parts of Archimedes' manuscripts. Rightfully, Kepler looked to Archimedes for inspiration:

> I had remembered that Archimedes, in seeking the ratio of the circumference to the diameter, once divided a circle into an infinity of triangles—this being the hidden force of his reductio ad absurdum. Accordingly, I instead of dividing the circumference, I now cut the plane of the eccentric into 360 parts by lines drawn from the point whence the eccentricity is reckoned.[71]

Figure 7.11 illustrates Kepler's approach. Equation (7.8) must be modified, for now the lengths of the arc segments are not equal. The modification becomes

$$T = \sum_{j=1}^{360} ka_j r_{sj} \qquad (7.9)$$

where a_j is the length of the jth arc segment.

From Figure 7.11, one recognizes, as Kepler did, that the term $a_j r_{sj}$ in Equation (7.9) is twice the area of the jth triangle of his partition and the summation of equation (7.9) is an approximation to the area enclosed by the orbit. Furthermore, Kepler intuits, as did Archimedes and Eudoxus before him, that

---

[70] Kepler, *New Astronomy*, p. 417.
[71] Ibid., p. 418.

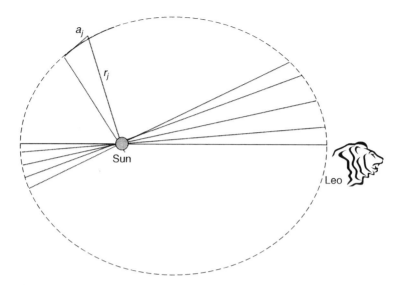

Figure 7.11. Partitioning the figure enclosed by the orbit into triangles, showing a few triangles in the partition along with the jth triangle.

refining the partition yields ever improving approximations that approach twice the actual area. Kepler applies this argument to the entire orbit, yielding the result that the period is proportional to the area enclosed by the orbit. Applying the same argument to any segment of the orbit yields the time required to travel between the endpoints of the segment. The argument's conclusion is that the corresponding time is proportional to the area swept out by the line adjoining Mars and the sun as Mars passes from the segment's initial point to the segment's endpoint. This is Kepler's second law.

The Second Law and Archimedes

The second law may appear as an additional constraint that is difficult to satisfy, but Kepler declares it sacrosanct, a guiding principle that does in fact guide Kepler to the correct result. Figure 7.12 illustrates the relation between the second law and Tycho's observations. Let S on the major axis denote the position of the sun, $M(0) = (m_1(0), m_2(0))$ denote Mars' initial position on the ellipse (the first observation in Table 7.1), and let $M(t) = (m_1(t), m_2(t))$ denote Mars' position at time t. The second law states that the area of the sector between the sun [i.e., $M(0)$], and $M(t)$ is proportional to the time difference between the observations. We will determine this area and ensure that parameters are set so that the area conforms to the observations.

Once more, for Kepler, the way forward was through the past. In his treatise *Conics*, Apollonius reviews Archimedes' work on the ellipse. Archimedes demonstrates how to find sector areas for the ellipse using a circle. These results

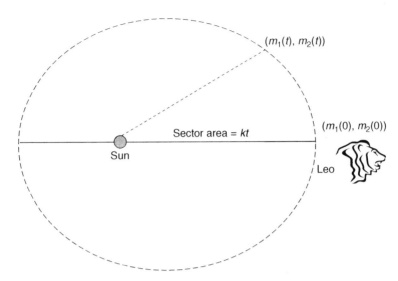

Figure 7.12. Time required to travel along arc determines area of accompanying sector.

allow Kepler to establish equations that satisfy the second law. As such, we begin with Archimedes.

Figure 7.13 illustrates the configuration of interest. A circle with diameter equal to the major axis of an enclosed ellipse circumscribes the ellipse. The figure associates points on the circle, $C(t) = (x(t), y(t))$, with points on the ellipse by

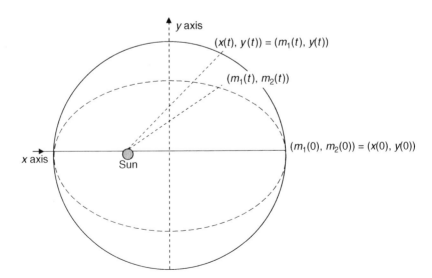

Figure 7.13. Association of Mars with its shadow point.

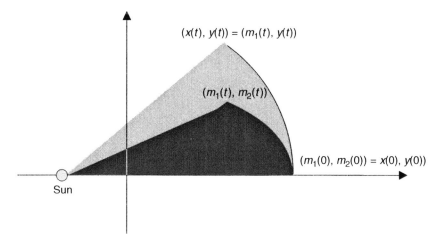

Figure 7.14. Association of points. Elliptic sector shown in black; circular sector, in gray, includes elliptic sector.

equating their x coordinates, $C(t) = (m_1(t), y(t))$. We refer to the corresponding points on the circle, $C(t) = (m_1(t), y(t))$, as *shadow points*. Accompanying the association of points is an association of sectors; Figure 7.14 depicts the associations. Archimedes demonstrates that the area of the elliptic sector is proportional to the area of the circular sector and the constant of proportionality is the ratio of the of the minor axis to the major axis. In equations form, we have

$$A_E(t) = bA_C(t) \qquad (7.10)$$

where $A_E(t)$ is the area of the elliptic sector, $A_C(t)$ is the area of the corresponding circular sector, and b is the length of the semiminor access.[72]

The following method, which establishes this result, is in the fashion of Archimedes. We first consider the case in which the initial position lies on the x axis at the point $(1, 0)$ as in Figure 7.15. The baseline for both the elliptic and circular sectors is the same; the segment that runs along the x axis from the sun S to the point $(1, 0)$. The cover of the elliptic sector consists of two components: (1) the line segment running from the sun to Mars and (2) the elliptic arc that runs from Mars to the point $(1, 0)$. Similarly the cover of the circular sector consists of two corresponding components, a line segment running from the sun to Mars' shadow and a circular arc that runs from Mars' shadow to the point $(1, 0)$.

Let $d_E(x)$, and $d_C(x)$ be the vertical distances from the baseline at the point x to the respective covers of the elliptic and circular sectors (see Figure 7.15).

---

[72]In general, the constant of proportionality is $\frac{b}{a}$, where a is the length of the semimajor access. We set $a = 1$.

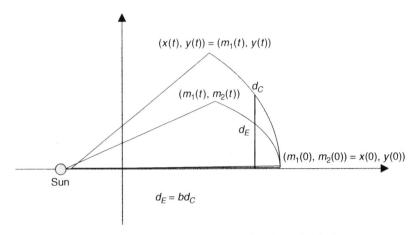

Figure 7.15. Distances from x axis to circular and elliptic arcs.

A critical observation is that the distances satisfy the following relation:[73]

$$d_E = bd_C \qquad (7.11)$$

Restricting equation (7.11) to the portion of the cover along the arcs yields the following relation:

$$bm_2 = y \qquad (7.12)$$

Here, once again, m_2 refers to the second coordinate on the ellipse, and y refers to the second coordinate on the circle.

A similarity argument applied to the triangles of Figure 7.16 illustrates that equation (7.11) also applies to the linear portions of the respective covers.

Archimedes knew that the areas of 2 two-dimensional objects sharing a common baseline and having proportional distances to their respective covers are in proportion to one another as the distances to their respective covers. In other words, since the elliptic sector and circular sector have the same baseline and

---

[73]For the basepoints with covers on the ellipse and circle, the relationship follows from the equations for the ellipse and the circle:

$$m_1^2 + \left(\frac{m_2}{b}\right)^2 = 1$$

$$m_1^2 + y^2 = 1$$

The top equation is the equation for the ellipse, and the bottom one is the equation for the circle. It is apparent that $y^2 = \left(\frac{m_2}{b}\right)^2$, and because Mars and its shadow share the same sign, $y = \frac{m_2}{b}$. But $\mid y \mid = d_C$ and $\mid m_2 \mid = d_E$. For the basepoints with covers on the corresponding line segments, this follows from the similarity between the triangles with vertices at the sun, basepoint, and cover, and the triangles with the cover point on both the segment and the arc.

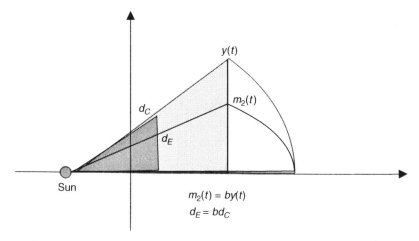

Figure 7.16. Distances from x axis to linear covers using similar triangles.

since equation (7.11) holds, equation (7.10) follows. The result is a special case of Cavalieri's theorem, named after a student of Galileo who lived in the seventeenth century. Archimedes had discovered the result nearly two millennia before Cavalieri.[74]

We have established equation (7.10) for the special case when the sector initiates at the point $(1, 0)$. This also establishes the general case using the following argument. If $M(t_1)$ and $M(t_2)$ are initial and final points along the ellipse, the corresponding sectoral area is obtained by subtracting the area of the sector initiating at the point $(1, 0)$ and terminating at $M(t_1)$ from the area initiating at the point $(1, 0)$ and terminating at $M(t_2)$.

The Shadow and the Second Law. The Archimedean result [eq. (7.10)] allows Kepler to establish a relationship between observations yielding Mars' shadow along its circular pathway and the time lapse between subsequent observations. Because the arc of the circular sector is proportional to the area of the elliptic sector and by the second law the area of the elliptic sector is proportional to the time, it follows that the area of the circular sector is also proportional to the time. Let us express this concept in equations.

Let the shadow of Mars be given by the point $C(t) = (x(t), y(t))$. To find the area of the circular sector initiating at the point $(1, 0)$, split the sector into two regions, a pie-slice-like region R, and a triangle T as illustrated in Figure 7. The area of the pie slice R is $A_R(t) = \frac{1}{2}\theta(t)$, where $\theta(t)$, the shadow angle the angle depicted in Figure 7.18. The area of the triangle T is $A_T(t) = \frac{1}{2}x_s$

[74]Cavalieri's proof (somewhat lacking) alongside an Archimedean argument are presented *Ellipse*. Archimedes successfully applied Eudoxus' method of exhaustion to several other p so it stands to reason that Archimedes was able to arrive at Cavalieri's theorem using method of exhaustion.

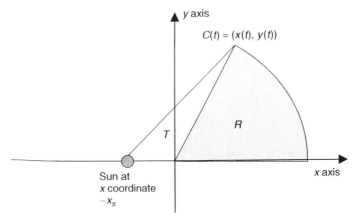

Figure 7.17. Splitting the circular sector into two regions.

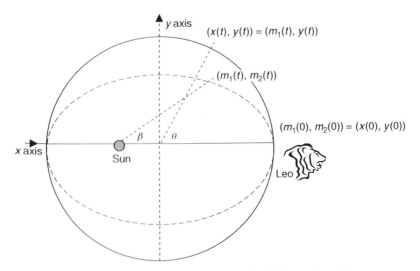

Figure 7.18. The shadow angle and the heliocentric angle.

the distance from the sun to the origin. Then the total area of the
· is given by the following:

$$A_C(t) = \frac{1}{2}\left(\frac{\pi\theta(t)}{180} + x_s y(t)\right)$$

$$= \frac{1}{2}\left(\frac{\pi\theta(t)}{180} + x_s \sin\theta(t)\right)$$

degree measurements for the angles, so the term $\frac{\pi}{180}$
'ement. Since the area is proportional to the time that

the shadow requires to pass from the point $(1, 0)$ to the point $C(t)$, we have the following equation

$$t = \frac{\kappa}{2} \left(\frac{\pi \theta(t)}{180} + x_s \sin \theta(t) \right) \tag{7.13}$$

where κ is the constant of proportionality.

Let us determine κ. The shadow completes one cycle in one Martian year and the Martian year is the unit time measurement: $\theta(1) = 360$. Placing this value into equation (7.13) yields the value of κ:

$$1 = \frac{\kappa}{2} (2\pi + x_s \sin 360)$$

$$\kappa = \frac{1}{\pi} \tag{7.14}$$

$$t = \frac{\theta(t)}{360} + \frac{x_s \sin \theta(t)}{2\pi}$$

Mars' Heliocentric Angle. Figure 7.18 illustrates the heliocentric angle $\beta(t)$ with respect to the x axis.

The angle $\beta(t)$ satisfies the following equation

$$\tan(\beta(t)) = \frac{m_2(t)}{x_s + m_1(t)} \tag{7.15}$$

where the position of Mars is $M(t) = (m_1(t), m_2(t))$.

One relates the shadow angle to the heliocentric angle using equations (7.12) and (7.15):

$$(m_1(t), m_2(t)) = (x_1(t), by(t))$$

$$= (\cos(\theta(t)), b \sin(\theta(t))) \tag{7.16}$$

$$\tan(\beta(t)) = \frac{b \sin(\theta(t))}{x_s + \cos(\theta(t))}$$

Kepler's Result

Kepler uses his Archimedes inspired method sparingly. Whereas he obtains his vicarious result for the entire Martian orbit, and then compares the theory with Tycho's data, Kepler's entire tabular results to demonstrate that the ellipse conforms to observations consists of only three points. What's more, the ellipse that Kepler uses is not even the final ellipse that he champions. Why does Kepler go through so much trouble, over 70 iterations, to test a theory that he dismisses on the basis of an 8-minute discrepancy, yet barely demonstrates that observations confirm his trophy? Kepler himself explains:

But given the mean anomaly, there is no geometrical method of proceeding to the equated, that is, to the eccentric anomaly.[75] For the mean anomaly is composed of two areas, a sector and a triangle, and while the former is numbered by the arc of the eccentric, the latter is numbered by the sine of that arc multiplied by the value of the maximum triangle, omitting the last digits. And the ratio between the arcs and sines are infinite in number. So, when we begin with the sum of the two, we cannot say how great the arc is and how great its sine, corresponding to this sum, unless we were previously to investigate this area resulting from the given arc; that is, unless you were to have constructed tables and to have worked from them subsequently.

This is my opinion. And insofar as it is seen to lack geometric beauty, I exhort the geometers to solve me this problem:

Given the area of a part of a semicircle and a point on the diameter, to find the arc and the angle at that point, the sides of which angle, and which arc, encloses the given area. Or, to cut the area of a semicircle in a given ratio from any given point on the diameter.

It is enough for me to believe that I could not solve this a priori, owing to the heterogeneity of the arc and the sine. Anyone who shows my error and points the way will be for me the great Apollonius.

Kepler wishes to use the relationships between the time, the area swept out by the line adjoining Mars and the sun, and the area as calculated using the associated shadow point, to determine Mars' position given the time of an observation. Then, for each observation time, he could compare the calculated position with the observed position. The complexity of the relations do not allow Kepler to perform the required calculation. Thus, the ellipse does not receive the scrutiny of the vicarious hypothesis.

Table 7.3 is the table where Kepler first performs calculations using an ellipse.[76] Relevant to the story of the ellipse's discovery, it is noteworthy that Kepler presents the table in Chapter 47 and performs the calculations long before writing Chapter 58, where he settles on the ellipse as the planetary pathway. It is in Chapter 47 where Kepler proposes the ellipse as a proxy for an ovum-shaped orbit that he previously describes and believes to be the true orbit.[77]

Kepler takes his vicarious hypothesis as a near representation of the truth, but for false reasons. Kepler determines column 3, the circular obit, using a value of $x_s = 0.09264$, whereby the radius of the orbit is 1. Following Ptolemy, Kepler refers to x_s as the eccentricity. Kepler's intention in performing this calculation is to demonstrate that a circular orbit obeying the second law does not comport with the observations and is thus wrong.

---

[75]Using the notation above, the mean anomaly is the time t, and the eccentric anomaly is the longitude β.

[76]Kepler, *New Astronomy*, Chapter 47, p. 477.

[77]While we continue to use the convention that one period represents a unit of time, Kepler divided the period into 360 units so that one unit of time represents $\frac{1}{360}$ th of a period.

TABLE 7.3. Kepler's Calculations Using an Ellipse

Common Mean Anomalies: Time from Aphelion	Longitude from Aphelion in Degrees		
	Vicarious Hypothesis	Circular Orbit	Elliptic Orbit
0.13542593	41.3425	41.48166667	41.23583333
0.26474383	84.70055556	84.70722222	84.66166667
0.38542593	131.12388889	130.99027778	131.23472222

For the ellipse as a proxy, Kepler uses the same eccentricity $x_s = 0.09264$. In the notation of the previous section, the length of the semiminor axis of Kepler's proxy ellipse is $b = 0.99142$, whereby the length of the semimajor axis is 1.[78]

A test of formulas (7.14) and (7.16) using Kepler's entries in columns 3 and 4 of Table 7.3 confirms Kepler's accuracy. As an example, consider the first entry of the elliptic orbit. Equation (7.14) is the relation between the time t and the shadow angle θ:

$$t = \frac{\theta}{360} + \frac{x_S \sin \theta}{2\pi}$$
$$0.13542593 = \frac{\theta}{360} + \frac{0.09264 \sin \theta}{2\pi}$$

Kepler's request of the geometers is to determine the value of θ for any given t. For this particular t, Kepler indicates that the value of the shadow angle is $\theta = 45$. Most likely, Kepler began with the shadow angle and used equation (7.14) to arrive at the time. With the shadow angle known, we can call on equation (7.16) to determine the longitude β:

$$\tan(\beta) = \frac{b \sin(\theta)}{x_s + \cos(\theta)}$$
$$= \frac{0.99142 \sin(45)}{0.09264 + \cos(45)}$$
$$= 0.876557$$

Taking the inverse tangent results in the value of the longitude, $\beta = 41.23707$. I performed my calculations on a spreadsheet. Kepler performed his using tables and a lot of arithmetic. Kepler's longitude, the first entry in the final column of Table 7.3, and the result of the spreadsheet differ by less than $\frac{13}{10,000}$ ths of a degree, a testimony to Kepler's diligence.

---

[78]The sun of the proxy ellipse is not on the focus of the ellipse. In *New Astronomy*, Kepler never relates the eccentric with the focus even though an ellipse configured from Table 7.3 does place the sun on the focus. After completing *New Astronomy*, Kepler does discover that the sun lies on the ellipse's focus and states so in reference.

Of more consequence than the accuracy of his computations are the realization of the strength of the method and the conclusions that Kepler draws. Concerning the method, Kepler writes the following:

> The investigation of the eccentricity in chapter 42 depends upon aphelial and perihelial distances, and in these there can be some slight error which is increased tenfold in establishing the correct eccentricity. Therefore, it should be noted in passing that if a perfectly reliable way of equating through the physical causes is finally found, a perfectly true eccentricity can afterwards be established, and through it the aphelial and perihelial distances can be entirely corrected.

What Kepler is stating is that with the correct equations for the pathway, one can find the eccentricity by matching the data to the equations. Kepler goes on to explain how one makes the correction for the ellipse and suggests that the true value of the eccentricity is $x_s = 9230$. Much later comes the epiphany of Chapter 57:

> My line of reasoning was.... The circle of chapter 43 errs in excess, while the ellipse of chapter 45 errs in defect.... But the only figure occupying the middle between a circle and an ellipse is another ellipse.

By "errors," Kepler means the discrepancies between the vicarious hypothesis and other entries of Table 7.3. There, at each observation, the circle and the ellipse are on opposite sides of the values from the vicarious hypothesis, which Kepler accepts as accurate for false reasons. The top and bottom values of the vicarious theorem are very close to midway between the circular and elliptic values, so Kepler proposes the ellipse between the two as the prize.

After leading the reader through a battlefield of equations and calculations, demonstrating the meticulous effort that he had undertaken, Kepler does something very un-Keplerian and declares victory with sparse confirmation from the observations.

> Nor was there any need to compute the equations anew from the ellipse. I knew they were going to perform their function without further prompting. But if this were to have happened, I had already prepared a refuge, namely the uncertainty of 200 units in distances.

Thus, Kepler presents no new tabulations of Table 7.3. Furthermore, once having seen that he had the right equations, he does not follow his excellent nose and pursue the scent that he describes in Chapter 47. He does not use the ellipse to improve estimates of eccentric, perihelial, and aphelial distances.[79] Instead,

---

[79] At least Kepler makes no claims to do so. Kepler's translator, William Donahue, claims that using the equations for the ellipse, Kepler does invent some data for his table at the end of Chapter 53. This table assists Kepler in finding distances between Mars and the sun at several points along Mars' orbit.

Kepler is less than forthright. The ellipse of Chapter 43 is of somewhat different dimension than the dimensions given in Chapter 53. Table 7.2 uses the results from Chapter 53, which prove to be quite accurate. Kepler must have noted the discrepancy. Was he exhausted by his travails and did not have the energy to sort out the final details? One indicator that this was so is that there is no indication in *New Astronomy* that the sun lies at one focus of the ellipse. It's hard to imagine that Kepler, a genius at distilling relations from disparate information, did not consider this possibility during his investigations. Years later, after recuperating from his battles with Mars, Kepler did sort out the details and placed the sun on one of the ellipse's foci.

Returning to *New Astronomy*, Kepler's cry to the geometers is legitimate. Using equations (7.14) and (7.16) to determine the longitude, given the time of observation, is untenable. However, a difficult yet tenable proposition is to go the other way; that is, given the observations, both longitude and latitude, determine the time of the observation. Table 7.1 gives the input data for the observations' positions. Using this table and the parameter of Table 7.2, one arrives at the results listed in Table 7.4.

The first column of Table 7.4 identifies the observations corresponding to those from Table 7.1. The second column gives the calculated time using the convention that the counting begins at the first observation. One unit of time represents a Martian year, the time it takes for Mars to go about its orbit. (Table 7.2 gives the conversion to Earth years; one Martian year is equivalent to about 1.88 Earth years.) The third column converts the times from Tycho's data (Table 7.1) to the conventions of column 2. The fourth column is the difference between the previous two columns.

The final column approximates a longitudinal difference with longitude measured in Mars' orbital plane. The error does not reflect an error in longitude, but an error in the difference of the longitudes between the first observation and

TABLE 7.4. Further Observations at Opposition

Date	Calculated Time	Actual Time	Time Error	Error in Longitude
November 18, 1580	0	0	0	0'
December 28, 1582	1.12092494022	1.12105402255	−0.00012908233	−2.79'
January 31, 1585	2.23394533898	2.23415111280	−0.00020577382	−4.44'
March 6, 1587	3.34687338208	3.34705916146	−0.00018577938	−4.01'
April 14, 1589	4.46776064749	4.46790392407	−0.00014327658	−3.09'
June 8, 1591	5.61087176798	5.61072603497	0.00014573301	3.15'
August 26, 1593	6.78908056659	6.78899496006	0.00008560653	1.85'
October 31, 1595	7.94825168483	7.94818584848	0.00006583634	1.42'
December 14, 1597	9.07603443654	9.07582903185	0.00020540469	4.44'
January 19, 1600	10.19104380358	10.19080845064	0.00023535294	5.08'
February 21, 1602	11.30325526843	11.30299066047	0.00026460795	5.72'
March 29, 1604	12.41987623598	12.41966033346	0.00021590252	4.66'

the associated observation. The values are in minutes, so that one can compare these differences with standards that Kepler uses throughout *New Astronomy*. Kepler seems to be fine with 2–3-minute differences. On occasion he gets up to 5-minute differences. Here, the maximum difference is 5.72 minutes or around 0.095 degrees.[80] Since the calculations are made in Mars' orbital plane, we have not considered errors in latitude.[81]

Since I have a computer program available, I tried to carry out Kepler's program of using the observations to calibrate the ellipse. Adjustments to the eccentricity and angle of inclination generally increased the error. I could not find a set of parameters that significantly improved the results. As all the parameters come from Table 7.2, this once more demonstrates the accuracy of Kepler's work.

If I could construct the table above (Table 7.7), so could Kepler. If the thought occurred to me, it certainly occurred to Kepler. True, I have a computer that performs the computations, and he doesn't. But Kepler calculates compulsively. There are others who are known to have had compulsive fits of calculations. Confronting insomnia, Euler (1707–1783) mentally computed the first six powers of the whole numbers up to 100.[82] Newton had computed tables of binomial interpolations to arrive at his binomial theorem. Ptolemy must have been another compulsive calculator, and Gauss (1777–1855) is said to have carried out the calculation of logarithms in his head. But of all these numbers enthusiasts, Kepler was in his own class. Kepler is the only one to have gone so far as to calculate his precise time of conception, May 16, 1571, and the time from conception until birth, 224 days, 9 hours, and 53 minutes. Considering the emotional investment that Kepler made in *New Astronomy*, and Kepler's predisposed need to quantify everything, it would be difficult for Kepler to keep himself from making a similar

---

[80]Column 3 (in Table 7.4) uses the incorrect assumption that Mars revolves about the sun at a constant angular speed of one degree per $\frac{1}{360}$ units of time. Multiplying the angular speed by the time error gives an estimate of the longitudinal error.

[81]The following three-step algorithm was used to produce the times of Table 7.4:

1. Choose a coordinate system in which the $x–y$ plane coincides with Mars' orbital plane and adjust Tycho's observations (Table 7.1) so that they all lie within the orbital plane. One accomplishes this by applying a rotational transformation to the data. The angle of rotation is given by the inclination angle of Table 7.2, and the axis of rotation is the intersection between the ecliptic and Mars' orbital plane, also given in Table 7.2. In the calculation above another rotation was performed so that the x axis coincides with the ellipse's major axis. Once again, Kepler's choices from Table 7.2 are employed. Once the rotations are performed, the longitudes of each observation are available in the new coordinate system.

2. For each observation, convert the longitudes β to the shadow angles θ. One accomplishes this by identifying each observation's coordinates using the ellipse with major and minor axes as Kepler provides in Table 7.2, and a coordinate system centered at the center of the ellipse. The tangent of the angle is then $\tan\theta = \frac{m_2}{bm_1}$, where $\frac{m_1}{m_2}$ are the coordinates of the corresponding observation. From this relation the angle θ may be determined.

3. Using equation (7.14), determine the times corresponding to each observation and make an adjustment so that the first observation occurs at time $t = 0$.

[82]Euler, functionally blind, had a photographic memory and memorized his table.

tabulation.[83] Either my estimation of Kepler is incorrect, or he did not disclose his calculations. If it is the latter, the results may not have been to his liking.

THE MENTOR

Copernicus displaced the earth from the center of the universe, but left the earth a resident on the perfect circle. Kepler deflated the circle and demonstrated that God's design does not rest on humankind's notions of perfection. Rather, God defines the universe on God's terms. Penetrating the mind of God would require humankind to put aside age-old prejudices. This is where Kepler, the hyper-self-critical analyst, makes his breakthrough.

In 1965 Richard Feynman gave his acceptance speech to the King of Norway and the Nobel Prize committee. It was a radical speech in that, unlike the acceptance speeches of previous Nobel laureates, it contained sparse technical information. Instead, the speech describes all the wrong turns, all the dead ends, and all the pitfalls that Feynman encountered on his journey that culminated in his prize-winning results. Many modern scientists believe that Feynman's speech ought to be required reading, for this is the true path of research and more importantly, Feynman informs us that the path to the truth must follow the corridor of integrity. Kepler was a mentor to Feynman. Perhaps it is the message of integrity that was Kepler's most important contribution.

---

[83] Kepler's numeric disposition explains the extensive calculations behind his 70 iterate vicarious methodology. He complains but could not keep himself from scaling that mountain.

CHAPTER 8

THE AUTHORITY

LOUIS XIV, THE SUN KING

Returning from the battlefront on France's northern border, Louis XIV entered the chamber of Madame Montespan, his mistress with whom he fathered seven children. Twenty-five years previously the charming, elegant, witty, beautiful, and married belle had captured Louis' heart, but that last adjective cast a pall over their relation. Kings were entitled to their dalliances, and Louis, who ascended the throne at age 12, who believed that the throne was his by divine right, who burdened his citizens with the construction of Versailles, who burdened his citizens with needless wars, and who executed opposition members on a whim, would not be denied his entitlement. Perhaps Louis recognized that accompanying divine right was a moral obligation and that adultery did not clear the hurdle. But a more likely explanation for Louis ending his 24-year affair is that a younger lady captivated him.

The king could legitimize his misbegotten children conceived on a bed shared by an adulterer and an adulteress, but as with most of Louis' perquisites, someone else paid the price. On this occasion it was Madame Montespan. To attenuate a source of gossip from the grist, Louis separated all the children from their real mother and placed them in the care of a younger unwed surrogate mother, Madame Maintenon. After many years, Madame Maintenon won over Louis' affections. On the day of his return from the battlefront, once within Madame Montespan's chamber, Louis greeted the chambermaids, and then with Madame

Shifting the Earth: The Mathematical Quest to Understand the Motion of the Universe,
First Edition. Arthur Mazer.
© 2011 John Wiley & Sons, Inc. Published 2011 by John Wiley & Sons, Inc.

Montespan he retired to a private room. Madame Montespan could not contain her grief as Louis informed her of his decision. Through her tears she foresaw a childless future cast away from a social circle that shunned her for her sin.

The year that Queen Anne-Theresa, Louis' legitimate wife and daughter of the Spanish king Philip, died, Louis married Madame Maintenon, the surrogate mother of Louis' now legitimate children. Louis' own lineage was in question. His predecessor, Louis XIII, a homosexual, and his mother had a chilly relationship, while Cardinal Mazarin and the queen mother were intimate.

Europe itself mirrored the tangled lives of its royalty; threaded into the web of religious and national passions that precipitated the Thirty Years' War was a family feud. Cardinal Mazarin's probable sin that provided Louis XIII with an heir was of service to Europe's royals as intermarriage had depleted the royal stock. Among Louis XIII's opponents in the Thirty Years' War were his father-in-law, the Spanish king, Philippe III, and later his brother-in-law Philippe IV. To reinforce royal bonds, the brother-in-law relationship was cast in a doubl hull. Phillipe married Louis' sister Elisabeth while Louis married Philippe's sister, Anne. Also a foe of Louis XIII was his wife's cousin and brother-in-law, Ferdinand III the Habsburg Holy Roman Emperor. The next generation inherited and continued the family feud as Louis XIV was both the son-in-law and nephew of his nemesis, King Philippe IV of Spain. Leopold I, successor to Ferdinand III, was Louis XIV's cousin.

With Spain defeated and its resources spent, Europe emerged from the Thirty Years' War, exhausted, in economic ruins, and no wiser. Europe's new conundrum was to determine the division of Spain's claims in Italy and the Spanish Netherlands as an enfeebled Spain would not be able to maintain these possessions in the face of more powerful predators. Europe freed itself of one tangle only to configure a new knot.

After the war, France was the preeminent power on the continent, and Louis marshaled France's population and economy in pursuit of personal glory through conquest. In 1667, an initial foray into the Spanish Netherlands ended in the Treaty at Aix-la-Chapelle, where Louis agreed to cease hostilities. Europe had been forewarned and the Elector of Mainz, Von Boinenberg, tapped the talents of a young, brilliant lawyer, Gottfried Leibniz, to act as his personal representative in the French court. Leibniz' mandate was to persuade the king to leave Europe alone. Was there a way to distract France from Europe other than offering her Egypt? Leibniz saw this as the best way to assuage Louis and proposed that a French army supported by other European contingents attack the infidel. Louis didn't bite. Instead, being as dismissive of treaties as he was toward his mistresses, Louis resumed his war on Europe in 1672.

Accompanying Louis' military force was a young Randolf Churchill, son of Winston Churchill,[1] who represented France's ally, king Charles II of England. The young Churchill witnessed firsthand the French policy of pillaging conquered

---

[1] The twentieth century's Winston Churchill son of Randolf Churchill is the progeny of the seventeenth century's Randolf Churchill, son of Winston Churchill.

territory and terrorizing noncombatant populations in the style of the Romans. Throughout European history, the threat of hegemony of one nation caused the threatened nations to congeal into an opposing alliance, regardless of the underlying enmity and distrust among alliance members absent the threat. The French policy of terror accelerated the congealing process. William III of Orange, cousin to France's ally by bribery, Charles II of England, united a disparate army of Netherlanders, Germans, and Dutch that was able to halt the French offensive. In 1674 the treaty of Nymegen codified the balance of power on the continent. Sentiment is stronger than treaties. The sentiment in Louis' court was that the job was unfinished. Meanwhile, an anti-French sentiment that even struck a chord in Britain stirred throughout the continent. The year 1681 saw Louis resume his offensive with the acquisition of Strasbourg and incursions north of the Rhine into the lands of German principalities. While staunchly Christian, Louis' hubris succeeded in offending all of Christian Europe to the point where his only ally was the Turks, who in the minds of Christian Europe had a long-term objective of Islamicizing the continent. Louis was delighted when in 1683, the Sultan amassed his Janissary army on the Balkans and sent it north. A Habsburg victory in the battle for Vienna gave Louis pause, and in 1684 Christendom agreed to a 20-year truce. For the Habsburgs the truce afforded the opportunity to consolidate gains to the east. For Louis, the truce ensured that a well-trained army of German states would not turn their attention on France and instead remain in eastern Europe. In England, Charles' brother James continued Charles' policy of accepting French bribes while acceding to public demand by providing no material support to the French cause. His decision to sign onto the truce was an easy one. The Dutch could not stand up to the French alone and also signed onto the truce.

This was not peace; it was truce, and Louis, who shredded a peace treaty when it suited him, had equally little regard for a truce agreement.[2] In 1688, Louis' army was once again on the move. The army encroached again on the Spanish Netherlands, Italy, and north of the Rhine, and eventually marched on Spain. William of Orange once again assembled an army of the indignant that encompassed his traditional allies. This time, however, he had an additional resource.

James of England had managed to antagonize his citizens to the point where they were in open revolt.[3] Having learned the Cromwellian lessons of rebellion the hard way; rather than resorting to a republic, British parliamentarians invited William of Orange to dislodge James and assume the British crown.[4] Against

---

[2]Peace would have required the Habsburgs and Dutch to recognize French gains on their territories, which was a nonstarter.

[3]Distrust between the Stuart house, who reacquired the throne on the death of Elizabeth, and the general population had brewed for decades. The Stuart kings were openly supportive of returning England to the Catholic Church against the wishes of the majority Protestant population.

[4]An earlier rebellion against the Stuarts resulted in the dictatorship of Oliver Cromwell who surprised the English by being even more unbearable than the Stuart kings. With the yoyo at its bottom, the English tugged on the string and restored the Stuarts by welcoming as their king Charles II, the son of Charles I whom Cromwell had beheaded.

French expectations, William succeeded and the superior British navy was now at his disposal.

In 1697, Eugene of Savoy (Savoy was the state of Kepler's final residence) terminally dispatched the Turkish threat with a victory at Zenta along the Danube. Eugene had earlier distinguished himself in the Battle for Vienna and would later demonstrate his military competence as well. Louis was exhausted by the weight of the alliance arraigned against him and worried by the implications of Eugene's victory. William of Orange's alliance was equally exhausted by Louis' powerful army. Once again, a weary Europe sat at the negotiating table, and once again a peace treaty resulted.

While the 1648 treaty of Westphalia marks the end of the Thirty Years' War, it is not a distinctive milestone. It does not mark the beginning of peace; it does not mark the beginning of respected borders; nor does it mark the beginning of cooperation. The only useful historical reference is that it identifies the moment that Spain sank. Over the next 50 years war continued through cycles of violence followed by hollow treaties, and in 1697, despite the latest treaty, all indicators pointed to continuing war. History affords many surprises; in this case the surprise is not that the peace did not last, but the manner in which the peace was broken.

In 1700, many years after the expected event, Louis' brother-in-law and cousin, King Carlos II of Spain, passed away without heir. Inbreeding must have given way to insanity, for the will of the king could only have been left by a man totally bereft of common sense. King Carlos named as his successor his grandnephew and Louis' grandson, Phillip of Anjou. It was the Catholic French under Louis XIII who connived with Protestant Europe to diminish Catholic Spain's authority, fomenting the Thirty Years' War. It was Louis XIII who pursued the Spanish after the treaty of Westphalia and tossed the Spanish beyond the Pyrenees. It was Louis XIV who violated every agreement that he signed and encroached on Spanish holdings in the Netherlands and Italy and again Louis XIV who sent his army across the Pyrenees and took control of Barcelona. King Carlos in one stroke willed to the Bourbons what two generations of Bourbon kings and 80 years of war had denied them: the rights to Spanish holdings throughout Europe. Louis would grasp at the holdings ordained to him through a divine contract, but an equally determined alliance of the indignant would never allow it. When Louis accepted the will and sent his grandson to Spain, he laid down the gauntlet and prepared for the oncoming war.

What was the balance of European power as it once again cycled into war? Although the old guard that lead his previous campaigns was long gone, their policies of professionalizing the military endowed Louis with the best army in Europe in terms of both manpower and training. On the other side, the alliance survived the death of William of Orange, and his successor, Queen Anne, continued William's policies. The alliance had two additional sources of strength. Whereas England had previously not contributed to the armies on the continent, the Dutch and English armies were united under the very capable command of the Duke of Marlborough, Randolph Churchill. Free of the Turkish threat, the Habsburgs were able to deploy a significant force commanded by their most capable general, Eugene of Savoy. The French did, however, lay a strategic fissure in the alliance. They cajoled Maximilian II, the Elector of Bavaria, into splitting with his father-in law, Leopold I, and allying with France. Beyond the manpower deprived to the alliance and added to the French cause, this netted Louis a contiguous path from French-held territories beyond the Rhine to Vienna, the seat of the Habsburg empire. With the Spanish crown on a French head, Louis could decimate the allies and reconstitute Spain's territories under French rule by taking Vienna.

In 1704, as Louis' army was en route to Vienna, the Duke of Marlborough marching his army from the west and Eugene of Savoy marching his army from the east linked and intercepted the French at Blenheim. In a seesaw battle for control of Europe, one that might have gone either way, the alliance prevailed and the French army returned to French soil. Europe caged itself in a stalemate of misery until after 13 years, an exhausted Europe grudgingly negotiated a peace settlement. The central tenet of European politics, a continental balance of power in which no nation was preeminent, had been restored.

NEWTON, THE MATH KING

In 1672, at age 29, Isaac Newton, the Lucasian professor of mathematics at Cambridge, wrote a letter to the British Royal Society that had enthusiastically welcomed him 2 years earlier:

> Sir, I desire to that you will procure that I may be put out from being any longer a Fellow of the Royal Society. For though I honor that body, yet since I see I shall neither profit from them, nor partake of the advantage of their assemblies, I desire to withdraw.[5]

The immediate cause of Newton's threat of resignation was his anger at the criticism leveled against his work on optics. Criticism came from others within the Royal Society, predominantly Robert Hooke, and continental scientists for whom the Royal Society was a forum for scientific debate. Henry Oldenburg, the president of the Royal Society, received letters on Newton's work, forwarded them to Newton, and Newton would respond. Two years of sallying forth in combat from behind his comfortable Cambridge cocoon had exhausted Newton and left him disenchanted.

Newton's threat to resign was all too real. While Newton enjoyed social praise, he would engage with society only on his own terms. In this case, the terms of the hypersensitive scientist were that he would credit nobody for any of their ideas that motivated his work and he would accept no criticism; and if society did not agree to his terms, Newton had the capacity to shun.

Newton developed the capacity to shun at a young age. Newton's father, an illiterate farmer, died prior to Newton's birth, and when Newton was age 3,

[5]Christianson, *Isaac Newton*, p. 47.

his mother, Hannah Ayscough, remarried. The groom, Barnaby Smith, lived in a town several miles away. The stepfather had no wish to have anything to do with the stepson, so Hannah left Isaac in the care of her parents in her home village of Woolsthorpe. The feeling left in Newton's heart was that of abandonment; he was alone. Seven years later, on the death of her second husband, Hanna returned to Woolsthorpe. Accompanying her were three children, Isaac's stepsiblings. Immediately on the mother's return, Hannah Smith arranged for Isaac to study at the Grantham School in the distant town of Grantham and board with strangers while Newton's stepsiblings remained with their mother. The decision may have been well intended, for this was the nearest school where Isaac could continue his education. Nevertheless, the decision most likely reinforced Isaac's insecurities and caused Isaac to fortify his already established emotional capacity to stand alone in the world.

It was when he was alone, not engaging the world, that Newton was at his best. The capacity to turn inward allowed Newton to focus his tremendous intellect without interference. This was evident in his schooldays when the friendless Isaac would busy himself with fashioning sundials, lanterns, kites, windmills, and watermills while other students engaged in their more kidlike social activities. His mastery of design and skill with tools, each far ahead of his years, were precursors of Newton's latter phenomenal mathematical and scientific achievements while in isolation. Another precursor, one with less positive consequences, was the fury that Newton unleashed on his fellow students as he directed his ire against any indignity that he imagined. There are records of many fistfights attesting to a disposition that brought out the worst not only in himself but also from those in his company. Isaac's physical prowess was no match for his intellectual prowess; nevertheless, he continued to get into fights despite generally being the one who came out the more bloody.

Newton's mother removed Isaac from school at age 17. She expected him to oversee the family farm. A brief stint on the farm demonstrated that Newton was completely ill-qualified for such an occupation. Common sense prevailed as Newton's schoolmaster, Henry Stokes, intervened on Newton's behalf and convinced Hanna Smith to allow her son to enter Cambridge University.

There is no record to indicate that Newton demonstrated his genius during his first 3 years at Cambridge. Then in 1664, as a precautionary measure against a plague that had decimated London, Cambridge closed its doors and dismissed its students. Cambridge had sufficiently fertilized Newton's intellect by introducing him to the scientific topics of the day so that his yet-unleashed intellect understood the more prescient issues. Newton's farm was the perfect space for Newton the loner to blossom. No experienced academic was available to guide Newton's intellectual development, nor was such an academic able to hinder him. Back at the family farm, Newton aged, 22, embarked on his own research agenda and accomplished miracles.

Isaac Newton, an unknown novice with no previous accomplishment in mathematics or sciences, took his first cut at the very frontiers of knowledge and slashed through barriers that allowed him to expand the frontiers significantly beyond their pre-Newtonian boundaries. In optics, Newton proposed the radical concept that white light is a composition of all the colors of the rainbow. In mathematics, Newton explored infinite series and generalized the binomial theorem to accommodate any real power.[6] Combining his investigations into mathematics with investigations of motion, Newton invented what he coined the "method of fluxions"; the more commonly known term for this branch of mathematics is *calculus*. Newton's research into motion initiated an effort to find laws by which bodies move. Interest in planetary motion caused Newton to think deeply about gravity and its relation to elliptic orbits. Overtaking Kepler, Newton outlined a program that would allow him to demonstrate that gravity alone caused the elliptic orbit of the planets. He made remarkable strides but did not complete this program for another 20 years. Any single one of these achievements would have been a mark of distinction for which the discoverer would rightfully have earned honor. But for these wide-ranging, penetrating, and highly creative discoveries to come from the mind of a 22-year-old youth while wandering about his farm in intellectual isolation seems mythical.

As we are concerned with the motion of a planet, let us peer into the thoughts of the 22-year-old Newton. Recall Kepler's difficulty in conceiving of a planetary orbit based on gravity. Kepler believed that an attractive force would result in a planet's falling into the sun. Kepler, along with Galileo, did state the law of inertia, the tendency of a body's motion to be unaltered unless a force acts on it, but neither grasped the full consequences of inertia. Newton did, and saw clearly that inertia was the counterbalance to gravity and the two together keep a planet in orbit. Newton created the following thought experiment. From a high mountain, launch a projectile at a tangent to Earth's surface. The projectile's inertia wills it to maintain its direction, while gravity wills the projectile to fall toward Earth's center. If the projectile's initial impetus is not that great, the projectile falls on the ground in a short time. Strengthening the initial impetus causes the projectile to travel further about Earth prior to its inevitable collision. Strengthening the impetus still more causes the projectile to travel even further. Finally, with sufficient impetus, the seemingly inevitable collision of the projectile does not occur; rather, the projectile returns to its original position and continues indefinitely to orbit Earth. The projectile is forever falling, yet at the same time, inertia counterbalances the fall, maintaining the orbiting projectile's distance from the larger attracting body. Figure 8.1 is a drawing based on one from Newton's notebooks.

Cambridge reopened in 1667, and one would think that on returning, Newton would let his discoveries be known. This was not the case. In fact, Newton kept all his discoveries to himself. It was not that Newton did not like attention or accolades; on the contrary, he enjoyed his share of praise. Newton's insecurity caused him to fear rejection, so he did not disclose his work. He would then become incensed as others received their due praise for unveiling works

---

[6]The binomial theorem provides a method for finding the coefficients of the polynomial $(a + b)^n$, in which n is a positive integer. Newton's generalization provides a method for expanding the expression into a power series of infinite terms for every real value of n.

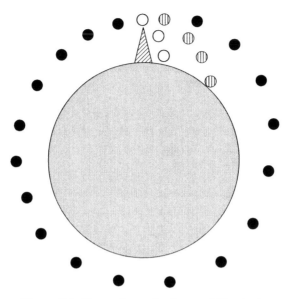

Figure 8.1. Newton's notebook; an orbiting body.

that Newton previously discovered but held secret. A pattern emerged whereby Newton would release his discoveries only after others had laid claim to them. In the case of calculus, this would not occur for another 40 years. However, the most immediate encroachment on Newton's world came from Nicholas Mercator (1620–1687), a very accomplished German mathematician. In 1668 Isaac Barrow (1630–1677), holder of the Lucasian chair in mathematics and Newton's mentor at Cambridge, passed Newton a paper from the German mathematician indicating a series method for determining logarithms. Only then did Newton reveal his work in infinite series that went beyond Mercator's paper. After some cajoling, Newton agreed to have his work published, earning him respect as a leading mathematician. This must have eased Newton's insecurities and boosted his confidence.

Optics fascinated both Barrow and Newton. The great Scottish scientist, James Gregory (1638–1675), had conceived of, but not constructed, the reflecting telescope and sent design specifications to Barrow. Barrow passed the design to Newton. Newton made improvements and used his skills honed by constructing windmills and the like as a schoolboy to construct a device. The telescope proved to be far more powerful than the conventional refracting telescope. In 1671, 2 years after Barrow resigned his position as holder of the Lucasian chair in mathematics, the next occupant of the Lucasian chair of mathematics, Newton, was persuaded to allow the Royal Society to examine the telescope. The awe-inspired members of the Royal Society invited Newton to become a member and further persuaded him to participate in their forums.

The accolades further boosted Newton's confidence, and Newton believed that he could engage the world on his terms. For the first time, Newton unveiled his earlier discoveries in optics. Newton, who as a boy was well acquainted and equally ill-equipped to cope with the rough and tumble of the schoolyard, was not in the least prepared for the rough and tumble of academic scrutiny. Newton's theories proved most controversial.[7] After more than a year of hostile exchanges, Newton penned his letter of resignation to the Royal Academy.

A flurry of diplomacy among Newton's supporters along with a promise to exempt Newton from paying dues averted Newton's formal resignation. However, the experience left Newton shell-shocked, and he reflexively did what was most natural to the boy abandoned by his mother; he relied on his emotional capacity to retreat from social discourse and stand alone. Newton stated his wish to keep his works unpublished during his lifetime. Newton's calculus and his work on motion and gravity remained his own secret.

Newton not only retreated from social discourse; he exited mathematics and physics altogether. For a period of roughly 10 years, from 1675 through 1684, alchemy absorbed Newton's attention. His dedication to this peculiar discipline was single-minded. Experiments in alchemy dictated Newton's time, and he considered other endeavors that are central to most humans, such as eating and sleeping, to be distractions. For all his efforts, Newton accomplished little.[8]

Central to Newton's successes in mathematics, mechanics, and optics was groundwork that had been laid by Newton's predecessors. For example, a short list of mathematicians who had tippy-toed around the field of calculus includes Archimedes, Kepler, Pascal, Fermat, Huygens, Gregory, and Newton's own mentor, Barrow. Additionally, René Descartes (1596–1650), by uniting algebra with geometry in the field of analytic geometry, changed the structure of mathematical communication and hence the way that mathematicians approach problems. This yielded an invaluable tool in the development of calculus. Calculus, mechanics, and Newton's contributions to optics were not the product of a lone genius, but the culmination of many contributors.[9] Perhaps the loner Newton did not develop the perspective to understand the influence of these men on his own thinking. Perhaps after his successes, Newton believed that he could be equally successful in any field. But this was not the case; the groundwork in chemistry was not prepared, and a breakthrough was still two centuries into the future. Newton must have been frustrated by his lack of progress.

Was it this frustration that caused Newton to welcome a visitor on a quest? Edmond Halley, a member of the Royal Society, had a fascination with comets. He, along with two other members of the Royal Society, Robert Hooke and

---

[7]The most controversial aspect of Newton's theories was his corpuscular theory of light in opposition to the wave theory of light. Although no participant in the debate could foresee three centuries ahead, this very same debate raged at the beginning of the twentieth century. Physicists have reached a compromise with one another and with nature, accepting that light has the properties of both particles and waves.

[8]In 1680, a sparring match with Hooke did motivate Newton to briefly address gravitation and planetary motion.

[9]This in no way diminishes Newton's contributions. It took his genius to add critical content to the disparate pieces laid by his predecessors.

Christopher Wren, had many discussions concerning the shape of a comet's pathway. Their consensus was that a comet's orbit is no different from those of planets and is accordingly a Keplerian ellipse. They also came to the same conclusion as Newton: that the sun's gravity was the locomotive force for comets as well as planets. Hooke made one additionally astute conjecture: that the force from the sun on a point in space is inversely proportional to the square of the distance from the sun to the point, specifically, the inverse-square law that Kepler had previously posed. Hooke, who frequently boasted far more than he could deliver, stated that he had a proof of the elliptic orbit based on the inverse-square law.

Having given Hooke his chance to deliver the proof, Halley sought out others who might have insight. At the suggestion of a friend, Halley traveled to Cambridge for a visit with Newton. When Halley posed his problem, without hesitation Newton responded that the shape of a comet's orbit is an ellipse. Dumbfounded by the quickness and certainty of Newton's response, Halley requested an explanation. Newton stated that he had proved this long ago and unsuccessfully attempted to locate his work. Then, very uncharacteristically, Newton promised to rework the solution and give it to this stranger, attesting to Halley's charm.[10] With that promise, Halley left.

History-altering moments are rare. We learn about these moments in school: Christopher Columbus' discovery of America, the Habsburg victory over Turkey, Waterloo, Hitler's invasion of Poland, Pearl Harbor, and not many more. The meeting between Halley and Newton was a private affair between two men. The public eye was far removed from the occasion. The topic of discussion was so arcane that few would have concerned themselves with it, and fewer would have considered it of any import. The moment is so modest that it appears to not be worthy of mention. But history is playing a Keplerian jest on those who pass by the moment on account of its modesty. For like Kepler, who, while bringing inspired genius to his work, feigns modesty, history disguises this pivotal moment in a veneer of modesty. The meeting between Halley and Newton unhermited the man whose mind possessed the ideas that underpin the science, technology, and economy of modern civilization.

Halley had a facility with men and could in fact bring out the best of those in his company.[11] Several months after Halley's visit, and much to his surprise, Newton furnished Halley with the solution. Newton took the time to reconstitute his result, which relied on his method of fluxions, using the mathematical conventions of his time. There was not a trace of calculus in the paper; all was Euclidean geometry. The work consisted of nine pages, and Halley, a well-trained

---

[10]My conjecture is that aside from deploying his charm, Halley left Newton with the impression that Hooke was closing in on the solution. Newton hated Hooke and would not let Hooke claim priority for that which he had discovered.

[11]Newton was not the only genius who Halley would assist, but he was the most outstanding. After Newton's death, as president of the Royal Society, Halley encouraged and provided assistance to John Harrison in his efforts to develop an accurate seaworthy timekeeper. This is one more demonstration of character in which Halley, of aristocratic background in a class-oriented society, assisted a commoner on account of merit. For a full account see *Longitude* by Dava Sobel (Sobel 1995).

scientist, recognized its value. The quest, stretching perhaps a millennium before Eudoxus and two millennia before Newton, to determine the governance of the motion of the heavens had come to an end. Newton bagged the prize.

An excited Halley requested permission from Newton to read the paper to the Royal Society. Newton carried grudges to his grave, and his previous experience at discourse with the Royal Society had caused much pain. But somewhere within Newton was a desire for recognition, and Halley had special interpersonal skills that overcame Newton's resistance. Universal praise from the Royal Society without a trace of controversy convinced Newton that he was dealing with the world on his terms, and so in December 1684, he allowed the work's publication.

An apple tree bearing one delicious apple most likely has many more. Halley sought to harvest more knowledge from Newton's mind. Together Newton and Halley conceived of a project in which Newton would give a fuller account of his ellipse-conquering analysis and then address further problems as well. Newton gave the work a Latin title, *Principia*. Halley convinced the Royal Society to fund the work's publication. The Royal Society agreed, but having suffered a financial setback due to the failure of a previous publication, the Royal Society never covered the gap between its promise and actually providing funds.[12] Rather than see the project go by the wayside, Halley financed *Principia's* publication from his own pocket.

After several revisions, the first volume of *Principia* was printed in 1687. One revision in particular deserves attention. Newton's original version somewhat surprisingly showed a conciliatory attitude toward Hooke and conferred a small degree of recognition on him. After a prereading, Hooke conveyed to Halley his belief that he was the discoverer of the inverse-square law and that *Principia* should give him this recognition. Halley passed Hooke's message on to Newton, which only had the effect of outraging Newton. It is true that Hooke had in the past shared the inverse-square law with Newton, but Newton claimed to have discovered it in 1665. In reality, Kepler deserves the credit for having first proposed the inverse-square law of gravity, although it is likely that neither Hooke nor Newton were aware of this.

The resolution of the latest Newton–Hooke conflict can be found in *Principia* as Newton grudgingly gives the following credit:

> The case of the 6th Corollary obtains in the celestial bodies (as Sir Christopher Wren, Dr. Hooke, and Dr. Halley have severally observed); and therefore in what follows, I intend to treat more at large of those things which relate to centripetal force decreasing in a duplicate ratio of the distances from the centers.

This resolution is a testimony to not only Halley's diplomatic skills but also Newton's sly style. By spreading credit for the inverse-square law on three individuals, Newton diffused the amount of credit conferred on Hooke. In addition, he scored points with both Halley and Wren.

---

[12]These guys were not savvy marketers. The failed publication that the Royal Society did finance was a work on bird species. They were unable to sell the copies from the first printing. On the other hand, *Principia* is still in print and earning money over 300 years after its initial publication.

The full work, *Principia*, occupies three volumes separately published between 1687 and 1689. The first volume contains Newton's laws of motion, after which Newton applies the laws to many different problems. The solution of orbiting bodies about an attractive source is one such problem. Newton also investigates the motion of the moon; formalizes the Keplerian explanation of the moon's gravity as the cause of the tides; demonstrates that Earth is not a perfect sphere, but bulges at its center; explains the dynamics of the precession of Earth's axis; and examines problems in fluid statics and dynamics. The scope of problems addressed is staggering and unprecedented. However, a central component to all the analysis that Newton performs is missing: his "theory of fluxions," or as we know it, *calculus*. Just as with his 1684 paper, Newton conveys all his mathematics using conventional geometry.[13]

Also in 1684 another genius for the first time published work that he had been holding for 10 years. Gottfried Leibniz (1646–1716), in a journal that he had founded, published an article on a new type of mathematics, calculus. Leibniz earned high praise on the continent, but the circulation of his work was most likely limited. Newton was unaware of Leibniz' work for perhaps a decade. In 1699 an admirer of Newton's, Nicolas Fatio, unleashed a scathing critique of Leibniz and accused him of plagiarizing Newton's work. The claim of plagiarism was far-fetched; Newton withheld his material on fluxions from all but a very small circle. Yet Newton believed with firm conviction that somehow Leibniz, living in Paris and then Hanover obtained his material, altered the notation, and published it as his own work.

Acting on his conviction, Newton traced the source of Leibniz' alleged plagiarism to Leibniz' 1675 visit to London. At that time the Royal Society invited Leibniz to demonstrate his calculating machine. The machine had not yet been perfected, and its performance was subpar; it was a performance that begged the ever-cantankerous Hooke to get his hooks into Leibniz. Also during this visit, John Collins, an admirer within Newton's inner circle with whom Newton shared some unpublished works, showed Leibniz a work of Newton's on infinite series. Exactly how much information Leibniz gleaned from a casual glance at Newton's paper on infinite series is uncertain, but even if he grasped the entire paper with full comprehension, its relation to Leibniz' results on calculus is quite remote. Leibniz made his own breakthrough and legitimately developed calculus independent of Newton. But that was not Newton's interpretation of Leibniz' private screening. For Newton, this was the evidence that convicted Leibniz of the crime. Newton encouraged Fatio to attack Leibniz and even penned attacks himself that Fatio subsequently signed and submitted as his own.

Newton's work on fluxions, motivated by his research into motion, intimately connected the new discipline with physics, and Newton's laws of motion may be presented in the mathematical language of calculus. Indeed, Newton most likely developed the results of *Principia* using his version of calculus and then

---

[13]The geometry and notation is conventional, but Newton does use limiting arguments of modern calculus.

translated the results into conventional geometric methods. Alternatively, Leibniz does not utilize physics to motivate his development of calculus. His development is more of a purely mathematical exercise. Leibniz' talents and interests were wide-ranging; aside from mathematics, Leibniz made contributions in many other fields: physics, history, philosophy, theology, law, technology, and linguistics.[14] Applying a special interest in languages and linguistics, Leibniz thought deeply about notational conventions and terminology for the new branch of mathematics that he coined *calculus*. History has been kind in offering these two geniuses, each arriving at calculus through his own path. Newton demonstrated how the new mathematics relates to our physical world. Leibniz developed superior notational conventions that are intuitive and facilitate communication of the new mathematics.

In 1700, as Europe prepared for the War of Spanish Ascension, Newton fought an ongoing war against Leibniz. The War of Ascension lasted 14 years. Newton's war against Leibniz lasted longer. Newton waged the war beyond Leibniz' death in 1716 until Newton's own death in 1727. During Newton's war, Newton secured his terms and conditions for his engagement with the world. In 1703 Hooke died and Newton became president of the Royal Society. Newton utilized the Royal Society as a weapon in his war against Leibniz. The Royal Society established its own independent committee to investigate the charge of plagiarism against Leibniz. What a surprise, when the committee's findings fully supported its president.

While Leibniz lived out his last few years under a cloud of suspicion, Newton attained celebrity status. In 1708 Queen Anne further honored Newton by knighting him.

Had the insecure Newton who feared controversy chosen as his first foray into publishing the method of fluxions instead of his work on optics, perhaps his career would have unfolded differently. While Newton's conclusion that white light is a blend of all colors of the rainbow was verifiable by experiment, the corpuscular claim was not, and it was almost inevitable that this would stir the controversy that Newton feared. By contrast, Newton's work on fluxions did not rely on experimentation and was demonstrable by argument in the form of proof. In 1671 Newton completed a treatise, *The Method of Fluxions*, but withheld publication.[15] Just as Leibniz initially received high praise for his publications on calculus, praise that later turned to acrimony as Newton's public relations campaign against Leibniz took effect, so would Newton have received such praise had he chosen to publish that work. So, why didn't Newton publish his work on fluxions?

When posed that question, Newton himself stated that he feared that publication would embroil him in further controversy, which he wished to avoid. Judging Newton's response from the bench of reason, one is skeptical. But it was not from the bench of reason that Newton made his life choices. It was

---

[14]Leibniz matriculated from Leipzig University with degrees in philosophy and law.
[15]This text was posthumously published in 1736.

from his mind that was seeded with an innate temperament and experiences of a difficult youth, followed by a difficult experience with his publication on optics. This is the mind where calculus first resided, and this mind chose to maintain the secrecy of those powers, for fear of ridicule.

In 1715 Louis XIV died at the age of 77. Newton died 12 years later at the age of 84. The offering that Louis XIV, the most dominant European political figure of his era, left to posterity was an obscenely opulent palace in a village on the outskirts of Paris. Newton's gift to posterity was a scientific revolution that changed humankind's relationship with nature and nurtured human technological and economic progress.

THE INFLUENTIAL *Principia*

Above it is claimed that *Principia* is the most influential book ever written. This claim deserves some explanation. What made *Principia* so revolutionary? Without diminishing the value of Kepler's contributions, it is necessary to look at a shortcoming of Kepler's work and demonstrate how Newton addresses this shortcoming. Kepler's work is descriptive; it is a beautiful mathematical representation of a single observed phenomenon; indeed, two of Kepler's three laws do nothing more than describe the phenomenon. The first law states that a planet orbits the sun following an elliptic pathway, and the third law states the relation between the length of a planet's year and the major axis of its elliptic orbit. As descriptive laws, they have no general application to other phenomena, and they have absolutely no predictive value. While the other law, Kepler's second law of conservation of angular momentum, turned out to be a more general statement, Kepler never realized it. The progress of science and technology would have proceeded excruciatingly slowly if all phenomena under investigation would have required their own independent descriptive laws that in turn have no predictive value.

Newton's laws are more far-reaching and general. They apply to the analysis of the motion of any body or system of bodies; they are equally applicable to the motion of an aircraft as to the motion of planets. Newton's laws also are predictive. They enable one to uncover a phenomenon before it is observed. Not only did Newton use his laws to demonstrate all three of Kepler's laws, reducing their stature from laws to propositions; he also applied them to problems other than the orbit of planets and made predictions about previously unobserved phenomena. As one example, Newton used his laws to demonstrate that Earth is chubby, it bulges at its equator, a phenomenon that nobody ever observed, but Newton correctly concluded.[16]

[16]The proposition of a chubby Earth caused another stir of controversy as Gian Cassini, a very capable Italian astronomer who worked for the French, took measurements of Earth's curvature and came to the opposite conclusion: Earth is skinny. The controversy prompted the French to send two expeditions, one to Peru, and one to Sweden, to take further measurements and settle the debate. In 1737 Maupertuis, the leader of the Swedish expedition and an enthusiastic disciple of Newton's, was the first to return with results that confirmed Newton's. Voltaire, also an enthusiastic disciple

A set of general laws with predictive capacity is invaluable to science and technology. We give two post-Newtonian examples. Astronomers were aware of a perturbation in the orbit of Uranus. The French astronomer Urbain Leverrier (1811–1877) applied Newton's laws to conclude that there must be another planet beyond Uranus and predict the position of the unobserved planet. In 1846, the same day that he received Leverrier's letter containing Leverrier's prediction of the position of the undiscovered planet, Johanne Galle of Berlin pointed his telescope in the direction of Leverrier's Newtonian calculation and discovered Neptune.

Among the performance requirements of an aircraft design are the weight that the aircraft can carry, the strength of the wings, and a stabilizing response to wind fluctuations. Engineers use Newton's laws to design aircraft to specification and predict the performance of any design. Modern engineering requires the predictive latitude of Newton's laws, without which technology that underlies our economy would not have been developed.

Principia also has an inspirational role. Its success boosted our collective confidence; we no longer resort to superstition to protect ourselves from natural phenomena that we don't understand. We attack the unknown with the intellect, confident in our capacity to find answers. It is with the confidence instilled by Newton's *Principia* that humankind closed the door on the Counter-Reformation and marched forward to the Enlightenment. *Principia* yields the means and inspiration to develop the science and technology that distinguish our modern lives. It is the most influential book ever written.

The Dry *Principia*

Principia's influence rests on its substance. It is a purely technical presentation that, like *New Astronomy*, was written in Latin. Stylistically, that's where the similarity with *New Astronomy* ends. *Principia* contains no nuggets of humor, there is no attempt to connect with the reader at a personal level, there are no philosophical interpretations, and there are no words of wonder; there is little that is Keplerian about *Principia*; it is 99% Euclidean. Just as Euclid in *The Elements* plunges right into technicalities and remains exclusively in that domain, so does Newton. A comparison between the two is apt.

Euclid's *The Elements* begins with definitions; the following is the first of 23.

Definition 1.   A point is that which has no part.

Newton's *Principia* also begins with definitions; the following is the first of eight.

Definition 1.   The quantity of matter is a measure of the same, arising from its density and bulk conjunctly.

of Newton's, took the opportunity to take a jab at Maupertuis, a former friend, with the quip that Maupertuis went to the ends of the earth to find that which Newton discovered at his desk.

After placing forth his definitions, Euclid presents his axioms, 10 in all. He divides the axioms into two categories: postulates and common notions. Euclid's first postulate reads as follows.

Postulate 1.   To draw a straight line from any point to any point.[17] A set of axioms also follows the definitions in Newton's *Principia*.

The axioms are stated in the form of laws, the first of which reads as follows.

Law 1.   Every body perseveres in its state of rest, or of uniform motion in a right line, unless it is compelled to change that state by forces impressed thereon.

With his definitions and axioms in place, statements of theorems and their proofs fill the remainder of Euclid's book. The first of Euclid's theorems, stated as a proposition reads as follows.

Proposition 1.   To construct an equilateral triangle on a given straight line.[18]

A proof that such a construction is possible follows the statement of this theorem.

In identical fashion, theorems follow Newton's definitions and axioms. Newton's first theorem, stated as a corollary, reads as follows.

Corollary 1.   A body by two forces conjoined will describe the diagonal of a parallelogram, in the same time that it would describe the sides, by those forces apart.

A proof of the statement follows.

Newton grounds *Principia* on the axiomatic–deductive process of Euclid. The success of *Principia* has bound physics within that framework.[19] But wait, Newton deviates from Euclid's technical style on the final four pages of his 443-page work.[20] In those final four pages, Newton gives a glimpse of his religious sentiment: "This most beautiful system of the sun, planets, and comets, could only proceed from the counsel and dominion of an intelligent and powerful Being."

Thus Newton ends *Principia* in the manner of Ptolemy, Copernicus, and Kepler—the work brings him nearer to an understanding of God.

---

[17]This is the translation. A schoolmarm would insist on an entire sentence such as the following: It is possible to draw a straight line from any point to any point.

[18]The objection of the school marm in footnote 16 applies here as well. Correcting that objection gives the following: It is possible to construct an equilateral triangle on a given straight line.

[19]Undoubtedly this is the correct style for presenting technical material. My only gripe is that, unlike Kepler's *New Astronomy*, *Principia* is nearly devoid of quotable material to place into this book.

[20]The book size refers to the translated copy.

WITHOUT FLUXIONS

Newton's Laws without Fluxions

As noted above, it is a certainty that Newton used his method of fluxions, alias calculus, to derive the results that he presents in *Principia*. How can we be so certain? Conventional mathematics of Newton's times are insufficient for deriving the results that Newton presents in *Principia*. Newton could have derived them only using new computational methods, and those methods can be nothing but calculus. However, the conventional mathematics of the day was sufficient to demonstrate the results once they had been derived. To avoid confrontation and present his stunningly original work in conventional terms, this is precisely what Newton chose to do. Here, we follow Newton's fluxionless path as far as possible and then demonstrate that the ellipse abides by Newton's laws of motion. We begin by placing forth some definitions and then giving mathematical expressions for Newton's second law of motion without explicitly using calculus.[21]

Definition 1.   The speed of an object at a point in time is the distance that the object would travel in a standard unit of time if its motion continued unaltered throughout the unit of time.

Definition 2.   The velocity of an object at a given time is the object's speed along with its direction of travel.

Definition 3.   The momentum of a body is the product of its mass and velocity.

While Newton identified the concept of momentum, he did not coin the term. Newton denoted the concept by the quantity of motion, and his definition reads as follows.

Definition 3′.   The quantity of motion is the measure of the same, arising from the velocity and quantity of matter conjointly.

Definition 4.   A force on a free body is anything that alters the body's momentum.

With these definitions we can state Newton's laws of motion directly from their translation into English.

Law 1.   Every body perseveres in its state of rest, or of uniform motion in a right line, unless it is compelled to change that state by forces impressed thereon.

---

[21]The knowledgeable scholar may object to this and the following section. Rigor is not complete, and sniffing the expressions yields an aroma of calculus. I concede this, but let the objector know that the purpose is not to be rigorous, but intuitive, so that the reader without knowledge of calculus may gain insight. The aroma emanating from the expressions should be familiar, not only to those familiar with calculus, but to those without exposure as well.

Law 2. The alteration of motion is ever proportional to the motive force impressed; and is made in the direction of the right line in which that force is impressed.

Law 3. To every action there is an opposite and equal reaction: or the mutual action of two bodies on each other are always equal, and directed to contrary parts.

Law 1 is somewhat of a puzzle to the logician. Laws as axioms ought to be independent of one another, and yet law 1 is a consequence of law 2. Indeed, according to law 2, the alteration of motion is proportional to the motive force; consequently, when there is no motive force, there is equally no alteration of motion, which is a restatement of law 1. Newton was a master of the axiomatic–deductive process and was aware that the first law follows from the second. He had a purpose for stating it.

The first law defines reference frames in which the remaining laws apply. A reference frame in which the first law does not apply disqualifies the use of the remaining laws. As an example, consider a reference frame that rotates. A body with no force acting on it, at rest in a nonrotating frame would appear to rotate about the coordinates of a rotating frame. The coordinates do not indicate that the body is at rest or moves in a right line even though there is no force acting on the body. We cannot directly use Newton's laws of motion to describe the body's motion in the rotating frame.[22] Reference frames where Newton's laws directly apply are known as *inertial reference frames*. They have the property that they move uniformly with respect to each other. This is significant in the special theory of relativity.

Another possibility for Newton stating the first law is that Newton wished to separate the force-free case from the case in which a force is impressed in order to first appeal to the reader's intuition and then build on that intuition. Examples that Newton gives after stating his first law do in fact have an intuitive appeal. Once the reader is comfortable with the first law, so that it is accepted as a fact of nature, the second law is easier to grasp and seems equally natural. For me, I read the first law, reflect on it, and relate it to my experience; a bowling ball once released continues in its motion rolling down the lane. This is convincing. Then I read the second law and there's an "aha" moment. "Sure," I say to myself, "only a force impressed upon a body could alter the body's motion and the greater the force, the greater its alteration." In this one instance, to ensure that the presentation is more readable, Newton was willing to forego the strict protocols of the axiomatic–deductive process.

It is the second law that yields the ellipse. While the words of the second law are convincing, an actual definition for the alteration of motion is lacking. After the initial pleasure of the "aha" moment wears off, my confidence sags. "I don't

---

[22]Using alternative descriptions of Newtonian mechanics, namely, Euler–Lagrange equations, one can express motion in rotating and other reference frames.

get it, what exactly is the alteration of motion?" This is where Newton separates himself from his predecessors. Newton knows a concise mathematical expression of the second law, one that allows him to calculate. But the concise mathematical expression is in terms of fluxions, which is Newton's private property. Having fenced fluxions off from the public, Newton proceeds to instill meaning using conventional mathematics.

Prior to giving a more rigorous meaning to the second law, let us consider a consequence that provides some insight. Galileo demonstrated the contra-Aristotelian result that bodies fall at the same rate regardless of their mass. Indeed, if we contemporaneously drop two bodies, one lighter and one heavier, from the leaning tower of Pisa, not only do the two bodies strike ground at the same time; they also have identical velocity profiles throughout their fall. Let us address the following question: What are the forces on the two bodies? Since the two bodies have identical velocity profiles, the alteration of their velocities, their acceleration, is identical. To alter the heavier object's velocity so that it matches that of the lighter body requires more force. How much force? According to Newton's second law, the force required to alter the body's velocity is proportional to the body's mass. In the case of the falling body, this is the body's weight. Newton exposes the flaw of Aristotle. The heavier body is able to fall at the same rate as the light body because its own weight provides the additional force necessary to accelerate at the same pace as the lighter body.

There are three concepts that Newton relates through the second law: a body's position, velocity, and the force impressed on the body. It is useful to represent these concepts as vectors. A *vector* is an arrow with both length and direction. In the case of a body's position, the length of the vector gives the distance of the body from a prescribed point, often the origin of a Cartesian coordinate system and the direction of the vector points in the direction of the body from the prescribed point. One also represents the velocity as a vector. The length of the arrow is the body's speed, and the direction that the arrow points toward gives the body's direction of travel. One finds that a vector is equally useful for representing a force. The length of the force vector is the magnitude of the force, and the direction of the force vector gives the direction in which the force acts on a body. Figure 8.2 illustrates position, velocity, and force vectors acting on a body moving along a given pathway.

While vectors may lie in spaces of arbitrary dimension, we restrict our vectors to two-dimensional space, as this is where the ellipse resides. Accordingly, the position, velocity, and force of a body at a particular time all have two components, one for each of two spatial dimensions. We denote a vector by a bold letter or by its two components as follows:

$$\mathbf{B} = \begin{bmatrix} b_1 \\ b_2 \end{bmatrix}$$

where b_1 is the component of the vector along one planar direction, usually the x direction; and b_2 is the component of the vector along the other planar direction,

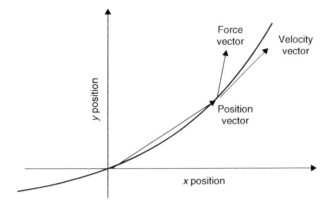

Figure 8.2. Position, velocity, and force vectors at a point in two dimensions. The body moves along the curve.

usually the y direction. For two vectors to be equal, each of their components must be equal.

Let us delve further into the concept of velocity and give a geometric interpretation of the velocity of a body from its known position. As a first effort, let us restrict motion to a single dimension, for example, along the x axis. In this case the position of a body, $x = x(t)$, is given as a function of time. The velocity of the object is the slope of the line tangent to the position curve. Figure 8.3 illustrates the geometry. The figure gives a position curve with respect to time. The figure also shows the tangent line to the position curve at two points: t_0 and t_1. The velocity of the body is the slope of the tangent line. In Figure 8.3, the velocity at time t_0 is greater than the velocity at time t_1.

How does this geometric definition of the velocity mesh with definitions 1 and 2 above, and Newton's laws of motion? According to law 1, if at time t_0 and beyond, no force is applied to the body, the body continues to move at uniform velocity. What would the graph of the position look like if one removed all forces beyond time t_0?

The body would move along the original curve until time t_0 and then follow the associated tangent line (see Figure 8.3). Using definitions 1 and 2, it follows that since the slope of the tangent line is the difference in position over the difference in time, the slope of the tangent line gives the body's velocity.

In two dimensions we have two position curves, and two associated velocities, one per direction. Let v_1 be the velocity along one direction and v_2 be the velocity along the complementary direction. Then the velocity vector is $\mathbf{V} = \begin{bmatrix} v_1 \\ v_2 \end{bmatrix}$. Using the Pythagorean theorem, one finds that the speed of the body in two dimensions is $v = \sqrt{v_1^2 + v_2^2}$. One also notes that the direction of motion is along the vector. Note that if the force causing a change in motion were to discontinue at time t_0 in accordance with law 1, the body would move in a right line as directed by the velocity vector at t_0, and this vector is tangent to the curve. Figure 8.4 illustrates the velocity in two dimensions.

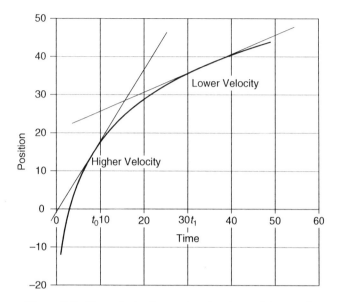

Figure 8.3. The velocity is equal to the slope of the graph.

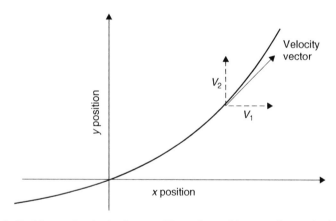

Figure 8.4. Position and velocity in two Dimensions. Absent a force, the body's path tracks the velocity vector.

Let us next consider the alteration of motion or momentum that Newton describes in his second law. There are two ways in which a force can alter motion. A force can increase or decrease the body's speed along its direction of motion. Additionally, a force can change the course of the body's pathway. We begin with the first alteration, a change in speed along the direction of motion. For this case, it is necessary to consider motion along only one dimension. Let the body move along the x axis so that its velocity vector also lies along the x

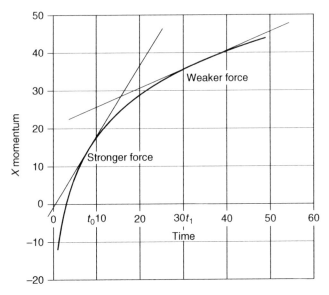

Figure 8.5. Momentum–force plot. The force is equal to the slope of the graph.

axis and is a one-dimensional value. We can plot the momentum, mass × velocity, as a function of time as in Figure 8.5. In that figure we assume that the force is changing continuously in time, causing the motion to change as well. Suppose that at the time t_0, the force ceases to change and instead stays at a fixed constant value. According to the second law, the change in motion from t_0 onward is equal to the fixed value of the force at time t_0.[23] The alteration in the body's motion is also constant, so the momentum curve is tangent to the original momentum curve at the point t_0. The slope of the tangent measures the force and is equal to the alteration in the body's momentum.

There is a natural analogy between the relationships of position to velocity and momentum to force. The velocity is the alteration of the position, and the force is the alteration of the momentum. When one holds the velocity constant, the graph of the body's position proceeds along a straight line that is tangent to its graph at the point where the velocity becomes constant. When one holds the force constant, the graph of the body's momentum proceeds along a straight line that is tangent to its graph at the point where the force becomes constant. Figures 8.3 and 8.5 both depict the relations. A point to note is that the curves in the figures are the same in order to emphasize the analogy. However, in general the momentum curve is very different from the position curve.

---

[23] The translation of Newton's second law is that the alteration of motion is "proportional to" the force. I make no claims to be a linguist, but it does appear that the phrase "proportional to" may at times be one and the same as "equal to." In the context of the second law it is evident that Newton took the two expressions as the same, for that is how he applies the second law throughout *Principia*.

In one dimension, it is common to express Newton's second law using the following notation. Let $x(t)$ represent the position of the body at time t, $v(t)$ represent the body's velocity, and $f(t)$ represent the body's force. Also, let $\dot{x}(t)$ and $\dot{v}(t)$ represent the corresponding alterations of position and velocity.[24] With this notation, the mathematical expressions for Newton's second law become

$$\dot{x}(t) = v(t) = s_x(t) \tag{8.1}$$

$$m\dot{v}_1(t) = f(t) = s_{mv}(t) \tag{8.2}$$

where $s_x(t)$ is the slope of the tangent line to the position curve at time t and $s_{mv}(t)$ is the slope of the tangent line to the momentum curve at time t.

It is sometimes convenient to divide equation (8.3) by the mass, yielding the following expression of the second law

$$\dot{v}_1(t) = \frac{f(t)}{m} = s_v(t) \tag{8.3}$$

where $s_v(t)$ is the slope of the tangent line to the velocity curve at time t.

Let us now apply these concepts to motion in two dimensions.[25] We can perform the same experiment that we considered in the one-dimensional case. Impress a two-dimensional force $\mathbf{F}(t) = \begin{bmatrix} f_1(t) \\ f_2(t) \end{bmatrix}$ on a body and graph its velocity $\mathbf{V}(t) = \begin{bmatrix} v_1(t) \\ v_2(t) \end{bmatrix}$ as a function of time. Then repeat the experiment hold the force fixed only at time t_0. At time t_0, the velocity travels on a tangent to the original curve; thus, each component of the velocity vector follows the tangent in accordance with equation (8.3) (see Figure 8.6). Newton's equations of motion then become

$$\dot{\mathbf{X}}(t) = \begin{bmatrix} v_1(t) \\ v_2(t) \end{bmatrix} = \begin{bmatrix} s_1(t) \\ s_2(t) \end{bmatrix} \tag{8.4}$$

$$\dot{\mathbf{V}}(t) = \begin{bmatrix} \dot{v}_1(t) \\ \dot{v}_2(t) \end{bmatrix} = \frac{1}{m} \begin{bmatrix} f_1(t) \\ f_2(t) \end{bmatrix} = \begin{bmatrix} s_{v_1}(t) \\ s_{v_2}(t) \end{bmatrix} \tag{8.5}$$

Note that in equation (8.5) division by the mass applies to both components of the force. The alteration of velocity is also known as the *acceleration*, and one frequently sees Newton's second law in the following vector form:

$$\mathbf{F} = m\mathbf{A}$$

where, as above, $\mathbf{F}$ is the force vector, m is the mass of the body, and $\mathbf{A}$ is the acceleration vector.

[24] This is Newton's notation that was later published, but not in *Principia*.
[25] We limit ourselves to two dimensions only for the purpose of focusing on the ellipse. One may generalize the concept to higher dimensions.

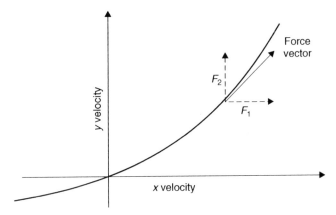

Figure 8.6. Force and velocity in two dimensions. Under a constant force, the velocity curve tracks the force vector.

Acceleration, that is, the alteration of velocity, in the one-dimensional case occurs only through a change in the body's speed. The two-dimensional case allows for a change not only in speed but also in direction. Indeed, by applying a force in a direction that is always perpendicular to a body's velocity, one can alter the direction without changing the speed. Applying a force so that it is neither perpendicular nor parallel to a body's velocity alters both its speed and its direction.

The Ellipse without Fluxions: Some Preliminary Considerations

Having presented Newton's equations of motion, we are one step closer to unveiling the ellipse as the planetary pathway. There are, however, some preliminaries that are necessary. This section gives two results that we use to construct the ellipse. The results are as follows.

Result 1. On an ellipse configured with semimajor axis a and semiminor axis b, the following relationship between the radial distance to a point on the ellipse r, and the x coordinate in a system with origin at the left focus of the ellipse holds

$$r = \varepsilon x + a(1 - \varepsilon^2) \tag{8.6}$$

where $\varepsilon > 0$ is the eccentricity of the ellipse, satisfying the equation, $\varepsilon^2 = \frac{a^2 - b^2}{a^2}$.

Result 2. The following equation defines the quantity angular momentum

$$J = m(x v_2 - y v_1) \tag{8.7}$$

where m is the mass of the body, x is the body's x coordinate, y is the body's y coordinate, v_1 is the body's velocity along the x coordinate, and v_2 is the body's velocity along the y coordinate.

The result is is that the angular momentum J is a conserved quantity; J doesn't change as the planet moves about its orbit. This means that everywhere along the planet's orbit J is the same.[26]

Conservation of angular momentum is equivalent to Kepler's second law, which states that the area swept out by a line between the sun and the planet is proportional to the time. Newton demonstrates result 2. This is known as *conservation of angular momentum*, using an argument having similarities to Kepler's; see equation (7.9) along with an explanation on the same page. The similarity is that both Kepler and Newton divide the orbit into sectors, make an approximation in each sector, and then give a limiting argument to demonstrate that the approximation yields the correct value as the number of sectors increases. Newton goes one step beyond Kepler by demonstrating that his argument is consistent with his laws of motion. Using a different method, we will demonstrate conservation of angular momentum in a later section. For now, we accept the principle and stock it in our arsenal.

The reader anxious to move on to the ellipse may skip to the next section. For the reader desiring to see a derivation of result 1, here it is.

We have the standard relation between r, x, and y:

$$r = \sqrt{x^2 + y^2}$$

Let the designated ellipse have its left focal point at the origin (the center of the sun) with semimajor and semiminor axes respectively given by a and b and with the focal distance given by c. Recall that the focal distance satisfies the relation $c^2 = a^2 - b^2$. Starting with the equation for the ellipse, one arrives at the following desired expression:

$$\left(\frac{x-c}{a}\right)^2 + \left(\frac{y}{b}\right)^2 = 1$$

$$\left(\frac{x-c}{a}\right)^2 + \left(\frac{y}{b}\right)^2 + \left(\frac{x}{b}\right)^2 = 1 + \left(\frac{x}{b}\right)^2$$

$$\left(\frac{x-c}{a}\right)^2 + \left(\frac{r}{b}\right)^2 = 1 + \left(\frac{x}{b}\right)^2$$

$$\left(\frac{r}{b}\right)^2 = 1 + \left(\frac{x}{b}\right)^2 - \left(\frac{x-c}{a}\right)^2$$

$$a^2 r^2 = a^2 b^2 + a^2 x^2 - b^2 (x - c)^2$$

---

[26]In addition to angular momentum, energy is also a conserved quantity. The energy has two components, kinetic and potential. The kinetic energy is $\frac{1}{2}mv^2$, where v is the planet's speed. The potential energy is $\frac{mM\Gamma}{r}$, where M is the mass of the sun, r is the distance between the sun and the planet, and Γ is the gravitational constant. See *The Ellipse* (Mazer 2010) for the derivation of the energy as a conserved quantity.

$$a^2r^2 = a^2b^2 + (a^2 - b^2)x^2 + 2b^2cx - b^2c^2 \quad (8.8)$$
$$a^2r^2 = c^2x^2 + 2b^2cx + b^2(a^2 - c^2)$$
$$a^2r^2 = c^2x^2 + 2b^2cx + b^4$$
$$a^2r^2 = (cx + b^2)^2$$
$$ar = cx + b$$
$$r = \frac{c}{a}x + \frac{b^2}{a}$$
$$r = \varepsilon x + a(1 - \varepsilon^2) \quad (8.9)$$

In the last expression the eccentricity $\varepsilon = \frac{c}{a}$ has been introduced. The eccentricity is a positive value ranging between zero and one; the ellipse becomes a circle with eccentricity zero and approaches a flat line as the eccentricity approaches one. From the definitions of ε and c, it follows that $b^2 = a^2(1 - \varepsilon^2)$. Summarizing, in addition to equations (8.6) and (8.7), we will be using the following equations:

$$c = a\varepsilon \quad (8.10)$$

$$b^2 = a^2(1 - \varepsilon^2) \quad (8.11)$$

Constructing the Ellipse, No Fluxions

We now construct the ellipse in the following sense. Given an ellipse as the shape of the planet's orbit, we wish to find a viable velocity profile for the planet at every point along the orbit, that is, find the components of velocity v_1 and v_2 at every point along the planet's orbit. The construction is set on Cartesian coordinates in which the left-handed focus of the ellipse, where the sun resides, is the origin.

As illustrated by Figure 8.7, the velocity vector is everywhere tangent to the ellipse and the ratio of the velocity components $\frac{v_2}{v_1}$ gives the slope of the tangent line. We then have two relationships, conservation of angular momentum and tangency of the velocity vector, to determine the two velocity components, v_1 and v_2. To arrive at the velocities, we follow a two-step process.

Step 1. Determine the slope of the tangent line at every point on the ellipse.

Step 2. Simultaneously solve the equations for conservation of angular momentum, equation 8.7 and tangency from step 1 to determine the two velocity components v_1 and v_2.

We are ready to execute.

Step 1. Determine the slope of the tangent line at every point on the ellipse.

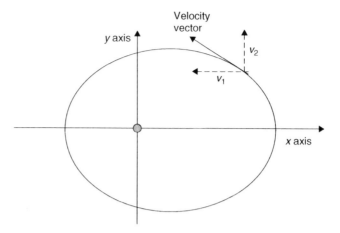

Figure 8.7. Components of the velocity vector along the ellipse.

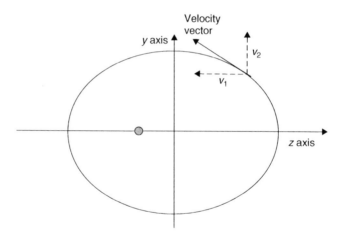

Figure 8.8. The ellipse in centered coordinates.

The calculation is simpler for the case where the ellipse is centered on the origin. Once we have found the result for the centered ellipse, the result for the shifted ellipse is obtained by shifting the value of x so that the origin once again lies on the left focus. To stress that we first work with the centered ellipse, we label the Cartesian variables (z, y) in which $z = x - c$ (see Figure 8.8)

A line is tangent to the ellipse when it intersects the ellipse at only one point. This is the property that allows one to find a tangent line. Let the centered ellipse be specified [eq. (8.12)] along with the tangent line [Eq. (8.13)]; in the following, a, b, μ, and β, are specified and the line is tangent to the ellipse. We wish to determine the associated values of z and y, yielding a relationship between the slope $\mu = \frac{v_2}{v_1}$ and a point on the ellipse (z, y):

$$\left(\frac{z}{a}\right)^2 + \left(\frac{y}{b}\right)^2 = 1 \tag{8.12}$$

$$y = \mu z + \beta \tag{8.13}$$

The intersection of the two graphs is found by substituting the equation for the line into that of the ellipse. On simplifying, one attains a quadratic in z:

$$\left(\frac{z}{a}\right)^2 + \left(\frac{y}{b}\right)^2 = 1$$

$$\left(\frac{z}{a}\right)^2 + \left(\frac{\mu z + \beta}{b}\right)^2 = 1$$

$$\left(b^2 + \mu^2 a^2\right) z^2 + 2a^2 \mu \beta z + a^2 \beta^2 - a^2 b^2 = 0$$

Applying the quadratic formula yields the solutions for z:

$$z = \left(-a^2 \mu \beta \pm \sqrt{(a^2 \mu \beta)^2 - (a^2 \beta^2 - a^2 b^2)\left(b^2 + \mu^2 a^2\right)}\right) \frac{1}{b^2 + \mu^2 a^2} \tag{8.14}$$

We are interested in the case when there is only one solution. This occurs when the value within the radical of equation (8.14) vanishes:

$$(a^2 \mu \beta)^2 - (a^2 \beta^2 - a^2 b^2)\left(b^2 + \mu^2 a^2\right) = 0 \tag{8.15}$$

Simplifying equation (8.15) yields a relationship between β and μ that must hold since the line is tangent to the curve:

$$\beta^2 = a^2 \mu^2 + b^2 \tag{8.16}$$

Continuing with the solution to the quadratic, as the radical in the quadratic formula is zero, we have the following solution:

$$z = -\frac{a^2 \mu \beta}{b^2 + \mu^2 a^2}$$

$$= -\frac{a^2 \mu \beta}{\beta^2} \tag{8.17}$$

$$= -\frac{a^2 \mu}{\beta}$$

Squaring the result [eq. (8.17)] substituting for β using equation (8.16), and simplifying yields the following solution:

$$z^2 = \frac{a^4 \mu^2}{\beta^2}$$

$$z^2 = \frac{a^4 \mu^2}{a^2 \mu^2 + b^2}$$

$$\left(a^2 \mu^2 + b^2\right) z^2 = a^4 \mu^2$$

$$(a^4 - a^2 z^2)\mu^2 = b^2 z^2$$

$$\mu^2 = \frac{b^2 z^2}{(a^4 - a^2 z^2)}$$

$$\mu^2 = \frac{b^4 z^2}{b^2(a^4 - a^2 z^2)}$$

$$\mu^2 = \frac{b^4 z^2}{a^4 b^2 \left(1 - \dfrac{z^2}{a^2}\right)}$$

$$\mu^2 = \frac{b^4 z^2}{a^4 y^2}$$

Replacing z by the x coordinate yields the result

$$\mu^2 = \frac{b^4 (x - c)^2}{a^4 y^2}$$

Taking the square root of the final value yields solutions for the slope $\mu = \frac{v_2}{v_1}$ as a function of the point (x, y) on the ellipse.

$$\mu = \frac{-b^2(x - c)}{a^2 y}$$

$$\frac{v_2}{v_1} = \frac{-b^2(x - c)}{a^2 y} \tag{8.18}$$

Equation (8.18) must hold everywhere along the planet's orbit.[27]

Step 2.   Simultaneously solve the equations for conservation of angular momentum [eq. (8.7)] and tangency from step 1 to determine the two velocity components v_1 and v_2:

$$x v_2 - y v_1 = \frac{J}{m} \qquad \text{[eq. (8.7)]}$$

$$\frac{v_2}{v_1} = \frac{-b^2(x - c)}{a^2 y} \qquad \text{[eq. (8.18)]}$$

[27] Taking the square root yields two solutions. From Figure 8.8 one notes that the negative solution is the one of interest because the tangent line as depicted has negative slope in the first quadrant.

$$v_2 = \frac{-b^2(x - c)}{a^2 y} v_1 \tag{8.19}$$

$$\left[-\frac{b^2(x - c)x}{a^2 y} - y\right] v_1 = \frac{J}{m} \quad \text{[substitute for } v_2$$

in first line of eq. of (8.19)]

$$\left[-b^2(x - c)x - a^2 y^2\right] v_1 = \frac{Ja^2 y}{m}$$

$$\left[-b^2(x - c)(x - c) - a^2 y^2 - b^2 c(x - c)\right] v_1 = \frac{Ja^2 y}{m}$$

$$\left[-b^2(x - c)^2 - a^2 y^2 - b^2 c(x - c)\right] v_1 = \frac{Ja^2 y}{m}$$

$$\left[-a^2 b^2 - b^2 c(x - c)\right] v_1 = \frac{Ja^2 y}{m} \quad \text{[eq.(8.8), ellipse]}$$

$$v_1 = \frac{-Ja^2 y}{mb^2 \left[a^2 + c(x - c)\right]}$$

$$= \frac{-Ja^2 y}{mb^2 \left[a^2 + a\varepsilon(x - a\varepsilon)\right]}$$

[eq. (8.10)]

$$= \frac{-Jay}{mb^2 \left[a + \varepsilon(x - a\varepsilon)\right]}$$

$$= \frac{-Jay}{mb^2 \left[a(1 - \varepsilon^2) + \varepsilon x\right]} \tag{8.20}$$

$$= \frac{-Jay}{mb^2 r} \quad \text{[eq.(8.6)]} \quad \tag{8.21}$$

Returning to equation (8.19), we solve for v_2 as follows:

$$v_2 = \frac{-b^2(x - c)}{a^2 y} v_1$$

$$= \frac{b^2(x - c)Jay}{ma^2 y b^2 r} \quad \text{[eq.(8.20)]} \quad \tag{8.22}$$

$$= \frac{(x - a\varepsilon)J}{mar}$$

We have successfully constructed a velocity profile on an elliptic orbit that conserves the angular momentum. This is the only possible candidate for a solution that respects Newton's second law. Below, we demonstrate that the candidate does, indeed, satisfy Newton's second law.

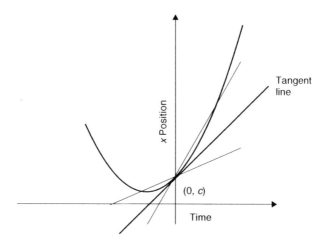

Figure 8.9. The tangent line.

Fluxions

There is one final step to take. It is necessary to demonstrate that the ellipse constructed in the preceding section is consistent with Newton's laws of motion. More specifically, we must demonstrate compliance with law 2; the force of the sun acting on the planet must alter the planet's motion as expressed by equation (8.5). We confront equation (8.5) in the next section. In this section, we prepare the tool that is necessary, fluxions. In particular, it is necessary to be able to determine the slopes of equation (8.5). The method of fluxions gives us the means of doing so. We start with an example and then give a general rule.

Let us find the slope of the tangent to the parabola, $x(t) = at^2 + bt + c$, at the time $t = 0$. We can use the method of the previous section where we determined the tangent to the ellipse. First, identify the point $x(0) = c$. Then, among all lines passing through the coordinates, $(x, t) = (c, 0)$, excluding the vertical line, identify the line that intersects the parabola only once. This is the tangent line (see Figure 8.9).

Rewriting the equation of the parabola and introducing the corresponding tangent line gives the following

$$x(t) = at^2 + bt + c \tag{8.23}$$
$$x(t) = \mu t + c$$

where μ is the slope that we seek.

Finding points of intersection yields the following identity:

$$at^2 + bt + c = \mu t + c \tag{8.24}$$
$$at^2 + (b - \mu)t = 0$$

Equation (8.24) has only one solution when $\mu = b$. In the case $\mu \neq b$, there are two solutions indicating two points of intersection between the parabola and the line. We may conclude that the slope we seek is $\mu = b$.

Let us interpret the result. In the vicinity of the point $t = 0$, where t is very small, the parabola behaves like the tangent line (as seen in Figure 8.9, the two graphs are very close) and the term at^2 is insignificant.[28] Accordingly, when finding the slope of the tangent line at $t = 0$, only the linear term matters.

Let us take this line of reasoning to polynomials of higher order. Consider the polynomial

$$x(t) = p_n t^n + p_{n-1} t^{n-1} + \cdots + p_2 t^2 + p_1 t + p_0 \tag{8.25}$$

where p_j are fixed coefficients.

What is the slope of the tangent line at the point $t = 0$? Following the reasoning of the preceding paragraph, in the vicinity of the point $t = 0$, where t is very small, the polynomial behaves like the tangent line and the sum $p_n t^n + p_{n-1} t^{n-1} + \cdots + p_2 t^2$ is insignificant. Accordingly, when finding the slope of the tangent line at $t = 0$, only the linear term matters. The slope we seek is $\mu = p_1$.

We next consider the problem of finding the slope of the tangent line to a polynomial as in equation (8.25) at an arbitrary point t_0. First return to the parabola in equation (8.23). We can reduce the problem to the known case above as follows. In equation (8.23), let $t = t_0 + \Delta$, where t_0 is fixed and Δ is a variable. The result is a parabola in Δ:

$$x(t) = at^2 + bt + c$$
$$x(t_0 + \Delta) = a(t_0 + \Delta)^2 + b(t_0 + \Delta) + c$$
$$\tilde{x}(\Delta) = a\Delta^2 + (2at_0 + b)\Delta + (at_0^2 + bt_0 + c)$$
$$= q_2\Delta^2 + q_1\Delta + q_0 \tag{8.26}$$

The q_j are the coefficients of the parabola in Δ, specifically, $q_1 = 2at_0 + b$. Equation (8.26) evaluated at Δ is identical to equation (8.23) evaluated at $t = t_0 + \Delta$. Figure 8.10 gives the graphs of the two parabolas coordinatized by their corresponding variables. The difference between the graphs is a horizontal shift by the quantity t_0. As the figure illustrates, the difference in the tangent lines is also a horizontal shift by the quantity t_0.

The horizontal shift has no effect on the slope of the shifted tangent line; accordingly, the slope of the tangent line to equation (8.23) at $t = t_0$ is the same as the slope of the tangent line to equation (8.26) at $\Delta = 0$. But we know the latter quantity, $\mu = q_1 = 2at_0 + b$, and so this is the slope that we seek.

---

[28] As t gets small, t^2 gets even smaller. For example, $\frac{1}{20^2} = \frac{1}{400}$.

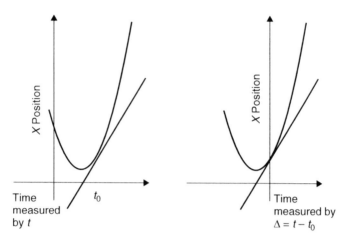

Figure 8.10. Plots of two parabolas coordinatized with respect to their corresponding variables.

Let us generalize to a polynomial of any order. Starting with equation (8.25), we replace t by

$$x(t_0 + \Delta) = p_n(t_0 + \Delta)^n + p_{n-1}(t_0 + \Delta)^{n-1} + \cdots$$
$$+ p_2(t_0 + \Delta)^2 + p_1(t_0 + \Delta) + p_0$$

This expression is a polynomial in the variable Δ

$$\tilde{x}(\Delta) = q_n \Delta^n + q_{n-1} \Delta^{n-1} + \cdots + q_2 \Delta^2 + q_1 \Delta + q_0$$

where the q_j are coefficients.

Applying the same arguments used in the case when $n = 2$, we find that the slope of the tangent line to equation (8.25) at $t = t_0$ is $\mu = q_1$. The following formula is obtained by use of the binomial theorem:[29]

---

[29] To find q_1, we use the binomial theorem

$$(t_0 + \Delta)^m = \sum_{j=0}^{m} C_j^m t_0^{m-j} \Delta^j$$

where $C_j^m = \frac{m!}{(m-j)!j!}$. Placing the expression into the expansion $x(t_0 + \Delta)$ and extracting the linear terms with $j = 1$ yields

$$q_1 = p_n C_1^n t_0^{n-1} + p_{n-1} C_1^{n-1} t_0^{n-2} + C_1^{n-2} p_{n-2} t_0^{n-3} + \cdots + p_1 C_1^1$$

Equivalence between this expression and equation (8.26) follows because $C_1^m = m$.

$$q_1 = np_n t_0^{n-1} + (n-1)p_{n-1}t_0^{n-2} + (n-2)p_{n-2}t_0^{n-3} + \cdots + p_1$$

Let us summarize what we have found. To find the slope of a polynomial at an arbitrary point, center the polynomial about the arbitrary point and extract the resulting linear coefficient. Newton went further. He found the slope for general functions using infinite series. Let the following equation hold at the point t_0 for an arbitrary function $x(t)$:

$$x(t_0 + \Delta) = q_0 + q_1\Delta + q_2\Delta^2 + q_3\Delta^3 + \cdots \tag{8.27}$$

The terms in this equation may continue indefinitely. Newton concluded that the slope to the function at $t = t_0$ is $\mu = q_1$.[30]

We demonstrate Newton's result by example. Let $x(t) = \frac{1}{t}$, with $t > 0$. To find the slope of the tangent line at $t = t_0$, we look for a series expansion as in equation (8.27)

$$x(t_0 + \Delta) = \frac{1}{t_0 + \Delta}$$

$$= \frac{1}{t_0} - \frac{1}{t_0^2}\Delta + \frac{1}{8t_0^3}\Delta^2 + \cdots \tag{8.28}$$

where the coefficients are using long division (see the footnote).[31]

Another method for determining the coefficients is to multiply the series expansion of $x(t_0 + \Delta) = \frac{1}{t_0 + \Delta}$ by its reciprocal, $t_0 + \Delta$, and equate the result to one. The calculation proceeds as follows. Let $x(t_0 + \Delta) = q_0 + q_1\Delta + q_2\Delta^2 + q_3\Delta^3 + \cdots$, where the elliptical dots indicate that the series never terminates.

$$1 = x(t_0 + \Delta)(t_0 + \Delta)$$

---

[30]Some analysis is necessary to demonstrate convergence of the infinite series. Consult a standard book on analysis such as Principles of Mathematical Analysis by Walter Rudin.

[31]

$$\frac{1}{t_0} - \frac{1}{t_0^2}\Delta + \frac{1}{t_0^3}\Delta^3 \ldots$$

$$t_0 + \Delta \quad | \overline{1.0000}$$

$$\underline{1 + \frac{\Delta}{t_0}}$$

$$-\frac{\Delta}{t_0}$$

$$-\frac{\Delta}{t_0} - \frac{\Delta^2}{t_0^2}$$

$$\frac{\Delta^2}{t_0^2}$$

$$= \left(q_0 + q_1\Delta + q_2\Delta^2 + q_3\Delta^3 + \cdots\right)(t_0 + \Delta)$$

$$= q_0 t_0 + (q_1 t_0 + q_0)\Delta + (q_2 t_0 + q_1)\Delta^2 + \cdots + (q_j t_0 + q_{j-1})\Delta^j + \cdots$$

Equality holds only when the constant term $q_0 t_0$ is 1, and the remaining coefficients are identically zero. This results in the following set of equations:

$$q_0 = \frac{1}{t_0}$$

$$q_j = -\frac{q_{j-1}}{t_0}$$

The solution to these equations is as follows

$$q_j = \frac{(-1)^j}{t_0^{j+1}}$$

for all integer values of $j \geq 0$.[32]

Here one sees the connection between Newton's work in infinite series and differential calculus. This is what arose Newton's suspicion that Leibniz plagiarized Newton's work after Leibniz' sneak preview of Newton's work in infinite series.

Newton was a master at finding series expansions of functions. For the purpose of finding the slope of the tangent line, it is unnecessary to expand a function out to all orders; it is necessary to determine only the first-order term, q_1.

There is one other interpretation that is useful. If we interpret $x(t)$ as the position of a body at time t, then the speed of the body at time t_0 allows one to find a good approximation to the position of the body around the time t_0 as follows

$$x(t_0 + \Delta) \approx x(t_0) + v_0\Delta \tag{8.29}$$

where v_0 is the velocity at time t_0.

The equation states that during the time interval Δ, the body moves an approximate distance of $v_0\Delta$. The value q_1, the slope of the position graph, is nothing more than the velocity, $q_1 = v_0$. We shall use this relation below.

The Ellipse and Newton's Second Law

We have assembled all of the elements necessary to demonstrate that the constructed ellipse satisfies Newton's second law. Recall that Newton's second law

---

[32]One may generate the solution one term at a time. Obtain q_0 from q_1. Then obtain q_2 from q_1 and so on. After a while one recognizes a pattern. One can check that the solution is correct using induction.

states that the mass times the acceleration vector must equal the force vector. Rewriting Newton's second law, we have the following:

$$\dot{v}_1 = \frac{F_1}{m} \tag{8.30}$$

$$\dot{v}_2 = \frac{F_2}{m} \tag{8.31}$$

Let us consider the force vector. Since the force acting on the planet is the sun's gravitational pull, and the sun lies at the origin, the planet must be pulled toward the origin. This requires that force act in opposition to the planet's position, which we express mathematically as follows:

$$\mathbf{F} = -f(x, y) \begin{bmatrix} x \\ y \end{bmatrix} \tag{8.32}$$

The expression $-\begin{bmatrix} x \\ y \end{bmatrix}$ ensures that the direction of the force is toward the origin where the sun lies. The choice of the function $f(x, y) > 0$ determines the strength of the force. We wish to apply the inverse square law. Newton specified the law in some detail as follows

$$F = \frac{mMG}{r^2} = \frac{m\Gamma}{r^2} \tag{8.33}$$

where F is the magnitude of the force, m is the mass of the planet, M is the mass of the sun, G is a constant known as the universal gravitational constant, and r is the distance between the sun and the planet. For notational convenience, we have set $\Gamma = MG$.

The law states that the force of attraction is proportional to the product of the masses of the two bodies and inversely proportional to the square of the distance between the bodies. The constant G is the constant of proportionality.[33]

Let us choose $f(x, y)$ so that the magnitude of the gravitational force is correct. Using the Pythagorean theorem, the force's magnitude is as follows:

$$F = \sqrt{f^2(x, y) \left(x^2 + y^2 \right)}$$
$$= \sqrt{f^2(x, y) r^2}$$
$$= f(x, y) r$$

Equating the result with Newton's force law yields the following:

$$f(x, y) r = \frac{m\Gamma}{r^2}$$

[33]Physicists have conducted experiments to measure G. Because it is impossible to isolate two bodies from external sources of gravitational pull, it is one of the hardest physical constants to determine.

$$f(x, y) = \frac{m\Gamma}{r^3} \tag{8.34}$$

$$= \frac{m\Gamma}{\left(x^2 + y^2\right)^{3/2}} \tag{8.35}$$

It is usually more convenient to give the expression in terms of r as in equation (8.34). We can now specify the components of the force:

$$F_1 = -\frac{xm\Gamma}{r^3} \tag{8.36}$$

$$F_2 = -\frac{ym\Gamma}{r^3} \tag{8.37}$$

It is next necessary to calculate the force components of our configured ellipse. Recall that in accordance with the second law, each component of the force is equal to the slope of the tangent line to that component's velocity, where time is the independent variable. For an arbitrary position on the ellipse (x_0, y_0) and an associated time t_0 when the planet's position is (x_0, y_0), it is necessary to expand each velocity component in the following form:

$$v_j(\Delta) = q_0 + q_1\Delta + \cdots \tag{8.38}$$

The meanings of the symbols Δ, q_0, and q_1 are as defined in the previous section, and j is the component of interest, $j = 1$ or $j = 2$.

If we can accomplish this, we have extracted the tangent line, and then, as shown in the previous section, the jth component of force at the point (x_0, y_0) is the coefficient q_1. We carry out the calculation for $j = 1$, the x component of the velocity. The reader may follow a similar argument for the y component.

We have already calculated v_1 in terms of its coordinate position [eq. (8.20)], $v_1 = \frac{-Jay}{mb^2[a(1-\varepsilon^2)+\varepsilon x]}$. If we can expand the x and y variables in a manner analogous to equation (8.27), then we obtain an expression for v_1 in Δ. Since we know the velocities of both the x and y variables, we have the following expansions:

$$x(\Delta) = x_0 + v_1^0\Delta + \cdots \tag{8.39}$$

$$y(\Delta) = y_0 + v_2^0\Delta + \cdots \tag{8.40}$$

Recall equation (8.29), which shows that the coefficient of the first-order term is the velocity. Accordingly, v_1^0 and v_2^0 are the known velocity components of the planet at time t_0. Placing the expansions for x and y into the expression for v_1 and simplifying allows one to extract the tangent line of v_1. In the calculations below we include only the constant and linear components of the series expansions as these are the only necessary terms:

$$v_1 = \frac{-Jay}{mb^2\left[a(1 - \varepsilon^2) + \varepsilon x\right]}$$

$$v_1(\Delta) = \frac{-Jay(\Delta)}{mb^2\left[a(1-\varepsilon^2)+\varepsilon x(\Delta)\right]}$$

$$\approx -\left(\frac{Ja}{mb^2}\right)\frac{y_0+v_2^0\Delta}{a(1-\varepsilon^2)+\varepsilon\left(x_0+v_1^0\Delta\right)} \tag{8.41}$$

$$\approx -\left(\frac{Ja}{mb^2}\right)\left(\frac{1}{r_0+\varepsilon v_1^0\Delta}\right)(y_0+v_2^0\Delta) \tag{8.42}$$

Note that we have simplified the expression by using the relationship between r and x, equation (8.6): $r = a(1-\varepsilon^2)+\varepsilon x_0$. We next simplify the fractional expression containing Δ.

Using equation (8.28), which gives the series expression, $\frac{1}{t_0+\Delta} = \frac{1}{t_0} - \frac{1}{t_0^2}\Delta + \cdots$, we find the following expansion for the second term of equation (8.42):[34]

$$\frac{1}{r_0+\varepsilon v_1^0\Delta} = \frac{1}{r_0} - \frac{\varepsilon v_1^0}{r_0^2}\Delta + \cdots \tag{8.43}$$

Substituting the constant and linear term of equation (8.43) into equation (8.42) and simplifying results in the following:

$$v_1(\Delta) \approx -\left(\frac{Ja}{mb^2}\right)\left(\frac{1}{r_0+\varepsilon v_1^0\Delta}\right)(y_0+v_2^0\Delta)$$

$$\approx -\left(\frac{Ja}{mb^2}\right)\left(\frac{1}{r_0} - \frac{\varepsilon v_1^0}{r_0^2}\Delta\right)(y_0+v_2^0\Delta)$$

$$\approx -\left(\frac{Ja}{mb^2}\right)\left[\frac{y_0}{r_0} + \left(\frac{v_2^0}{r_0} - \frac{\varepsilon v_1^0}{r_0^2}\right)\Delta\right] \tag{8.44}$$

Once again we consider only the linear term and discard all terms of higher order; thus the term Δ^2 does not appear. The constant term of equation (8.44) gives the x component of the velocity at time t_0, $-\frac{Jay_0}{mb^2r_0}$. By Newton's second law [eq. (8.5)], the coefficient of the linear term, $-\left(\frac{Ja}{mb^2}\right)\left(\frac{v_2^0}{r_0} - \frac{\varepsilon v_1^0}{r_0^2}\right)$, is the component of the force, F_1. Simplifying the coefficient yields the following result:

$$F_1 = -\left(\frac{Ja}{mb^2}\right)\left(\frac{v_2^0}{r_0} - \frac{\varepsilon v_1^0}{r_0^2}\right)$$

---

[34] Replace t_0 with r_0 and Δ with $\varepsilon v_1^0\Delta$.

$$= -\left(\frac{Ja}{mb^2}\right)\left(\frac{\dfrac{(x_0 - a\varepsilon)J}{ar_0}}{r_0} + \frac{\varepsilon\dfrac{Jay_0}{b^2r_0}}{r_0^2}\right) \qquad \text{[eqs.(8.21), (8.22)]}$$

$$= -\left(\frac{J^2a}{mb^2r_0^3}\right)\left(\frac{(x_0 - a\varepsilon)r_0}{a} + \frac{\varepsilon ay_0^2}{b^2}\right)$$

$$= -\left(\frac{J^2a}{mb^2r_0^3}\right)\left(\frac{(x_0 - a\varepsilon)r_0}{a} + a\varepsilon\left[1 - \left(\frac{x_0 - a\varepsilon}{a}\right)\right]^2\right)$$

[eq. (8.8), ellipse]

$$= -\left(\frac{J^2a}{mb^2r_0^3}\right)\left(\frac{(x_0 - a\varepsilon)\left(a(1 - \varepsilon^2) + \varepsilon x_0\right)}{a}\right.$$

$$\left. + a\varepsilon\left[1 - \left(\frac{x_0 - a\varepsilon}{a}\right)\right]^2\right) \qquad \text{[eq.(8.6)]}$$

$$= -\frac{J^2ax_0}{b^2r_0^3} \qquad\qquad (8.45)$$

Getting to the final result from the previous expression does take some work. However, it is nothing more than standard algebraic operations that simplify the expressions in the second set of parenthesis to x_0. Comparing (8.36) and (8.45) gives the value of the angular momentum that equates the forces:

$$J^2 = \frac{m^2\Gamma b^2}{a} \qquad\qquad (8.46)$$

We have shown that a planet moving along the constructed ellipse with velocities given by equations (8.21) and (8.22) having angular momentum given by equation (8.46) respects Newton's laws of motion, and accordingly the planet follows the prescribed pathway.[35] A question arises. Is it possible that the planet could deviate from the constructed ellipse once it is on that ellipse's pathway? The answer under our assumption of a fixed position for the sun is that the planet may not deviate from its elliptical orbit without an external force. To deviate from the ellipse, one would have to alter the planet's velocity from that given by equations (8.21) and (8.22). But then one would need to change the force from equation (8.36) to something else. Since the sun's force acts according to the inverse-square law of equation (8.36), only an external force could alter the planet's pathway.

---

[35]It is possible to place the value of J as perscribed by equation (8.46) into the equations for velocities, (8.21) and (8.22). If the velocities and focal distance are known, the unknown values for the semimajor and semiminor axis may be determined.

Odds and Ends; Conservation of Angular Momentum

Since we have expended some effort toward finding the tangent, we might as well take advantage of our ability and demonstrate some things of interest. First we demonstrate conservation of angular momentum under the assumption that the force is given by equation (8.32). If the alteration of the angular momentum is everywhere zero, that is, if the slope of the tangent against the time variable is zero, then the angular momentum never changes. Below we express the angular momentum as a series and extract the linear term. Let (x_0, y_0) be an arbitrary point in space and t_0 the time at which a body is at the following position:

$$J = m\,(xv_2 - yv_1)$$

$$J(\Delta) = m\,(x(\Delta)v_2(\Delta) - y(\Delta)v_1(\Delta))$$

$$\approx \left(x_0 + v_1^0\Delta\right)\left(mv_2^0 + F_2^0\Delta\right) - \left(my_0 + v_2^0\Delta\right)\left(v_1^0 + F_1^0\Delta\right)$$

$$\approx m\left(x_0v_2^0 - y_0v_1^0\right) + \left(x_0F_2^0 + mv_1^0v_2^0 - y_0F_1^0 - mv_1^0v_2^0\right)\Delta$$

$$\approx J_0 + \left(x_0F_2^0 - y_0F_1^0\right)\Delta$$

$$\approx J_0 + (-x_0 f(x_0, y_0)y_0 + y_0 f(x_0, y_0)x_0)\,\Delta \qquad \text{[eq.(8.32)]}$$

$$= J_0$$

The linear term vanishes with the result that the angular momentum does not change at the point (x_0, y_0). Since this point is arbitrary, the result holds everywhere and the angular momentum remains constant. One point of note is that the result holds for any force following equation (8.32), including a force that follows the inverse-square law but not precluding others.[36]

NEWTON'S LAWS, KEPLER'S LAWS, AND CALCULUS

Proving the elliptic orbit without calculus requires quite an effort. Alternatively, if calculus is in one's toolbag, one can extract all of Kepler's laws from Newton's rather handily. Let's get down to business.

Using Leibniz' notation with m as the mass and $\mathbf{v}$ the velocity vector, Newton's laws are as follows:

Law 1: $\quad \dfrac{d}{dt}m\mathbf{v} = 0 \quad$ (whenever there is no force on the body)

Law 2: $\quad \dfrac{d}{dt}m\mathbf{v} = \mathbf{F} \quad$ (where $\mathbf{F}$ is the force applied to the body)

---

[36]Conservation of angular momentum follows from the symmetry of the force about the origin; the force on the body is constant along circles centered at the origin. David Hilbert (1862–1943) and later Emmy Noether (1882–1935) generalized the relation between symmetries and conserved quantities. The result in known as *Noether's theorem*.

One may also express law 2 as $\frac{d^2}{dt^2}m\mathbf{x} = \mathbf{F}$, where $\mathbf{x}$ is the position vector.

Law 3 applies to a closed multibody system in which the bodies may impress forces on each other:

Law 3: $\mathbf{F}_{ij} = -\mathbf{F}_{ji}$ (where $\mathbf{F}_{ij}$ is the force that body i impresses on body j)

As above, we consider the planet to be a single body in which the sun sitting at a fixed position at the origin applies a force in accordance with the inverse-square law expressed by equation (8.32).

Conservation of Angular Momentum; Kepler's Second Law

Just as Kepler discovered his second law, conservation of angular momentum, prior to discovering the ellipse, we shall first establish the second law and use it to derive the elliptic pathway. Kepler did not state his law in terms of conservation of angular momentum. Rather, he expressed it in terms of the area swept out by a line connecting the planet with the sun. For this purpose, it is convenient to use polar coordinates that are related to standard Cartesian coordinates by the following identities:[37]

$$r = \sqrt{x^2 + y^2}$$

$$\theta = \arctan \frac{y}{x}$$

$$x = r \cos \theta$$

$$y = r \sin \theta$$

We need to express the planet's velocity in terms of the polar coordinates. The variable v_r is the planet's radial velocity outward from the sun: $v_r = \frac{d}{dt}r$. The variable v_θ is the planet's velocity in a direction perpendicular to the radial velocity. The convention is that v_θ is positive when the movement is counterclockwise and negative when the movement is clockwise. Figure 8.11 illustrates the relation between (v_1, v_2) and (v_r, v_θ). A velocity vector $\mathbf{V}$ may be expressed in either its standard Cartesian form or polar form. The transformation between the polar and radial velocities is given by the following equations:[38]

$$v_\theta = -\sin(\theta)v_1 + \cos(\theta)v_2 \tag{8.47}$$

$$v_r = \cos(\theta)v_1 + \sin(\theta)v_2 \tag{8.48}$$

$$v_1 = \cos(\theta)v_r - \sin(\theta)v_\theta \tag{8.49}$$

$$v_2 = \sin(\theta)v_r + \cos(\theta)v_\theta \tag{8.50}$$

---

[37] See *The Ellipse* (Mazer 2010) for a derivation of the identities.
[38] For a derivation of the transformations, see *The Ellipse* (Mazer 2010).

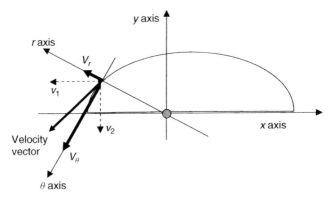

Figure 8.11. Components of the velocity vector along the ellipse.

It is helpful to note that the the velocity component v_θ is proportional to the angular velocity $\frac{d}{dt}\theta$ and the radius r:

$$v_\theta = r\frac{d}{dt}\theta \tag{8.51}$$

With these preliminaries, we can establish Kepler's second law.

The area A swept out by the line connecting the planet and the sun between two angles, θ_0 and θ, is given by the following integral (see Figure 8.12):

$$A(\theta) = \int_{\theta_0}^{\theta} r^2(\vartheta)d\vartheta$$

$$A(t) = \int_{t_0}^{t} r^2(\tau)\frac{d\theta}{d\tau}d\tau$$

$$= \int_{t_0}^{t} r(\tau)v_\theta(\tau)d\tau$$

where we have changed the variable of integration from the angular variable θ to the time variable τ. The value t_0 is the time when the body is at θ_0 and t is the time when the body is at θ.[39]

Kepler's second law states that $A(t) = \frac{J}{m}t$, or equivalently $\frac{d}{dt}A = \frac{J}{m}$, where J is a constant. Applying the fundamental theorem of calculus to the last integral

---

[39] Note that we have implicitly demanded that r be a function of the angular variable.

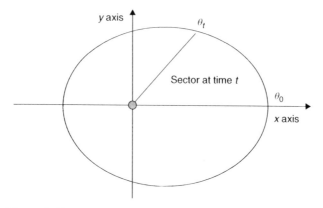

Figure 8.12. The area of the sector is proportional to the time.

and then converting to Cartesian coordinates yields the following expression:

$$\frac{d}{dt}A = rv_\theta$$

$$= -r\sin(\theta)v_1 + r\cos(\theta)v_2$$

$$= -yv_1 + xv_2$$

In the section "Odds and Ends" section [following eq. (8.46) and related text] we showed that the last expression is a constant. The same argument using more conventional notation follows. Define $J = (-yv_1 + xv_2)m$. We must show that applying Newton's laws with the force as given by equation (8.32), yields the result $\frac{d}{dt}J = 0$:

$$\frac{d}{dt}J = m\frac{d}{dt}(-yv_1 + xv_2)$$

$$= m\left(-v_1\frac{d}{dt}y - y\frac{d}{dt}v_1 + v_2\frac{d}{dt}x + x\frac{d}{dt}v_2\right)$$

$$= m\left(-v_1v_2 - y\frac{d}{dt}v_1 + v_2v_1 + x\frac{d}{dt}v_2\right)$$

$$= -yF_1 + xF_2 \qquad \text{(Newton's second law)}$$

$$= yf(x,y)x - xf(x,y) \qquad \text{[eq.(8.32)]}$$

$$= 0$$

Using equations (8.50) and (8.51) to convert the expression for angular momentum into polar coordinates results in the following identity:

$$J = -yv_1 + xv_2$$

$$= mrv_\theta \qquad (8.52)$$

The Ellipse; Kepler's First Law

Using calculus, we will be able to generate the ellipse as the planetary pathway. This is in contrast to the previous case, where, once given the ellipse as the pathway, we could only verify that the pathway is in accord with Newton's laws. As a preliminary step, it is necessary to introduce another constant of motion, the energy:

$$E = \frac{1}{2}m\left(v_1^2 + v_2^2\right) - \frac{m\Gamma}{\sqrt{x^2 + y^2}} \tag{8.53}$$

The first term on the right, half of the square of the planet's speed, is the kinetic energy.[40] The next term is the potential energy. One may apply the same method used to demonstrate the conservation of angular momentum to the energy and demonstrate that the energy is, indeed, a fixed quantity. In polar coordinates the energy is as follows:

$$E = \frac{1}{2}m\left(v_r^2 + v_\theta^2\right) - \frac{m\Gamma}{r}$$

We can express the energy completely in terms of radial variables by substituting for the variable v_θ using equation (8.52), $v_\theta = \frac{J}{mr}$. It is then possible to determine a relation between v_r and r:

$$E = \frac{1}{2}\left(mv_r^2 + \frac{J^2}{mr^2}\right) - \frac{m\Gamma}{r} \tag{8.54}$$

$$v_r = \pm\sqrt{\frac{2E}{m} + \frac{2\Gamma}{r} - \frac{J^2}{(mr)^2}} \tag{8.55}$$

With these preliminaries we can generate the ellipse. Let r be a function of θ, $r = r(\theta)$:

$$\frac{dr}{d\theta} = \frac{\dfrac{dr}{dt}}{\dfrac{d\theta}{dt}}$$

$$= \frac{v_r}{\dfrac{J}{mr^2}} \qquad \text{[eq. (8.52) and definition of } v_r \text{ and } v_\theta]$$

$$= \frac{mr^2}{J}\sqrt{\frac{2E}{m} + \frac{2\Gamma}{r} - \frac{J^2}{(mr)^2}}$$

---

[40] The concept of energy was developed following publication of Principia. Contributors include Leibniz, Euler, Lagrange, and Coriolis.

$$d\theta = \frac{J\,dr}{mr^2\sqrt{\dfrac{2E}{m} + \dfrac{2\Gamma}{r} - \dfrac{J^2}{(mr)^2}}}$$

$$\theta = \int \frac{J\,dr}{mr^2\sqrt{\dfrac{2E}{m} + \dfrac{2\Gamma}{r} - \dfrac{J^2}{(mr)^2}}}$$

Using substitution of variables

$$s = \frac{J}{mr} \tag{8.56}$$

one arrives at the following:

$$\theta = -\int \frac{ds}{\sqrt{\dfrac{2E}{m} + 2\dfrac{m\Gamma}{J}s - s^2}} \tag{8.57}$$

Completing the square within the radical and simplifying results in the following:

$$\frac{2E}{m} + 2\frac{m\Gamma}{J}s - s^2 = -\left(s - \frac{m\Gamma}{J}\right)^2 + \left(\frac{m\Gamma}{J}\right)^2 + \frac{2E}{m} \tag{8.58}$$

Substitution of equation (8.58) into (8.57) and simplifying yields the following:

$$\theta = -\int \frac{ds}{\sqrt{\dfrac{2E}{m} + 2\dfrac{m\Gamma}{J}s - s^2}}$$

$$= -\int \frac{ds}{\sqrt{-\left(s - \dfrac{m\Gamma}{J}\right)^2 + \left(\dfrac{m\Gamma}{J}\right)^2 + \dfrac{2E}{m}}} \tag{8.59}$$

For simplicity, we use the following notation:

$$\alpha^2 = \left(\frac{m\Gamma}{J}\right)^2 + \frac{2E}{m} \tag{8.60}$$

Using trigonometric substitution, one arrives at the following:

$$s - \frac{m\Gamma}{J} = \alpha \cos u \tag{8.61}$$

$$\theta = -\int \frac{ds}{\sqrt{\alpha^2 - \left(s - \frac{m\Gamma}{J}\right)^2}}$$

$$= \int \frac{\alpha \sin u \, du}{\sqrt{\alpha^2 - \alpha^2 \cos^2 u}}$$

$$= \int \frac{\alpha \sin u \, du}{\alpha \sqrt{1 - \cos^2 u}}$$

$$= \pm \int \frac{\alpha \sin u \, du}{\alpha \sin u}$$

$$= \pm \int du$$

$$= \pm u \tag{8.62}$$

A relation between θ and r follows by tracing through the variable changes:[41]

$$\alpha \cos u = s - \frac{m\Gamma}{J} \qquad [\text{eq.}(8.61)]$$

$$\alpha \cos \theta = s - \frac{m\Gamma}{J} \qquad [\text{eq.}(8.62)]$$

$$\alpha \cos \theta = \frac{J}{mr} - \frac{m\Gamma}{J} \qquad [\text{eq.}(8.56)]$$

$$\left(\alpha \cos \theta + \frac{m\Gamma}{J}\right) r = \frac{J}{m}$$

$$r = \frac{J}{m\left(\alpha \cos \theta + \frac{m\Gamma}{J}\right)}$$

$$r = \frac{\frac{J^2}{m^2\Gamma}}{1 + \frac{\alpha J}{m\Gamma} \cos \theta} \tag{8.63}$$

Equation (8.63) is the polar form of the equation for an ellipse. This can be seen by returning to equation (8.6), which relates the variables r and x on the

---

[41] In equation (8.62) we have set the constant of integration to zero. After some effort, one finds that the constant determines the orientation of the final result without affecting the result's shape.

ellipse, and substituting for the variable x:

$$r = \varepsilon x + a(1 - \varepsilon^2) \qquad \text{[eq.(8.6)]}$$

$$r = \varepsilon r \cos \theta + a(1 - \varepsilon^2)$$

$$(1 - \varepsilon \cos \theta) r = a(1 - \varepsilon^2)$$

$$r = \frac{a(1 - \varepsilon^2)}{1 - \varepsilon \cos \theta} \tag{8.64}$$

Using the following identities, equations (8.63) and (8.64) are the same:

$$\varepsilon = -\frac{\alpha J}{m \Gamma}$$

$$= \sqrt{1 + \frac{2E J^2}{m^3 \Gamma^2}} \qquad \text{[eq.(8.60)]} \tag{8.65}$$

$$a(1 - \varepsilon^2) = \frac{J^2}{m^2 \Gamma} \tag{8.66}$$

$$a\left[1 - \left(1 + \frac{2E J^2}{m^3 \Gamma^2}\right)\right] = \frac{J^2}{m^2 \Gamma}$$

$$-a\frac{2E J^2}{m^3 \Gamma^2} = \frac{J^2}{m^2 \Gamma}$$

$$a = -\frac{m \Gamma}{2E} \tag{8.67}$$

Equation (8.67) gives the semimajor axis of the ellipse, while equation (8.65) gives the ellipse's eccentricity, which we take to be positive.[42] Recall equation (8.11) relating the semiminor axis with the eccentricity, $b^2 = a^2(1 - \varepsilon^2)$. Using equations (8.66) and (8.67) to simplify the expression for b^2 sets the semiminor axis: $b^2 = \frac{-J^2}{2mE}$. A point of note for the champions of perfection is that when the eccentricity is zero, the solution is the perfect circle. But any small perturbation causes a nonzero eccentricity, and an ellipse emerges.

Although we have generated the ellipse as the planetary pathway, further specification is possible. In particular, we can find a relation between the constants, Γ, E, and J. Note that the Γ character of is different from those of E and J. Whereas the constant Γ is universal in that it applies to every possible ellipse, both E and J apply to a specified ellipse. Accordingly, for a specified ellipse, one in which the values a and ε are known, let us determine the associated energy and angular momentum. Let us calculate the energy of the planet at the point where the planet is closest and farthest from the sun: the points on the x axis given by Cartesian coordinates, $(-a(1 - \varepsilon), 0)$ and $(a(1 + \varepsilon), 0)$.

---

[42]The positive value places the sun at the left-handed focus, while the negative value places the sun at the right-handed focus.

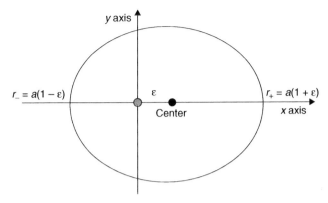

Figure 8.13. Radial distances at apogee (right) and perigee (left) from the sun.

In polar coordinates, we have the radius of the nearest position, $r_- = a(1 - \varepsilon)$ and the radius of the farthest point $r_+ = a(1 + \varepsilon)$ (see Figure 8.13). Using equation (8.54), we can express the energy at both these positions as follows

$$E = \frac{1}{2}\left(mv_{r_\pm}^2 + \frac{J^2}{mr_\pm^2}\right) - \frac{m\Gamma}{r_\pm} \tag{8.68}$$

where the symbol $\pm$ indicates that the equation holds at both the nearest and farthest points from the sun.

At the positions of interest, the radial velocity vanishes: $v_r = 0$. One way to see this is through a continuity argument; the radius increases as the planet approaches its farthest position from the sun at r_+ and then decreases once it has passed its farthest position. The radial velocity then goes from positive to negative at its farthest position from the sun and accordingly must pass through zero at that point. A similar argument at the closest position shows that the planet's velocity passes from negative to positive and must be zero there.[43] Aside from the continuity argument, one may explicitly differentiate equation (8.63) with respect to time to determine v_r. Doing so yields a factor $\sin\theta$ that is zero at the points of interest, $\theta = 0$ and $\theta = \pi$.

Setting the radial velocities to zero and equating the values of the energy at the nearest and farthest positions in equation (8.68) yields a relationship between J and Γ:

$$\frac{J^2}{2ma^2(1 - \varepsilon)^2} - \frac{m\Gamma}{a(1 - \varepsilon)} = \frac{J^2}{2ma^2(1 + \varepsilon)^2} - \frac{m\Gamma}{a(1 + \varepsilon)}$$

---

[43]The reader familiar with optimization theory may note that the vanishing of the velocity implies that the radius is at a critical point. Indeed, the radius achieves a maximum as the planet crosses the positive x axis and a minimum as the planet crosses the negative x axis.

$$\frac{J^2}{2ma^2(1-\varepsilon)^2} - \frac{J^2}{2ma^2(1+\varepsilon)^2} = \frac{m\Gamma}{a(1-\varepsilon)} - \frac{m\Gamma}{a(1+\varepsilon)}$$

$$\frac{2\varepsilon J^2}{ma^2(1-\varepsilon^2)^2} = \frac{2m\varepsilon\Gamma}{a(1-\varepsilon^2)}$$

$$J^2 = m^2\Gamma a(1-\varepsilon^2) \qquad\qquad (8.69)$$

$$= \frac{b^2}{a}m^2\Gamma$$

Note that equation (8.69) is the same as equation (8.46), which was obtained using a completely different method. Once the angular momentum is known, the energy follows by substituting equation (8.69) into the energy. At the farthest point where $v_r = 0$, we have the following:

$$E = \frac{1}{2}\left(mv_{r_+}^2 + \frac{J^2}{mr_+^2}\right) - \frac{m\Gamma}{r_+}$$

$$= \frac{J^2}{2ma^2(1+\varepsilon)^2} - \frac{m\Gamma}{a(1+\varepsilon)}$$

$$= \frac{1}{2a^2(1+\varepsilon)^2}\left[\frac{J^2}{m} - 2a(1+\varepsilon)m\Gamma\right]$$

$$= \frac{1}{2a^2(1+\varepsilon)^2}\left[\Gamma ma(1-\varepsilon^2) - 2a(1+\varepsilon)\Gamma m\right] \qquad [\text{eq.}(8.69)]$$

$$= -\frac{m\Gamma}{2a(1+\varepsilon)^2}\left[-(1-\varepsilon^2) + 2(1+\varepsilon)\right]$$

$$= -\frac{m\Gamma}{2a(1+\varepsilon)^2}[1+\varepsilon]^2$$

$$= -\frac{m\Gamma}{2a} \qquad\qquad (8.70)$$

For a given elliptic orbit, equations (8.69) and (8.70) determine the angular momentum and energy, respectively.

The Ellipse; Kepler's First Law by a Differential Equation

Another perhaps more direct method for generating the ellipse utilizes differential equations. The differential equation one obtains directly from Newton's laws is a second-order differential equation with time as the independent variable. The approach shown below is to use polar coordinates and express the radial variable as a function of the angle $r = r(\theta)$. We start with Newton's second law in Cartesian coordinates. Using the inverse-square law [eqs. (8.36) and (8.37)] as the force in Newton's second law results in the following equations:[44]

$$\frac{d^2x}{dt^2} = -\frac{x\Gamma}{r^3} \tag{8.71}$$

$$\frac{d^2y}{dt^2} = -\frac{y\Gamma}{r^3} \tag{8.72}$$

Next, differentiate equation (8.49) with respect to time taken to find Newton's law in polar coordinates:[45]

$$v_r = \cos(\theta)v_1 + \sin(\theta)v_2$$

$$\frac{dr}{dt} = \cos(\theta)v_1 + \sin(\theta)v_2$$

$$\frac{d^2r}{dt^2} = \frac{d}{dt}(\cos(\theta)v_1 + \sin(\theta)v_2)$$

$$= (-\sin(\theta)v_1 + \cos(\theta)v_2)\frac{d\theta}{dt} + \cos(\theta)\frac{dv_1}{dt} + \sin(\theta)\frac{dv_2}{dt}$$

$$= v_\theta\frac{d\theta}{dt} + \cos(\theta)\frac{dv_1}{dt} + \sin(\theta)\frac{dv_2}{dt} \qquad [\text{eq.}(8.47)]$$

$$= r\left(\frac{d\theta}{dt}\right)^2 + \cos(\theta)\frac{dv_1}{dt} + \sin(\theta)\frac{dv_2}{dt} \qquad [\text{eq.}(8.51)]$$

$$= r\left(\frac{d\theta}{dt}\right)^2 - \cos(\theta)\frac{\Gamma x}{r^3} - \sin(\theta)\frac{\Gamma y}{r^3} \qquad [\text{eqs.}(8.71), (8.72)]$$

$$= r\left(\frac{d\theta}{dt}\right)^2 - \Gamma\frac{\cos^2(\theta) + \sin^2(\theta)}{r^2}$$

$$= r\left(\frac{d\theta}{dt}\right)^2 - \frac{\Gamma}{r^2}$$

$$= \frac{J^2}{m^2r^3} - \frac{\Gamma}{r^2} \qquad [\text{eqs.}(8.51), (8.52)] \tag{8.73}$$

The same substitution that assisted with the integration above, $r = \frac{1}{ms}$, assists in this calculation. Applying the chain rule and changing variables results in the

---

[44] Recall that the magnitude of the force is $m\sqrt{\left(\frac{x\Gamma}{r^3}\right)^2 + \left(\frac{y\Gamma}{r^3}\right)^2} = \frac{m\Gamma}{r^2}$.

[45] The following question may have occurred to the reader: Why not directly apply Newton's second law in polar coordinates. The answer is that polar coordinates are not at rest, but rotate with the system. In fact, since polar coordinates rotate, the polar coordinate description of a point at rest away from the origin is a rotation in the opposite direction of the rotating polar coordinates. This description violates Newton's first law that a particle at rest without a force acting on it remains at rest. The example shows that Newton's laws don't directly apply to a description in polar coordinates.

following equations:

$$\frac{dr}{dt} = \frac{dr}{d\theta}\frac{d\theta}{dt}$$

$$= \frac{dr}{ds}\frac{ds}{d\theta}\frac{d\theta}{dt}$$

$$= -\frac{1}{ms^2}\frac{J}{mr^2}\frac{ds}{d\theta} \qquad [eq.(8.52)]$$

$$= -J\frac{ds}{d\theta} \qquad\qquad\qquad (8.74)$$

$$\frac{d^2r}{dt^2} = -J\frac{d}{dt}\left(\frac{ds}{d\theta}\right)$$

$$= -J\frac{d}{d\theta}\left(\frac{ds}{d\theta}\right)\frac{d\theta}{dt}$$

$$= -mJ^2s^2\frac{d^2s}{d\theta^2} \qquad [eq.(8.52)] \qquad (8.75)$$

On substituting the value $r = \frac{1}{ms}$ into equation (8.73) and then equating the results with equation (8.75), one obtains the following equation:

$$\frac{J^2}{m^2r^3} - \frac{\Gamma}{r^2} = -mJ^2s^2\frac{d^2s}{d\theta^2}$$

$$mJ^2s^3 - m^2\Gamma s^2 = -mJ^2s^2\frac{d^2s}{d\theta^2}$$

$$mJ^2s^2\frac{d^2s}{d\theta^2} + mJ^2s^3 - m^2\Gamma s^2 = 0$$

$$ms^2\left(J^2\frac{d^2s}{d\theta^2} + J^2s - m\Gamma\right) = 0 \qquad (8.76)$$

There are two solutions, a trivial solution ($s = 0$) and the solution obtained by setting the term in parenthesis to zero. The trivial solution demands an unbounded radius for all times, which we do not consider. Let us solve the differential equation that results when the term in parentheses is set to zero:

$$J^2\frac{d^2s}{d\theta^2} + J^2s - m\Gamma = 0$$

$$\frac{d^2s}{d\theta^2} + s = \frac{m\Gamma}{J^2} \qquad (8.77)$$

The standard solution method follows three steps:

Step 1.   Determine a particular solution, s_p.
Step 2.   Determine a solution to the homogeneous equation, s_h. In this instance the homogeneous equation is $\frac{d^2 s}{d\theta^2} + s = 0$.
Step 3.   Add the solutions from steps one and two and then match the constants of the homogeneous solution to additionally known conditions, that is, initial conditions.

We now carry out these steps.

Step 1.   Determine a particular solution, s_p. By inspection, one notes a particular solution is $s_p = \frac{m\Gamma}{J^2}$.
Step 2.   Determine a solution to the homogeneous equation, s_h. In this instance the homogeneous equation is $\frac{d^2 s}{d\theta^2} + s = 0$. The general solution to the homogeneous equation is $s_h = A \sin\theta + B \cos\theta$, where A and B are constants to be determined by additional information.[46]
Step 3.   Add the solutions from steps one and two and then match the constants of the homogeneous solution to additionally known conditions, that is, initial conditions. The solution is of the form $s = A \sin\theta + B \cos\theta + \frac{m\Gamma}{J^2}$. To specify the constants, additional information is necessary. We consider the initial conditions: $s(0) = s_0$ and $\frac{ds}{d\theta}|_{\theta=0} = 0$. We can demonstrate that the initial condition places the planet at either aphelion or perihelion.[47]

The solution at $\theta = 0$ is $s(0) = B + \frac{m\Gamma}{J}$. To match the initial condition requires that $B = s_0 - \frac{m\Gamma}{J^2}$. The first derivative at $\theta = 0$ is $\frac{ds}{d\theta}|_{\theta=0} = A$. To match the initial condition requires that $A = 0$. With A and B established, we can express the planet's orbit in (s, θ) variables:

$$s = \left(s_0 - \frac{m\Gamma}{J^2} \right) \cos\theta + \frac{m\Gamma}{J^2}$$

In polar coordinates, the planet's orbit is as follows:

$$\frac{1}{mr} = \left(\frac{1}{mr_0} - \frac{m\Gamma}{J^2} \right) \cos\theta + \frac{m\Gamma}{J^2} \tag{8.78}$$

---

[46]See a standard text on ordinary differential equations.
[47]For those familiar with optimization theory, the initial condition sets the radius at a critical point. That critical point is a maximum provided that r_0 is large enough to ensure that ε is positive. We have $r_0 = a(1 + \varepsilon)$, which is the maximum distance from the left-hand focus to any point on the ellipse. With a smaller value of r_0 so that ε is negative, the origin of the coordinate system (i.e., the sun's position) lies at the ellipse's right-sided focal point.

$$r = \cfrac{1}{\left(\cfrac{1}{r_0} - \cfrac{m^2\Gamma}{J^2}\right)\cos\theta + \cfrac{m^2\Gamma}{J^2}}$$

$$= \cfrac{\cfrac{J^2}{m^2\Gamma}}{\left(\cfrac{J^2}{m^2\Gamma r_0} - 1\right)\cos\theta + 1} \tag{8.79}$$

In equation (8.79) one notes that the planet's orbit is an ellipse with the following parameters:

$$\varepsilon = 1 - \frac{J^2}{m^2\Gamma r_0} \tag{8.80}$$

$$a(1 - \varepsilon^2) = \frac{J^2}{m^2\Gamma} \tag{8.81}$$

$$a = \cfrac{J^2}{m^2\Gamma\left(1 - (1 - \cfrac{J^2}{m^2\Gamma r_0})^2\right)}$$

$$= \cfrac{J^2}{m^2\Gamma\left(\cfrac{2J^2}{m^2\Gamma r_0} - \left(\cfrac{J^2}{m^2\Gamma r_0}\right)^2\right)}$$

$$= \cfrac{r_0}{2 - \cfrac{J^2}{m^2\Gamma r_0}}$$

$$a = \frac{r_0}{1 + \varepsilon} \tag{8.82}$$

We remark that the ellipse of this section is identical to that of the previous section. Some algebra demonstrates equivalence between the parameters ε and a of equations (8.65), (8.66), and (8.67) with equations (8.80), (8.81), and (8.82).

The Planet's Periodicity, Kepler's Third Law

Recall that Kepler's third law states that the square of the planet's period is proportional to the cube of the semimajor axis: $T^2 = Ca^3$, where T is the period, a is the semimajor axis, and C is a constant that we determine below. For Kepler, the third law was an empirical observation of fact. The statement is a testimony to Kepler's brilliance in discerning complex patterns from obscure data. For Newton, Kepler's third law is a result of Newton's laws of motion as demonstrated below.

A restatement of conservation of momentum in integral form is

$$A(t) = \int_0^t r(\tau)v_\theta(\tau)d\tau$$

$$= \frac{J}{m}t$$

where $A(t)$ is the area swept out by the line adjoining the planet and the sun in time t.

Setting the time t to the period T gives the area of the ellipse:

$$\frac{J}{m}T = \pi ab$$

$$T = \frac{m\pi ab}{J}$$

$$= \frac{m\pi a^2\sqrt{1-\varepsilon^2}}{J} \tag{8.83}$$

Using equation (8.81), one finds

$$\frac{m\sqrt{1-\varepsilon^2}}{J} = \frac{1}{\sqrt{\Gamma a}} \tag{8.84}$$

On substituting equation (8.84) into equation (8.83) and simplifying, one obtains Kepler's third law with the constant of proportionality fully determined:

$$T = \frac{m\pi a^2\sqrt{1-\varepsilon^2}}{J}$$

$$= \frac{\pi}{\sqrt{\Gamma}}a^{3/2}$$

$$T^2 = \frac{\pi^2}{\Gamma}a^3$$

One point of note is that the only planet-related parameter that affects the period is the semimajor axis. Two planets with different eccentricities and different masses have the same period as long as their semimajor axes are the same.

Newton and Apollonius

In the methods above, we have implicitly assumed that the magnitude of ε is less than 1. With this assumption, the solution is, indeed, the ellipse, and the magnitude of ε is its eccentricity. What if we relax the assumption and allow ε to assume a value of 1 or greater? What solutions result?[48]

---

[48] We assume a positive value for ε, as the negative solution is a reflection of the positive solution about the y axis. For the ellipse, the sun lies at the right-handed focus.

The change does not affect the calculations. The solution in equation (8.79) is correct. If we transpose this back to Cartesian coordinates with $\varepsilon > 1$, we get an equation that has the same form as the equation for an ellipse

$$\frac{(x - c)^2}{a^2} + \frac{y^2}{b^2} = 1$$

with $c = a\varepsilon$ and $b^2 = a^2(1 - \varepsilon^2)$.

The crucial difference is that b^2 is negative whenever $\varepsilon > 1$. (Don't be put off by the square in b^2. The squaring of the term was arbitrary. Think of b^2 as any other number that may be positive or negative.) This yields the hyperbola as a solution to Newton's equations of motion. Indeed, with the correct initial condition, which is given in Table 8.1, a body such as a space probe will branch away from the sun along a hyperbolic trajectory. Figure 8.14 illustrates a hyperbolic trajectory.

When $\varepsilon = 1$, one attains a parabola as follows:

$$\frac{1}{r} = m\left(\frac{1}{mr_0} - \frac{m\Gamma}{J^2}\right)\cos\theta + \frac{m\Gamma}{J^2} \qquad [\text{eq.}(8.78)]$$

$$= -\frac{m\Gamma}{J^2}\left(1 - \frac{J^2}{\Gamma m^2 r_0}\right)\cos\theta + \frac{m\Gamma}{J^2}$$

$$= \frac{m\Gamma}{J^2}(1 - \varepsilon\cos\theta) \qquad [\text{eq.}(8.80)]$$

$$= \frac{m\Gamma}{J^2}(1 - \cos\theta) \qquad (\varepsilon = 1)$$

$$1 = \frac{m\Gamma}{J^2}(1 - \cos\theta)r$$

$$\frac{J^2}{m\Gamma} + r\cos\theta = r$$

$$\frac{J^2}{m\Gamma} + x = \sqrt{x^2 + y^2}$$

TABLE 8.1. Using the Energy Sign to Determine Trajectory Shape

ε	E, J	Shape
$\varepsilon = 0$	$E = -\dfrac{m^3\Gamma^2}{2J^2}$	Circle
$0 < \varepsilon < 1$	$E < 0, J^2 > 0$	Ellipse
$\varepsilon = 1$	$E = 0, J^2 > 0$	Parabola
$\varepsilon = 1$	$J^2 = 0$	Line
$\varepsilon > 1$	$E > 0, J^2 > 0$	Hyperbola

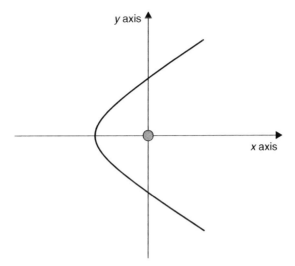

Figure 8.14. Hyperbolic trajectory of a space probe.

$$\left(\frac{J^2}{m\Gamma}\right)^2 + 2\frac{J^2}{m\Gamma}x + x^2 = x^2 + y^2$$

$$2\frac{J^2}{m\Gamma}x = y^2 - \left(\frac{J^2}{m\Gamma}\right)^2$$

$$x = \frac{m\Gamma}{2J^2}y^2 - \frac{J^2}{2m\Gamma}$$

The other possibility corresponding to $\varepsilon = 1$ is $J = 0$; see equation (8.80). In this case, the conservation of angular momentum in the form of equation (8.52) states that the angle remains a constant ($v_\theta = 0$) and the trajectory moves in a line along the radial direction. The line is a degenerate conic section.

As demonstrated thus far, one can determine the nature of the solution by the parameter ε. With $\varepsilon = 0$, the solution is the circle. Increasing the eccentricity up to but less than 1 yields an ellipse. At the value of 1, the solution is either the parabola or a line depending on whether the angular momentum is zero (line) or nonzero (parabola). Finally, when $\varepsilon > 1$, a body's trajectory is the hyperbola. All of the conic sections that Apollonius studied are Newtonian pathways, a fact that would have startled Apollonius, the originator of the epicycle.

Aside from examining the parameter ε, one can determine the shape of the trajectory in terms of the sign of the energy. Using relationship (8.65) $\varepsilon = \sqrt{1 + \frac{2EJ^2}{m^3\Gamma^2}}$, one may draw the conclusions listed in Table 8.1.[49]

---

[49] The table may be extended to the case of negative ε using symmetry of solutions. One replaces ε by its absolute value.

Note that the energy may be negative because the gravitational energy is a negative value; that is, the term $-\frac{m\Gamma}{\sqrt{x^2+y^2}}$ in the expression for energy, $E = \frac{1}{2}m\left(v_1^2 + v_2^2\right) - \frac{m\Gamma}{\sqrt{x^2+y^2}}$, is negative. The first term in the expression for the energy is the kinetic energy. The kinetic energy is always positive and depends on the body's speed. The second term in the energy expression, the gravitational term, is the potential energy. Table 8.1 states that when there is sufficient speed so that the kinetic energy dominates the potential energy, the sun is unable to confine a body within a finite region. Such a body with nonzero angular momentum follows a hyperbola. Conversely, when the potential energy dominates the kinetic energy, the sun's gravity traps the body within a finite orbit. Such a body with nonzero angular momentum follows an ellipse or a circle. Finally, when the kinetic energy and potential energy are in balance, a body with nonzero angular momentum follows a parabola.

In *Principia*, Newton demonstrates that all of the conic sections are pathways for bodies influenced by a force abiding by the inverse-square law. This is further evidence that Newton privately called on his invention of fluxions to assist with his calculations. It is difficult to imagine generating these solutions by any other process.

NEWTON'S AGENDA

It is fitting to allow Newton to conclude this chapter with the final paragraph from *Principia*. There Newton sets the agenda that has occupied humankind since Newton:

> And now we might add something concerning a certain most subtle Spirit which pervades and lies hid in all gross bodies; by the force and action of which Spirit the particles of bodies mutually attract one another at near distances, and cohere, if contiguous; and electric bodies operate to greater distances, as well repelling and attracting the neighbouring corpuscles; and light is emitted, reflected, refracted, inflected, and heats bodies; and all sensation is excited, and the members of animal bodies move at the command of the will, namely by the vibrations of this Spirit, mutually propagated along the solid filaments of the nerves, from the outward organs of sense to the brain, and from the brain into the muscles. But these are things that cannot be explained in a few words, nor are we furnished with that sufficiency of experiments which is required to an accurate determination and demonstration of the laws by which this electric and elastic Spirit operates.

CHAPTER 9

RULE BREAKERS

BEGINNINGS

As Louis XVI, grandson of Louis XIV, knelt before the chopping block with his head beneath the blade of the axman, what crossed his mind? What would happen the moment that the blade severed his head from his body, with the head rolling to the fore and his body slinking to the ground aft of the chopping block? On which side of the eternal divide would Louis find himself? Did Louis feel that God, who had graced his lineage with divine authority to rule over France, had abandoned him, or did he comfort himself believing that God would right this injustice and grant eternal salvation? Perhaps he cursed his advisers for convincing him to support the American Revolution, whose success inspired the French Revolution and in turn brought Louis to the chopping block.[1] Had he so cursed, the leaders of the French Revolution would be most obliging, for soon all associates of the crown would face the executioners of the Revolution.

The second half of the eighteenth century saw two revolutions. One led to the establishment of a democratic nation governed by a respected constitutional process. This system of governance served the new nation well, as the United States

---

[1]The tax burden to support this foreign venture without any tangible benefit to the French caused disenchantment with Louis' policies, whereupon the French drew the obvious conclusion; that is, if the Americans can rebel against their sovereign, so can the French.

Shifting the Earth: The Mathematical Quest to Understand the Motion of the Universe,
First Edition. Arthur Mazer.
© 2011 John Wiley & Sons, Inc. Published 2011 by John Wiley & Sons, Inc.

is the only nation on the planet to have transitioned through the industrial revolution with its system of governance intact. As for the other revolution, internal squabbling and a settling of old scores virally infected the French revolutionary government along with all of France. On top of their internal difficulties, the monarchies of Europe sought to destroy the revolution from its inception. Unlike the American Revolution, whose distance from Europe's monarchies allowed it time to mature after its difficult birth, the French Revolution never had a chance.

When hell visits a nation, chaos reigns, social conventions are nonexistent, everyone is a suspected enemy of the state, and on top of this foreign enemies await their opportunity to cross the boarder, a single overarching hope unites the populace: a hero. The hero is above and beyond ordinary people. Rules that apply to you and me do not apply to the hero. Indeed, the hero can break the rules, work miracles, and on behalf of the common good, change the course of the nation. With his military victories that not only defended France but also brought territorial acquisition, Napoleon was the anointed hero—and so the great accomplishment of the French Revolution was to displace the divine authority of the king with the divine authority of the emperor. None other than God's single point of contact with humankind, the Pope himself, Pope Pius VII, presided over the coronation, lending the Pope's authority and by extension God's authority to Napoleon's ascension.

Napoleon has both his defenders and detractors. The detractors ascribe Napoleon's military ventures to a flawed character that caused him to seek personal glory. Defenders argue that Europe's obsession with snuffing out all traces of the French Revolution narrowed Napoleon's military options; his campaigns were defensive responses to external threats. One point that detractors and defenders agree on is that, unlike Alexander, Napoleon showed himself to be an able administrator over his conquered territories. Even as good French administration assisted with an economic upturn that enriched Europe, the various nationalities that make up Europe seethed at French protectorship with resentment.

While Italian, German, Austrian, and Dutch pride took a French fist on their chin, Napoleon altered the status of a stateless people who had been disenfranchised for nearly two millennia. In 1807 Napoleon gave audience to a Council of French Jews. Napoleon asked the Council if, given citizenship, their religion would interfere with allegiance to the state. When the Council ensured Napoleon of their allegiance, Napoleon granted citizenship. Citizenship in several nations across Europe followed. In 1871, as the architect of the new state of Germany, Otto Bismark capped the process by extending citizenship to Jews within his domain.

The film *The Frisco Kid* opens with Gene Wilder acting the role of Avram, a midnineteenth-century yeshiva graduate in Poland, ice skating across a frozen pond along with his classmates.[2] Avram is all dressed up in shtetl garb, a black

---

[2]A yeshiva is a school that provides religious training to Jewish students. Prospective rabbis also received training at the yeshiva.

cap tops his long hair, a beard adorns his face, and underneath his black, robelike coat is a prayer shawl. The overall image portrays a devotion to Orthodox Jewish tradition, but then there are those very nontraditional ice skates. Avram's clumsiness in his negotiations with the ice skates collide with his uncontained joy, producing a very memorable scene. One might take the scene as a metaphor for the Jew's entry into the modern world. He enters while maintaining his traditions, finds it hard to keep his balance, but without hesitation plugs onward. The metaphor has some appeal, but is all wrong.

Two thousand years prior to Napoleon, during the Greek era, Jews flirted with the superior Greek intellectual culture. A civil war among Jewish traditionalists and assimilationists ensued. When the Greeks confronted the unrest with a heavy hand, traditionalist sentiment prevailed and for the most part, Jewish scholars remained within the confines of their traditional sources of inspiration, Jewish law—that is, until Napoleon. Then once again Jews flirted with an external intellectual culture. Once again there was internal discord within the Jewish community. But this time it was different. The Jews did not reside in a single country where leadership could stamp its authority on the populace. They were scattered throughout the diaspora. When the opportunity to escape the confines of the ghetto presented itself, a considerable number snatched that opportunity, raced toward modernity with an unimaginable ferocity, and tossed tradition aside lest it encumber them.

Among the states that Bismark dissolved and reassembled into a united Germany was Swabia. Bismark's united Germany caused mixed emotions among the citizens of this newly founded nation. While their victory over France[3] and supremacy of Austria established the new nation as the continental superpower, a point of national pride, there was also noticeable resentment of the manhandling that Bismark subjected the states to as he bullied them into subservience. However, within the Jewish community, such resentment never took hold. Families such as the Einstein family of Swabia never felt beholden to states that had for centuries not even allowed them to assume the position of the lowest rung on the ladder of citizens. They only saw the advantages of being citizens of the strongest nation in a newly defined Europe, identified themselves as loyal Germans, and appreciated Bismark for giving them citizenship. Culturally, they exited the temple with its old hymnal tradition and celebrated Mozart, Beethoven, and Bach. Many, such as the convert Mendelssohn, completely relinquished their Jewish identities to contribute to German culture as Germans. Albert Einstein was born in 1878, to a family that embraced assimilation.

---

[3]Bismark's French counterpart was Emperor Louis Napoleon, a nephew of the first emperor who did not possess his uncle's military genius. Louis Napoleon's ascension came in the aftermath of staged rebellions in France that sent King Louis Philippe running lest he suffer the same fate as Louis XVI. Louis XVIII was the last progeny of Louis XIV to reside in Versailles. In the aftermath of Waterloo, Metternich and his cohorts at the Conference of Vienna restored the French throne to the Bourbon dynasty in the personage of Louis XVIII. As for Louis Napoleon, he gained ascendancy not for his deeds but because the populace longed for a hero and the Napoleon name was the closest thing available. Unfortunately, Louis Napoleon did not measure up. France became a republic on the death of Louis Napoleon.

In 1880 Herman Einstein joined his brother Jakob at Jakob's electrical supply business, giving Jakob access to the wealth of Herman's in laws. Herman, his wife Marie Koch Einstein, and their son Albert, moved to Munich, where the Einstein's company competed with Siemens to electrify municipalities in southern Germany. Not only had the Einsteins given their son a good Prussian name;[4] the assimilationist Einsteins sent their son to a Catholic school that had a good reputation. This was the optimal pathway for merging into the mainstream of German society.

In 1835, the *Beagle*, a vessel using newly developed steam technology to propel itself across the oceans, carried aboard a 26-year-old whose observations would revolutionize scientific thought. Charles Darwin (1809–1882) spent only 5 weeks in the Galapagos archipelago. There, the evidence of the diversity of lifeforms through evolution revealed itself. Darwin had been introduced to the concept of evolution during his university days. With Copernican deliberation, over a 20-year stretch, Darwin transformed the concept into a more concrete theory. In his 1858 publication *On the Origin of Species*, Darwin expounds his theory. There he explains the evolution of lifeforms through a process of adaptation to the environment, endowing nature with the capacity for natural selection. Encompassed within evolution was a competitive viewpoint, lifeforms adapt to seek competitive advantages, ensuring their survival over the less adept.

While Darwin confined himself to the scientific realm, the power of his ideas soon found expression among social theorists. The wealthy were quick to use Darwin's theories as an explanation of their success—they were superior, as demonstrated by their wealth, and so they came out on top of a Darwinian competition. Racists also usurped and perverted Darwinism to demonstrate the superiority of their particular race; it was evident that nature's process of self-selection had endowed a particular race with superior qualities. At least it was self-evident among the racists, and although they were never able to demonstrate the presumed superior qualities, their ideas resonated with a certain set of people who needed some reason to feel superior. Both the social elites and the racists stressed the competitive aspects of Darwin's theory and declared themselves to be the winners.

"You don't talk about what to do with parasites and bacilli. Parasites and bacilli are also not reared. They are as quickly and fully as possible destroyed."[5] These are the 1887 writings of the Prussian intellectual and anti-Semite Paul de Lagarde addressing a solution to the Jewish problem. When incubated in tolerance, a country blessed with rich ethnic diversity can engender bursts of national creativity. This was not the case in 1888 throughout Austria, where the sentiments of the racist de Lagarde felt their presence. Linz, Austria, the town where Kepler more than two and a half centuries earlier resided after completing

[4]The name of the Prussian duke who saw political advantage in aligning himself with the then young Lutheran movement, expanding his domains and warring with his uncle, the Polish King Sigismond, was Albert of the house of Hohenzollern. Copernicus successfully defended the town of Frombork against Albert's forces.
[5]From Ian Kershaw's biography of Hitler (Kershaw 1998, 2000).

New Astronomy, is where Adolf Hitler grew up. His father, Alois Hitler,[6] with only an elementary school education, climbed considerably up the ranks of the Austrian civil service bureaucracy. Hitler's father, like Kepler's, was abusive, but unlike Kepler, who found solace in religion and later science, Hitler's heart never knew solace, while hatred established permanent residency.

In 1908, after the death of his mother, Hitler moved to Vienna, the capitol of Austria. Austria was a brew of nationalities: Slavs, Hungarians, Magyars, Germans, and ethnic minorities, including Jews and gypsies. It is within this brew that German racism fomented with the highest toxicity. Openly anti-Semitic politicians such as Schonerer and Lueger found success among the electorate. During Hitler's years in Vienna, Lueger was elected to the office of the mayor using a race-baiting campaign with Jews as targets. The Jew became the source of all social problems and people connected on a personal level. If only the Jews weren't destroying economic opportunity through usury, they would all be better off. If only the Jews did not tempt morality with their prostitution rings, the moral fiber of society would not be in decay.

Hitler had gone to Vienna as an aspiring artist. He wished to gain admission to Vienna's Academy of Fine Arts, but failed. There was a period when Hitler was very much down on his luck. Perhaps it is while he was at rock bottom, listening to Lueger's anti-Semitic raves that Hitler, too, blamed the Jew for his personal circumstances.[7] The Viennese experience also yielded a political education. Hitler noted that among a not-too-modest portion of the populace, racist ideas had appeal that a skillful politician could harness.

Thirteen years prior to Hitler's attempts to enter the Academy of Fine Arts, Albert Einstein also received a rejection notice, but one with promise. At 16, after dropping out of high school for a period, Albert Einstein took the entrance examinations to the Swiss Polytechnic in Zurich. While his overall scores did not cut the grade, Einstein's scores in mathematics impressed Professor Henrich Weber. Weber arranged for Albert's entrance if Albert agreed to complete his high school studies. Albert finished his high school studies in Switzerland, residing at the home of a warm family, the Wintelers.[8] Afterward, the Polytechnic kept its promise and allowed Einstein to enter its physics program despite the fact that Einstein showed little improvement in his entry exams.

Once within the Polytechnic, Einstein did not live up to the expectations of Professor Weber. He skipped classes and ridiculed professors for their backward thinking that did not incorporate modern ideas. Applying the physicist's affinity

---

[6] Alois had changed his family name from Schicklgruber to Hitler.
[7] Oddly enough, Hitler did have somewhat good relations with the few Jews whom he actually encountered. In 1909 he wrote a note of appreciation to Dr. Eduard Bloch, the Jewish physician who attended to Hitler's mother during her fatal illness. Hitler also supported himself through his paintings of famous Viennese landmarks. Hitler's paintings would find their way to paying customers through predominantly Jewish business partners, including Siegfried Loffner and Jacob Altenberg.
[8] The Winteler daughter, Marie, was Einstein's first but not last romantic interest. Later, Maja Einstein, Albert's sister, married Anna's brother, Paul, and one of Albert's future friends, Michelle Besso, married Marie's younger sister, Anna.

for the concept of symmetry, Einstein seemed to be indifferent to the impression that he created because the polytechnic did not make much of a positive impression on him.

Einstein would pay a price for his insolence. On graduation, while classmates readily found employment, no professor was willing to recommend Einstein to perspective employers. Strains between Einstein and Weber had reached the point where Einstein believed that Weber was actively undermining his prospects. It was anything but apparent that Einstein would amount to much, and for his part it took not only genius but also tenacity to overcome obstacles, some of which were Einstein's own making.

In the end, while the lack of support from the Polytechnic faculty did not restrain Einstein, there were three individuals who, for better or worse, impacted Einstein throughout his life. It is difficult to order their influence—who among the three had the most impact in making Einstein who he turned out to be? Let us start at the personal level. Einstein had a healthy male attraction to the opposite sex and carried on relations with women throughout his life. He was handsome, and when he had the premonition, romantic ideas could flow forth alongside mathematical equations. Also, he had a wicked wit. Mileva Maric, a Serbian physics student 7 years older than Einstein, caught his eye. Einstein, with a counterculture bent, found this woman, who bucked convention by entering into a man's profession, fascinating. During their student days, the fascination turned to romance. In 1901, a year after Einstein's graduation, while he was both unemployed and without any job prospects, Einstein committed to marrying Mileva. It may well have been a commitment compelled not by love but by obligation, for Mileva carried Einstein's child.

Mileva returned to Serbia and gave birth to a daughter. Einstein's initial response to be a good father did not pan out. While he promised to visit Mileva and his daughter in Serbia, Einstein broke that promise. With little choice, the still-single Mileva gave the child up for adoption; an out-of-wedlock upbringing would stigmatize not only the mother but the daughter as well. The separation of the daughter from the mother weighed heavily on Mileva's heart, and Einstein could not escape his role in her pain. Although Einstein kept his promise of marriage, the strain of the experience intruded into their marriage, gathering strength with time. Mileva could not find joy with Einstein. While there would be two more children, the efforts of Einstein to do the right thing and endure the hapless circumstances were doomed, and the trajectory of their marriage followed its initial momentum toward divorce. Mileva and Albert finalized their divorce settlement in 1916 after 14 years of marriage. But the entangled relationship between divorced parents and children affected Einstein throughout his life.

The other two influences from Einstein's college years were both positive and have been somewhat historically underrated. Among the professors who did not think highly of Einstein was Hermann Minkowski (1864–1909), a Jewish emigré from Lithuania and professor of mathematics. Einstein regularly ditched Minkowski's class, and his scores were not impressive. Minkowski viewed Einstein's later achievements as unexpected for, as Minkowski recalled,

Einstein was a somewhat unmotivated student. Minkowski's way of putting it was more colorful: "It came as a tremendous surprise, for in his student days Einstein had been a lazy dog."[9] In a seminal lecture, Minkowski is far more generous. Concerning the realization that time measurement is not universal and nature designates no preferred measurement, Minkowski writes: "the credit of first recognizing that the time of one electron is just as good as that of the other...belongs to A. Einstein."[10]

Minkowski himself was a genius with impressive creativity. Einstein's hallmark achievement, general relativity, was the culmination of years of effort. In this effort, Einstein was not alone. He was not Atlas, single-handedly holding up the relativistic world. Rather, there were others who made significant contributions. Among the others is Minkowski. It was Minkowski, by unveiling the four-dimensional reality of the universe, who endowed relativity with its underlying mathematical and geometric structure. The structure assisted Einstein in developing the general theory of relativity from the special theory of relativity.[11]

Another influence on Einstein was a student with whom Einstein formed a strong friendship, Marcel Grossman (1878–1936), the son of a Jewish industrialist. Grossman's initial assessment of Einstein was in stark contrast to Minkowski's. In a letter to his father, Grossman sized up his new friend Einstein with a remarkably accurate forecast. Grossman wrote that Einstein would make his mark in the world. While Einstein ditched Minkowski's class, Grossman took diligent notes and kindly passed them on to Einstein. Having a greater appreciation for Minkowski than Einstein, on graduation, Grossman entered a doctoral program in mathematics and specialized in non-Eulidean geometry. Einstein later turned to Grossman and obtained his assistance in incorporating Minkowski's work into a non-Euclidean structure that culminates in the general theory of relativity. Specifically how did Minkowski and Grossman's work influence Einstein? Let us restrain the tempo of this narrative and for now note that Einstein achieved his signature accomplishment with the assistance of associates from the Polytechnic.

We also note that beyond the intellectual realm of physics, Grossman proved to be an invaluable friend netting Einstein his physical sustenance. In 1902, 9 years before Hitler would hit rock bottom, Einstein discovered the bottom's empty yet hard surface. Marcel Grossman prevailed on his father to use his connections and attain for Einstein employment at the Swiss patent office. Not unlike Halley's

---

[9]This quote is from Isaacson's biography of Einstein (Isaacson 2007).
[10]Excerpt from Minkowski's speech entitled "Space and Time," presented at the University of Cologne (Köln) in 1908.
[11]It is a curiosity that these two towering intellects passed by one another at rather close distance without recognizing the other's talents. One can only surmise that Einstein's immaturity in dealing with authority was so unattractive that not only Minkowski but also the entire ETH (Eidgenössische Technische Hochschule, Zürich) faculty disregarded Einstein. As for Einstein's disregard of Minkowski, Minkowski introduced Einstein to rigorous methodology, which Einstein's still undisciplined spirit found unappealing. Only later would Einstein come to appreciate the power and the creativity behind the rigor.

intervention with Newton, the history books generally pass by this event—and yet it would be momentous.

DEBUNKING THE ETHER

In ancient Greece Aristotle intuited that the material composition of heavenly bodies was unlike that of Earth and coined the heavenly matter ether. The Aristotelian universe did not survive Copernicus, Kepler, and Newton, but ether reconstituted itself in a different context. Newton's corpuscular theory of light was the target of nearly two centuries of objections; each objection was a bullet in an apparently dead horse. James Maxwell's (1831–1879)[12] eloquent equations describing the electromagnetic fields and demonstration that light was an electromagnetic wave was the final, biggest, and most indisputable bullet. With light confirmed beyond doubt to be a wave, space required a medium through which the wave could propagate. Ether was reborn as the medium that transmits light waves across the universe, just as air is the medium that transmits sound waves across distances on Earth. As a concept based on the physicist's understanding of light, ether was fine. As a physical reality, the only problem with the ether was that there was not a physical trace of it. This nagged at the physicists, and many embarked on an experimental expedition to find the elusive ether.

The most promising of the experiments was one designed by the American physicist, Albert Michelson (1852–1931).[13] Michelson's experiment rested on the belief that just as a bicyclist's motion causes a headwind, the motion of the earth through the ether would cause an ether wind, and the wind would alter the relative speed of light. Michelson designed an apparatus known as an *interferometer* that partitions incoming light into two slivers. One sliver would penetrate a beamsplitter (a partially coated mirror) and traverse longitudinally down the axis of motion into the ether wind. Meanwhile, the other sliver would bounce off the beamsplitter in an orthogonal direction. Each sliver would travel down its apportioned arm of Michelson's interferometer until reaching arm's end, where a mirror would reflect the sliver back to its point of separation from its Siamese twin. At the point of separation, Michelson's interferometer recombined the slivers into a single beam. In the absence of an ether wind, the recombined beam would be indistinguishable from the original beam. This is not the case in the presence of the ether wind because the time required by each sliver to return to the separation point is not the same. The ether wind would cause the sliver traveling along the wind-oriented arm to return at a later time than the sliver traveling up the complementary arm.[14] The difference in return times

---

[12]The Scottish physicist James Clerk Maxwell proposed equations that describe electromagnetic fields. This discovery is central to relativity and further discussed later in this section.

[13]Michelson was born in Poland and, seeking opportunity along with an escape from anti-Semitism, emigrated to the United States.

[14]For a mathematical calculation of the return times, see Feynman's book, *Six Not So Easy Lectures* (Feynman 1997).

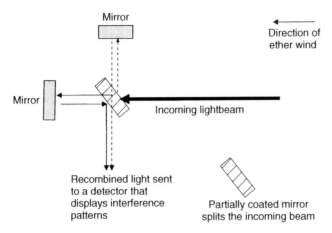

Figure 9.1. Michelson's interferometer.

would result in the waves of each sliver being out of sync with one another. The experimenter could read the signature of the ether wind in the interference patterns of the out-of-sync waves (see Figure 9.1).[15]

In 1881 Michelson rode his apparatus into the ether wind and found nothing. The experiment was a dud firecracker. There were only two ways to interpret the result; either there was no ether or the ether moved in unison with the earth, leaving no ether wind. The latter explanation—no ether wind—had already been debunked by a French experimentalist, Armand Fizeau (1838–1923), so Michelson, confident that there was an ether wind, redoubled his efforts to detect it. A colleague, Edward Morley (1838–1923), and Michelson enhanced the design of Michelson's interferometer, giving the apparatus greater sensitivity and accuracy. Throughout 1887 Michelson and Morley repeated the experiment on different days with different orientations of the interferometer. Each time, as with Michelson's very first effort, it was a dud. In frustration, they gave up. Both Michelson and Morley viewed their experiments as failures, which they nevertheless reported. Michelson would later receive the Nobel Prize for conceiving and executing the experiment. It was not a failure, but a point of departure from the ether.

In 1892 the Dutch physicist Hendrik Lorentz (1833–1928) at the University of Leiden puzzled over the Michelson–Morley results. To explain why motion through the ether had not been detected, Lorentz conjured up a fantastic process.

---

[15]The reader might think it rather fantastic that one could ensure that the lengths of each arm were identical. In fact, such skepticism is warranted. Michelson knew that the lengths would not be the same and that the slivers would be slightly out of sync because of minute differences in the arm lengths. These minute differences would cause a diffraction pattern. He knew that if he altered the orientation just slightly, the diffraction pattern would change because the alignment of the apparatus with the ether wind would also change. It is the dependence of the diffraction pattern on the orientation that would be the true signature of the ether wind.

Lorentz proposed that motion against the ether causes a physical contraction of matter along the direction of the motion, be it a measuring rod or the apparatus used by Michelson and Morley. Such a contraction would negate the effects of the ether wind; thus, nature somehow reset the length of the wind oriented arm so that the returned light from both arms arrives at the point of separation at the exact same time. Lorentz noted that a contraction previously proposed by the German physicist Woldemar Voigt (1850–1919) would yield Michelson and Morley's null result.

Another towering intellect, Jules-Henri Poincaré (1854–1912), took an interest in the problem and became familiar with Lorentz' work. Among mathematicians, Poincaré is known with due respect as a "monster of mathematics." Nobody has influenced more fields of mathematics than has Henri Poincaré. While Poincaré occupies many floors of the ivory tower, he surprisingly also assumed functions that other residents might consider mundane, and he applied himself in these endeavors with all the diligence that he did in the academic arena. Poincaré was a member and later director of the Bureau of Longitudes. It was that agency's mandate to accurately align the abstract longitudes of a map with actual locations on Earth.

The means for determining longitude had been known to Eudoxus, who was aware of a round planet. Calculations on simultaneous measurements of the angle between the North Star, two separate locations on Earth, and a designated star yield the difference in longitude between the locations. The plan fell victim to an inability to coordinate simultaneity of the observations at separate locations. Ptolemy's mentor Hipparchus suggested using lunar eclipses to coordinate observation times of different observers, a workable but limited solution. With the advent of the telegraph and undersea cables, there was a more robust solution. A telegraph signal would be used to coordinate clocks. Once clocks assumed a common standard, the astronomical measurements would be taken at a designated time, guaranteeing simultaneity. For practical purposes the solution worked well. But the problem inspired Poincaré to think of ways to account for the travel time of the emitted telegraph signal.

While the travel time of the telegraph signal was negligible and had little impact on the measurements of longitude, Poincaré pondered this problem. He considered the case when the emitting and receiving stations are not stationary, but in motion relative to one another. The research trail that started from the Bureau of Longitudes was not at all mundane, but went to the heart of defining not only time measurement but also time itself. Poincaré summarized his reflections in his paper "The Measure of Time." The paper demonstrates that coordination of time between moving and stationary observers results in the two measuring time differently. Poincaré also questions the very idea of simultaneity, and demonstrates that two events that are simultaneous to one observer may not be simultaneous to another.

Poincaré recognized that a redefinition of time must result in a new model of mechanics and broadened his investigations, bringing him in contact with Lorentz' works. In 1902, Poincaré published "Science and Hypothesis," where he discusses the limitations of classical Euclidean geometry, the limitations of the classical concept of time, and limitations with classical Newtonian mechanics. Poincaré sees in Lorentz' work some promise, but also presents challenges that Lorentz had not yet addressed. Two challenges in particular surface. One follows Poincaré's 1900 paper, "The Theory of Lorentz and the Principle of Reaction," in which Lorentz' line of reasoning contradicts Newtonian physics. The other is a challenge to demonstrate that Maxwell's equations satisfy the principle of relativity. *Science and Hypothesis* is a discussion of issues that lie at the core of relativity.[16]

Lorentz rose to one of Poincaré's challenges. He attacked the relativity conundrum, something that nagged many physicists and Poincaré in particular. In 1861 the Scottish mathematician James Maxwell (1831–1879) published his equations describing the dynamics of electromagnetic fields. A fundamental tenet of physics is that the physics of two systems in steady motion with respect to one another are identical. This being the case, equations describing the physics of phenomena from the perspective of either system are also identical. Poincaré coined the tenet, the principle of relativity. Indeed, Newton's laws of motion comply with relativity—the equations of a moving object are the same in all reference frames that are in uniform motion with respect to one another: $F = mA$.[17] Common belief was ditto for Maxwell's equations, but nobody could demonstrate that this was so.

Let us become more familiar with the concept of relativity with a concrete example. Let us say that two spacecraft are in uniform motion with respect to each other and are outside the influence of any gravitational field. The scientists aboard each spacecraft have radio communication and agree to perform the same experiment aboard their spacecraft. Aboard each craft is a programmable robot; both are manufactured on the same assembly line and are identical in all respects. The scientists on one spacecraft devise an algorithm that allows their robot to mimic an acrobat. While balancing a spinning plate atop a stick that is set on the robot's hand, the robot navigates a circular path about the craft. After successfully demonstrating the algorithm aboard their spacecraft, the scientists communicate the identical algorithm to their counterparts aboard the sister spacecraft. On programming their robot with the identical algorithm, the scientists in the sister spacecraft report that their robot is also an acrobat. Although the spacecraft are in motion with respect to one another, the physics aboard the two spacecraft are the same, so the algorithm works identically on both spacecraft. Of course, isn't that obvious.[18]

The principle of relativity is so intuitive that we almost feel that there is no need to mention it. But the intuitive principle presents a challenge when

---

[16]Poincaré and Lorentz developed a strong friendship along with mutual respect and admiration for each other's work. Poincaré stated that Lorentz was the preeminent physicist of his day and urged the Nobel committee to award Lorentz the Nobel Prize. In 1902, Lorentz bagged the prize.

[17]The equation is the familiar result that force is equal to the product of the mass and acceleration, and this is valid in all inertial reference frames.

[18]Stephen Hawking has a nice way of stating the principle of relativity. He says that pingpong players on a train moving at constant speed play the same game as those inside a gymnasium.

specific equations describing physical phenomena are put to the test. In the case of Maxwell's equations, it requires deep insight and a connection to the Michelson–Morley experiment in order to demonstrate that the principle of relativity holds.

There is a revery afforded to Maxwell's equations that equals the revery afforded to Newton's. All experimental evidence confirms the equations. They express nature with such elegance that nature herself seems pleased; through Maxwell's equations she reveals a secret or two—one secret is the speed of light. So it was with great consternation that physicists were unable to demonstrate that Maxwell's equations comply with the principle of relativity. Lorentz had previously worked on the problem and in 1895 gave an incomplete solution. Poincaré lauded Lorentz' approach but noted the shortcomings of his solution. Lorentz rose to Poincaré's challenge and in 1904 finished the job. Using the Lorentz transformations, Lorentz demonstrated that Maxwell's equations do, indeed, satisfy the principle of relativity. Alongside the spatial contractions that Lorentz had calculated for the specific purpose of explaining the Michelson–Morley experiments and time dilations that Lorentz introduced in his 1895 confrontation with Maxwell's equations, Lorentz introduced variants in Maxwell's waves that miraculously eliminated all of the troublesome terms in previous efforts. This was a staggering intellectual feat. It also exposed another weakness in Newton's mechanics, for anyone willing to take note.

Lorentz was one of many inspired by Poincaré. Who has not latched on to a song that would not exit the mind? Since his days as an undergraduate, Einstein had thought considerably about problems that Poincaré later posed in *Science and Hypothesis*. Einstein consumed Poincaré's book, and the ideas stirred his imagination. Einstein had befriended a circle of like-minded Bohemians who, as rejects from conventional social organizations, began their own. They jokingly called themselves the Olympia Academy, and Einstein was its center. Einstein's right-hand man was Michele Besso (1873–1955), a man whose awkwardness and inability to focus on personal goals masked a keen intelligence. Einstein incessantly brought up the topic of Poincaré's work with other members of the academy and Michele Besso in particular.

Einstein began his job at the Swiss Patent Office in 1902. The job brought much needed security, and the newlywed phase of his marriage masked the strains that would later surface. Besso turned out to be an insightful sounding board. The Olympia Academy of Bern housed in Einstein's tiny apartment with an additional desk at the Swiss Patent Office was an active research center of physics that rivaled those at Oxford, Cambridge, Leiden, Gottingen, Berlin, Paris, and Harvard. Einstein was far from a slouch at work and performed admirably. Einstein's efficiency afforded him the opportunity to utilize some of his worktime to reflect on physics. In 1905, these reflections culminated in the publication of five works, all significant, and three of which would change the direction of physics.

The eighteenth and first half of the nineteenth centuries showcased Newtonian physics and mathematics. His laws of motion yielded the answer to all problems that were put to the test, and his calculus allowed great minds such as Maxwell to describe other physical phenomena. Every experiment and observation confirmed Newton's supremacy, and in the first half of the nineteenth century there was no reason whatsoever to believe that the universe operated differently from Newton's formulation. Yet, from the imagination of one of the most outstanding minds in science, the possibility of a different reality emerged. George Bernhard Riemann (1826–1866) had extended notions of non-Euclidean geometry that his fellow German mathematician, Carl Gauss (1777–1855), first proposed. The geometry of Riemann and Gauss allows lines to curve, parallel lines to intersect, bounded lines to be of infinite length, and among other oddities, the angles forming a triangle to sum to greater than 180 degrees.

For Gauss and others who studied non-Euclidean geometry, these oddities were abstractions that were possibly useful for describing curved objects, like a globe. Nobody dreamed to challenge the notion that the structure of the universe itself was non-Euclidean. The success of Newton, constructed on the foundation of Euclidean geometry, precluded such nonsense. But then, with no physical evidence whatsoever, Riemann offered the non-Euclidean alternative. With uncanny insight, Riemann suggested that our limited perceptions only allowed us to see a Euclidean approximation of a non-Euclidean universe. He further forecasted that a non-Euclidean, non-Newtonian reality would be discoverable through investigations of phenomena existing on exceedingly large and exceedingly small spatial scales. So it was around 1860 that Riemann forecasted the twentieth-century revolutions in physics, general relativity, and quantum mechanics.

Shortly after Riemann's unfortunate death at age 40, experimental evidence confirming Riemann's insight surfaced. Concerning large spatial scales, Michelson and Morley's unsuccessful hunt for ether was a riddle that could not be cracked by Newtonian mechanics in a Euclidean framwework. Concerning small spatial scales, experiments designed to address Kirchhoff's challenge to describe blackbody radiation produced results that did not comport with classical theory.[19] Neither the experimenters nor the theorists of the nineteenth century had any notion that Riemann's insight was the key to the explanation of these results.

In 1900, 2 years prior to the establishment of the Olympia Academy, Max Planck (1858–1947), a proud Prussian, had formulated an explanation of Kirchhoff's problem on the basis of quantization of energy levels. The explanation was somewhat reminiscent of Lorentz' explanation of the Michelson–Morley results. Planck himself had difficulty believing his formulation as he could not understand a mechanism that caused quantization of energy levels. Einstein's 1905 publication "On a Heuristic Viewpoint Concerning the Production and Transformation of Light" furthered the work of Planck. Einstein

---

[19]In 1859 the German physicist Gustav Kirchhoff (1824–1887) challenged the physics community to determine the frequency distribution of radiation emited from a blackbody. Experimentalists rose to the challenge and made measurements. By 1895, the experiments of the German physicists Wilhelm Wien (1864–1928) and Otto Lummer (1860–1925) gave a good description of the phenomena. The distribution of the frequencies as measured by Wien and Lumer differed from the theoretical distributions based on classical methods.

proposes the bold theory that against much evidence, light was not a wave, but as Newton stated two and a half centuries earlier, a stream of particles later called *photons*. Einstein further proposes that photons interact with other particles to exchange energy only at discrete energy levels and gives a formula that quantifies the allowable energy exchange levels. Einstein further relates the energy levels to the frequency of the light when described as a wave.[20] Planck initiated Riemann's small-scale program, and Einstein accelerated it. Einstein discovered essential relations that lie at the heart of quantum mechanics. While Einstein would later bail out, others followed and finished the job.

The other end of the Riemannian program, investigations of large-scale phenomena, is where Einstein defined himself as a physicist and where he captured the imagination of the public. Einstein's 1905 paper "On the Electrodynamics of Moving Bodies" begins his 11-year quest for the general theory of relativity. This paper has remarkable similarities with Lorentz' 1904 paper. Einstein uses the Lorentz transformations to demonstrate that Maxwell's equations are in accord with the principle of relativity. This seems to not be an advance in science, because Lorentz accomplished this one year prior to Einstein. But there are differences that separate Einstein from both Lorentz and Poincaré.

Reminiscent of Newton's *Principia*, where Newton first states the laws of motion and then derives consequences, Einstein begins his paper by stating two propositions inherent in God's fundamental design of the universe and then derives consequences. Einstein's two propositions are (1) the principle of relativity and (2) the speed of light, which is a constant that is independent of the motion of the observer or light source. As noted above, the first proposition had been a tenet of physics since Galileo and is embedded in Newton's laws. The second proposition is where Einstein disembarks from Newtonian reality and strays toward Riemann. Einstein begins that path using Newton's axiomatic methodology to strike down Newton.

As the second proposition contains the point of departure from Newton, let us discuss it further. Let us imagine two observers measuring the speed of light that emanates from an observed source, where one observer is stationary with respect to the source and the other is in motion with respect to the source—let us say moving away from the source in the same direction that the incoming light moves. In Newtonian physics, the stationary observer's measurement of the light's speed would be different from that of the moving observer. As the moving observer moves along with the light, the relative speed of the light is lower than the speed as recorded by a stationary observer. While I bicycle down the road alongside the cars, their relative speed is slower than the speed when viewed by a stationary pedestrian. Of course, this is common sense.

Einstein says that when he developed theories, he pretended to be God. He would ask himself if God would in fact create something that accords with his

---

[20]The experiment of Frank Lenard, a German physicist whose rabid anti-Semitism caught the attention and earned the praise of Hitler, guided Einstein's investigation. Lenard reported that the energy level of electrons emitted from a plate bombarded by light depended only on the frequency and not the intensity of the light.

theory. It was Einstein's own call, based on his own beliefs and prejudices. Concerning the second proposition, Einstein believed that God would in fact conjure up a process that causes the first and second observers to have the same measurements for the speed of light. Why would anyone state as a fundamental principle a principle that so resoundedly violates common sense? Because the second proposition would account for the Michelson–Morley results. No experimental evidence was available to Riemann, so he could not formulate such a proposition. But Einstein had the advantage of the Michelson–Morley experiments. Whereas Michelson and Morley viewed their experiment as a failure, Einstein viewed it as evidence in support of his second proposition.

With these two propositions, Einstein derives the Lorentz transformations. He neither dithers nor dallies, nor obscures the results with obsolete notions. It is a very lucid paper in which Einstein demonstrates that so long as the two propositions are reality, the Lorentz transformations as a consequence of the propositions must also be reality. The spatial and time dilations really do exist. What doesn't exist in Einstein's reality is the ether. Einstein mentions it in a passing sentence that dispenses with the notion and then never returns to it again:

> The introduction of a "luminous ether" will prove to be superfluous inasmuch as the view here to be developed will not require an "absolutely stationary space" provided with special properties.

This distinction separates Einstein from Poincaré and Lorentz. It enables Einstein to make contributions to relativity in addition to those already placed forward by Poincaré and Lorentz, and it provides a fruitful perspective for future development. We further explore the contributions of Lorentz, Poincaré, and Einstein in the next section.

It is unfortunate that Einstein gives no bibliography, nor does he discuss sources that influenced his work.[21] With Poincaré and Lorentz addressing similar issues, there has always been an uncertainty concerning their role in his thinking. Einstein does credit one individual, Michele Besso. Of Besso, Einstein states:

> In conclusion I wish to say that in working with the problem here dealt with I have had the loyal assistance of my friend and colleague M. Besso, and I am indebted to him for several valuable suggestions.

In this way, an unemployed vagabond became Einstein's collaborator in science. Besso and Einstein continued their friendship for the remainder of their lives, and Besso's role as collaborator would continue as Einstein developed his theory of general relativity.

It was natural for Einstein and Besso to have developed a deep friendship. Both were Jewish with similar views on assimilation. Additionally, they shared

[21] In the introduction, Einstein does presents a result that comes from Lorentz' 1895 paper indicating that he knew of the Lorentz transformations and a first attempt at reconciling Maxwell's equations with the principle of relativity.

common interests in music, philosophy, and, of course, physics. Einstein assisted Besso with his employment problem by recommending Besso to the Swiss Patent Office, from which Besso retired decades later. Einstein also introduced Besso to his wife, Anna Winteler, of the same Winteler family who hosted Einstein as he completed high school. After Einstein's divorce, Besso looked after Mileva and became a father figure to Einstein's boys.

One other point of note concerning Einstein's life is that one of his 1905 papers, A "New Determination of Molecular Dimensions," served as Einstein's doctoral dissertation. Einstein received his degree in 1905 from the University of Zurich.

THE ESTABLISHED AND THE UNKNOWN

One of the most famous scientific quotes comes from the pen of Hermann Minkowski, demonstrating that in addition to his formidable mathematical talent, he had a gift for language. Minkowski begins a 1908 address to the Assembly of German Natural Scientists and Physicians, in the following manner:

> The views of space and time which I wish to lay before you have sprung from the soil of experimental physics, and therein lies their strength. They are radical. Henceforth, space by itself and time by itself are doomed to fade away into mere shadows, and only a kind of union between the two will preserve an independent reality.[22]

More remarkably, Minkowski accomplishes precisely the ambitious goal of the quote. While Einstein recognized that his laws were a point of departure from Newtonian mechanics, Minkowski penetrated deeper. Minkowski replaced the underlying framework of time and space with a new invention.

The universe where Newtonian mechanics resides has three spatial dimensions augmented by an independent time dimension. In this universe it is possible to designate a particular spatial reference frame as a universal reference frame and coordinate clocks across the universe. The time and coordinates of any event with respect to the universal reference frame and clock would identify the event for every being in the universe; beings observing two events would record the same time and spatial differences between the events.

Einstein's mechanics reside in a more bizarre environment. As Minkowski states, time, space, and the observer are inseparable. Minkowski intuited that the universe is a four-dimensional blob and there are no universal time and spatial coordinates that identify events in this blob. Even if two beings initially coordinated clocks and measuring rods at the same point within the universal blob, then, once they separated, their measurements of the distance between two events would be different. (In this context distance is a measurement of the spatial and time differences.) Minkowski specifies the underlying geometry that gives

[22]Excerpt from Minkowski's speech, "Space and Time" (see full see footnote 10, above).

structure to the blob. It is somewhat abstract, but without Minkowski's specification of the blob's non-Euclidean structure, special relativity has no scaffolding, and a further generalization of the theory is not possible.

It would take Einstein, with the assistance of Besso, Grossman, and Minkowski's friend and colleague, the phenomenal mathematician David Hilbert (1862–1843), 6 years from the time of Minkowski's publication to apply Minkowski's concepts in a more general setting and complete the general theory of relativity. Minkowski, who was perfectly situated in his four-dimensional universe to further contribute to the general theory, departed in 1909 shortly after this famous talk. This event was sadly noted by the scientific community.

Perhaps a few issues remained, but for the most part, Minkowski's unveiling of a new universal geometry wrapped up the subject of special relativity. Of course, special relativity set the stage for general relativity and as such, the theory was not complete. But the distinction between special and general relativity is not artificial, and one can identify components of special relativity that are significant contributions to physics. Let us return to the works of two contributors, Lorentz and Poincaré, and give them their due credit.

If one were to make a checklist of all the contributions of special relativity, Lorentz and Poincaré would fill out much of that list:

1. Explanation of the Michelson–Morley result. Check. Lorentz proposed the initial theory in 1891 for precisely that purpose.

2. Development of the complete set of Lorentz transforms. Check. Lorentz made the initial proposal. Poincaré generalized the set of all possible transformations in his 1905 paper, "On the Dynamics of the Electron."

3. Reconciliation between electromagnetic dynamics with the principle of relativity. Check. Lorentz and Poincaré demonstrated that Maxwell's equations are invariant under the Lorentz transformations.

4. The introduction of a radical concept of time. Check. Lorentz introduces time dilations dependent on the motion of the observer into the Lorentz transformations. Poincaré, as a result of his work at the Bureau of Longitudes, gives the time dilations physical meaning.

5. Ambiguity of simultaneity. Check. Poincaré is the first to point out simultaneity is a relative concept. Events recorded as simultaneous by one observer are not in general simultaneous events for another observer.

6. Endowing space and time with a non-Euclidean structure. Check. Poincaré discusses non-Euclidean geometries in his book, *Science and Hypothesis* and then in his paper "On the Dynamics of the Electron" presents a non-Euclidean structure consistent with the Lorentz transformations. Minkowski endows his universal blob with Poincaré's structure.

7. Reconciliation of dynamics with the principle of least action. Check. In *Science and Hypothesis*, Poincaré dwells at length on this principle and deems it a foundational concept. Demonstration that Lorentz' theory

comports with the principle of least action is central to Poincaré's paper "On the Dynamics of the Electron."[23]

Outside of a small academic community, few have heard of Lorentz and Poincaré, let alone know of their contributions to relativity. By contrast, Einstein is a household name, and he garners credit for ideas that stemmed from Lorentz and Poincaré. The checklist above redresses the Einstein myth that apportions all the credit for relativity to him alone. But the checklist is incomplete, and Einstein's contributions toward special relativity's completion rightfully earn him celebrity status.[24] Einstein was able to take relativity in a more fruitful direction than its original authors. Why did he find paths that eluded both Poincaré and Lorentz?

The general perception of scientists is that they are distilleries, distilling theories from observations by using their intellect and logic. There is an alternative view that I find more persuasive. Scientists germinate their theories not from the intellect, but from the gut. The scientist uses the intellect to develop supporting arguments in favor of a theory, but the theory itself comes from the gut. What were the compositions of Lorentz' and Poincaré's guts, and how did they differ from Einstein's?

The nutshell answer speaks volumes. Lorentz' gut guides him to maintain the ether and seek relations that explain experimental results. Poincaré's gut guides him to demonstrate consistency with the principle of relativity and the principle of least action. Einstein's gut tells him to establish the fundamental tenets of God's design and then deduce consequences, whatever they may be, from those tenets. All three men reveal their perspectives in their papers. Lorentz' 1904 paper "Electromagnetic Phenomena in a System Moving with any Velocity Less than that of Light" displays Lorentz' instinct. While the paper addresses Poincaré's criticism that the Lorentz theory must fully account for the principle of relativity, Lorentz raises his true concern in the opening paragraph, where he discusses the need to explain experimental results:

> The first example of this kind is Michelson's well known interference experiment, the negative result of which led Fitzgerald and myself to the conclusion that the dimension of solid bodies are slightly altered by their motion through the ether.

Later, Lorentz gives other experiments that need accounting for, such as a failed effort by the team of Rayleigh and Brace and another failed effort of Trouton and Noble to find the ether. The body of the paper utilizes the Lorentz

---

[23]The original version of the principle of least action applied to a setting free of potential energy. It states that a body follows a path that minimizes the summed momentum along the pathway. Recall that Newton identifies momentum with motion. Accordingly the principle states that the path minimizes the motion. In the absence of a potential that applies force, the path between two points that minimizes a body's motion is a straight line. In 1834, the great Irish mathematician Hamilton (1805–1865) incorporated potential energy into the principle and equated it with Newtonian mechanics.

[24]My gripe is not against the public acclaim afforded to Einstein; it is that Poincaré and Lorentz deserve to be universal celebrities as well.

transformations to describe the dynamics of an electron as it moves through ether. Lorentz then gets to the heart of his concern and applies his description of electron dynamics to explain the failed experiments.

In 1904 Poincaré traveled as an invited guest to the St. Louis World's Fair. On American soil, Poincaré gave a speech outlining his perspective on the current state and future of physics. It was culled from his 1902 publication *Science and Hypothesis*, and Poincaré submitted his views to The Congress of Arts and Science Universal Exposition in a paper, "Principles of Mathematical Physics." There Poincare presents physics not as an index of equations but as consequences of principles. There are six principles of import:

> The principle of the conservation of energy, or the principle of Mayer, is the most important, but it is not the only one; there are others from which we are able to draw the same advantage. These are: The Principle of Carnot, or the principle of degradation of energy. The principle of Newton, or the principle of equality of action and reaction. The principle of relativity, according to which the laws of physical phenomena should be the same whether for an observer fixed, or carried along in uniform translation; so that we would not and could not have any means of discerning whether or not we are carried along in such a motion. The principle of the conservation of mass, or principle of Lavoisier. I would add the principle of least action.

Poincaré discusses the status of the principles, stating that recent experiments threaten to topple what has been held sacred. The paper relays an uneasy mix of concern and opportunity. In his 1905 paper, "On the Dynamics of the Electron," Poincaré takes his measure of Lorentz' theory. Poincaré summarizes his analysis in the introduction: "*the postulate of relativity can be established in any rigour. It is what I show by a very simple calculation founded on the principle of least action.*"

After all the concern that physics is in such a state of disequilibrium that it may at any moment topple over, Poincaré sees in Lorentz' 1904 work some stability. The Lorentz transforms salvage the principle of relativity, and an auxiliary term in what is known as the *Lagrangian* proves the utility of the principle of least action. Both Lorentz and Poincaré demonstrate admirable acumen, but they limit their investigations. Their approach has taken them to the end of their trail.[25]

---

[25]Lorentz' path was a continuation of Michelson's. While Michelson could not find a physical trace of the ether, Lorentz felt that it would be central to any physical theory. For Lorentz, interactions between ether, matter, and energy would explain Lorentz' proposed spatial contractions as well as contradictions that Poincaré pointed out. Poincaré's ideas are a bit more ambiguous. He certainly was no conservative. He did redefine time and foresaw the eventual downfall of Newton. While he flirted with an etherless universe, in the end he held fast to the notion. In both Lorentz' and Poincaré's works the ether defines a special rest frame, and they seek out how perceptions in a moving frame manifest themselves in the special rest frame. It must also be noted that Poincaré's 1905 paper also contains a first stab at bringing gravity to the theory of special relativity. Poincaré shows keen insight in placing a speed of light limit on gravity; the gravitational field of a body can propagate no faster than the speed of light. Otherwise there is no further development along the lines considered by Poincaré.

Whereas Lorentz and Poincaré sought to explain experiments and maintain the integrity of principles, Einstein was an explorer. His two postulates form the exploratory vehicle through which Einstein could make further discoveries. In Einstein's 1905 paper, "On the Electrodynamics of Moving Bodies," after presenting his two postulates, Einstein's first discovery is the Lorentz transformations. There he discovers that measurements of time and space are dependent on the motion of the observer. Yes, Lorentz and Poincaré knew this, but Einstein's perspective causes him to use this discovery differently from his predecessors. Einstein intuits that critical to any physical theory is the relationship of perceptions among different observers moving at different speeds. Using his two postulates and Lorentz transforms, it is these relationships that Einstein pursues.

Einstein targets Maxwell's equations and discovers the relation between measurements of electric and magnetic fields among observers moving at different relative speeds. Yes, Lorentz and Poincaré beat Einstein to this discovery, but it is worthwhile to repeat the different perspectives. Einstein's predecessors calculated differences in electromagnetic fields between the observers in motion and an ether-centered observer in order to demonstrate the principle of relativity. For Einstein, the principle of relativity needs no demonstration; it is a given fact. What is critical is the discovery of relations in measurements between different observers, the discovery of which results from the principle of relativity.

It was inevitable that given Einstein's different purpose and perspective, he would discover relations that eluded Poincaré and Lorentz. Einstein discovers the shift that describes the frequency of light in terms of the relative velocities of the emitter and the receiver. This gives what is known as the *Doppler shift*, an effect used by astronomers to determine the relative motion of a star with the earth, and used by the friendly highway patrol to determine when you've pushed the gas peddle above 65 mph. Einstein also discovers that the intensity of light varies with relative velocity and specifies the relation. The faster you move toward an emitting light source, the quicker you'll get a tan, not only because you will move closer to the emitting source but also as your very motion itself increases the intensity of the light. Another critical discovery of Einstein's is that energy measurements of a light source depend on the relative motion of the receiver and emitter, and Einstein specifies this relationship. This final relationship becomes Einstein's gateway to his famous equation $E = mc^2$.

In 1905, Einstein presents a version of the energy mass relation in a follow-up paper, "Does the Inertia of a Body Depend upon Its Energy Content?" Einstein's demonstration of the equation is limited; it is based on a specific experiment on an electric charge. Einstein informs the reader not to constrain the result to an electron because any charged body could be used. He has in mind that the formula has general application. Einstein also uses a mathematical approximation that limits its application. The drawbacks of Einstein's presentation are worth mentioning. Einstein, fully aware of these drawbacks, knows that he is onto something big. Despite the fact that it has not been fully developed, he risks it and puts it out there, anyway. In 1906, Einstein refines his argument and eliminates some of the concerns.

Poincaré had the chance to beat Einstein to this famous formula. In his 1900 paper "The Theory of Lorentz and the Principle of Reaction," Poincaré presents a thought experiment in which the results according to Lorentz' theory are not in accord with the principle of equality of action and reaction. Poincaré restores the principle by introducing a fictive fluid that has the same mass energy relationship as that discovered by Einstein. Rather than seizing on the relation, Poincaré dismisses the fluid as not real and attributes no physical significance to the relationship between energy and mass. Here, Poincaré both discovered a hole and outlined the dimensions of a plug. He failed to realize the possibilities embedded in his fictive fluid and, rather than looking for physical significance in the fictional fluid, continued to use the example as a demonstration that fundamental principles of physics were under attack.

The equation $E = mc^2$ is arguably the most famous equation in science. It has an association with the atom bomb that etches its presence in the mind years after one has graduated high school and the more mundane formulas of trigonometry and the like have escaped one's memory and moved to the ether. While relevant to the atom bomb, the equation was central to special relativity. It stitches up a loose strand that would otherwise have unraveled the theory.

Lorentz and Poincaré were successful in roping Maxwell into the framework of relativity. In doing so they tossed Newton out. Were they any better off than the French were before and after their revolution? Just as the French exchanged a monarch for an emperor, Lorentz and Poincaré exchanged Newton for Maxwell. While there was an electromagnetic theory, there was none for mechanics. This was unsatisfactory and in 1904, Poincaré expressed his concern in *Principles of Mathematical Physics*:

> From all these results, if they were confirmed, would arise an entirely new mechanics, which would be, above all, characterized by this fact, that no velocity could surpass that of light, anymore than any temperature could fall below absolute zero, because bodies would oppose an increasing inertia to the causes which would tend to accelerate their motion; and this inertia would become infinite when one approached the speed of light.

Einstein's postulates yield the result that mass is not a constant but dependent on its motion. Among some strange yet true aspects of the theory, a body at rest has less mass than the same body in motion. Einstein's relativistic energy formula combined with the formula $E = mc^2$ gives the measurement of a body's mass as its speed alters. This expands the theory of special relativity to mechanics. One starts with Newton's formula relating the force on a body to its acceleration, $F = mA$. Here m is not a fixed mass, but a relativistic mass that varies in accord with Einstein's equations. Just as Poincaré predicts, the mass increases as the speed of the body does, and a body can never exceed the speed of light. Poincaré had the insight to foresee this reality. Einstein discovered the reality.

TROUBLE

Imagine Johannes, or Jean, or Giovanni, leaving his village for the opportunities of the city. He bids a tearful farewell to his family and promises to write. On arriving at the city, he stays in a dormitory where there are nameless others in similar circumstances. There is no running water in the dormitory. He has no privacy, as a single room is packed like a box to hold as many people as possible at a low cost. He confronts not only the strangers' noise and physical presence but also their stench. By day he follows a crowd down to a factory gate. There he is among dozens who hope to get selected to enter the factory for a meager wage that will get him through the day.

On days that he doesn't go down to the factory, he might see wondrous things. There are people in fine dress riding in fancy cars, returning to their mansion homes. There are cinemas, stage shows, restaurants with fine foods, and stores with fancy gadgets. But all of this is inaccessible to him. He finds himself homesick and recognizes that partly out of embarrassment for his situation, he has not kept his promise to write back home; and so with his embarrassment intact, he spends a few precious coins on a pen and paper and writes, "Everything is fine. I have a good job. There are wondrous things in the city." The parents are thrilled to receive this letter. They share it with the neighbors whose boy listens and knows he will also go when he is old enough.

Phillip Blom describes the beginning of the twentieth century as the vertigo years. The industrial revolution brought new lifestyles and broke traditional social units of family, farm, and village. The mass movement of individuals from the country side to the city continued to increase exponentially. Dynamos electrified the city. Mass production of goods changed economic activity. Trains moved at dizzying speeds, making distant lands accessible. Everything was in flux; even the fundamental workings of God as expounded by Einstein seem to have changed. It was a heady time with high expectations of the future.

There are times when expectations and reality clash with dangerous consequences. Mass migration to the city meant unemployment, exploitation, and in many cases emotional isolation from family and community. In Austria the different nationalities were in greater communication and contact, decreasing rather than increasing social cohesion. Jealousy of the success of a neighbor or someone from an ethnic minority found new expression among politicians who would capitalize on that jealousy. In Germany, resentment of Prussian domination didn't subside—and if Bismark had left Germans in angst, imagine the embitterment of France. Not only had she lost Alsace and Lorraine, but the realization that she was no longer the preeminent European nation was a stain on her national honor.

Social forces stirred the individual and the state in a whirlwind. The antidote to internal angst and divisiveness was war. War would bind individuals with the state in a common cause. From England to Russia, from the lower classes to the leadership, there was popular support for war. A minority of individuals stood against prevailing sentiment. In 1914, Albert Einstein, the newly appointed director of physics at the Kaiser Wilhelm Institute, joined a pacifist organization that circulated a petition decrying the calls for war. Their pleas drowned in the excited clamor of those demanding war. The demand was met.

Hitler never liked Vienna. He had a distinctively racist perspective of the world. Being among other races, dealing with those who should not be of concern, and watching other races awash in success while he mattered not a whit to anyone offended Hitler's German sensibilities. Those offended sensibilities and a desire to escape mandatory military service prompted Hitler to move to Munich in 1914. Cooperation between Austrians and Germans allowed German authorities to net Hitler, the draft evader. The Germans were ready to extradite Hitler to Austria, but then World War I intervened. Hitler enlisted in the German army to be with his brethren. Germany obliged and sent him not east to Austria, but in the opposite direction, west to the Belgian front.

Prior to the war Hitler was a failure, a nobody without family, without a home, and without social connections. The army became Hitler's home and brought him social connections. Hitler was a courier, shuttling communiqués between the front line and military command behind the line. It was not a job without dangers as there was a considerable casualty rate among the couriers. Hitler performed his job admirably and was respected among his fellow comrades. In *Mein Kampf*, Hitler recalls that the day he received his iron cross, a medal awarded for uncommon heroism, was the proudest day of his life. Hitler received the award for for protecting his commanding officer while under fire; it was an honor that Hitler genuinely earned. The image of Hitler at war is that he enjoyed the comradery arising from the common sense of purpose and was completely insensitive to the human suffering that was all about him.

While on the front line, Hitler became aware of a battle that had ominous implications for the future of warfare. In April 1915, after many days of waiting, winds were favorable for Fritz Haber (1868–1934) to launch an attack against French and Belgian positions using a secret weapon. The French and Belgian soldiers were defenseless; indeed, they never foresaw the means by which death would be delivered. A cloud of poisonous gas entered the lungs of over 5000 men. The chemically active chlorine disassembled human lung tissue and transformed it into a functionless jelly compound.

Who directed this ghastly yet historic attack, the possible means of German victory? Fritz Haber was the director of chemistry at the Kaiser Wilhelm Institute and a friend of his colleagues Max Planck, Max Born, and Albert Einstein. Haber's most noted achievement, one that later earned him the Nobel Prize, was the development of a process to synthesize ammonia from the air's nitrogen. It is Haber's process that launched the agricultural revolution and allows humankind to surpass the Malthusian limit. The same process yields explosives that arm nations for war. With the advent of war, Haber more pronouncedly aimed his talents toward the manufacture of death.

Born a Jew, Haber converted to Catholicism and became a fierce German nationalist. Like the majority of professors, including Max Planck, Haber supported the war and showed his support by rushing to enlist—a significant step further than the other nationalistic professors of the institute. Because of his

Jewish heritage, the army denied Haber a commission. His passionate German nationalism proved to be stronger than the snub, so Haber accepted a position as scientific adviser to the government. As the developer of the killing agent, Haber was most qualified to lead the attack.

The course of history is affected by inspiration as well as blunder. In this case, the German's blunder cost them victory. The decision to call on Haber to launch his attack was premature. Had the Germans built up an inventory of Haber's poison so that they could quickly launch follow-up attacks, the results on the allies would have been devastating, and Germans would have marched through to Paris. Instead, the Allies had a respite during which they developed a defense against the poison. The speed with which the English manufactured and deployed gas masks is a testimony to human ingenuity.

The English also had their brilliant chemists. Preeminent among them was the Russian-born Jew Chaim Weizmann (1874–1952). Weizmann is the father of industrial fermentation processes, the use of bacterial agents in the manufacturing of chemicals. One of Weizmann's processes yields acetate, a necessary ingredient in the manufacture of cordite. Cordite was the explosive that the English used as gunpowder. As acetate was in short supply, Weizmann's invention was critical to England's war effort. Unlike Haber, Weizmann did not aim to assimilate. Weizmann became a leader of the Zionist movement, had a role in the establishment of the state of Israel, and became Israel's first president. Historical events well beyond either human control would later cause Haber and Weizmann to cross paths.

What was going through the minds of the European populace when in 1914 they craved war? What sort of madness gripped them and swept them toward disaster? On all sides, the commonly accepted sentiment was that the war would last perhaps not more than a month with victory ensured. Four years and over 16 million casualties later, with economies in ruin and food supplies short, with amputees returning home to confront a limbless future, the dire consequences of war were self-evident. In Germany, even the most patriotic citizen could not help but take note of the stark contrast between the war propaganda and actual circumstances. On November 9, 1919, German citizens took it upon themselves to end the war. Individual acts of protest swelled into a movement that overthrew the Kaiser's rule and established a republic. Although Kaiser Wilhelm had recognized that defeat was imminent and initiated efforts to negotiate a settlement, it would be the republic that assumed the burden and blame of the final surrender.

Morale on the front reflected the discontent of the German citizens. In the summer of 1918, Lieutenant Hugo Gutmann, one of the few Jewish officers in the German army, summoned Adolf Hitler and a fellow courier. He promised the iron cross to the couriers if they would pass on a message to the front, an indication of the level of discipline within the army. Previously an order would have guaranteed delivery, now this bribe was necessary. Lieutenant Gutmann was true to his word. Despite pushback, Gutmann continued to badger his senior officers until they authorized what Gutmann had promised. On August 4, Hitler received his second iron cross.

On October 13, 1918, Hitler fell victim to a mustard gas attack launched by the British against German positions. The attack blinded Hitler, and comrades evacuated him to a hospital. While recuperating, Hitler learned of the overthrow of the Kaiser and the impending German surrender. Rage consumed Hitler. He believed that there was no price too high for a German victory and with equal certainty, that good Germans were all behind the war effort. This was a treasonable act that required assignment of blame and eventual punishment. For Hitler, the culprits were clearly identifiable. They were the same vermin that de Lagarde had identified and Lueger railed against. Hitler blamed the Jews as leaders of a Marxist conspiracy. The following passage from *Mein Kampf* reveals Hitler's state of mind:

> Kaiser Wilhelm II was the first German Emperor to hold out a conciliatory hand to the leaders of Marxism, without suspecting that scoundrels have no honor. While they still held the imperial hand in theirs, their other hand was reaching for the dagger. There is no making pacts with Jews; there can only be the hard: either-or.[26]

EXITING THE QUANTUM UNIVERSE

"The young man who came to meet me made so unexpected an impression on me that I did not believe he could possibly be the father of the relativity theory so I let him pass. . . . During the first two hours of our conversation he overthrew the entire mechanics and electrodynamics."[27] These are the recollections of Max von Laue, an assistant of Max Planck. Einstein's work had attracted Planck's attention, and Planck himself intended to visit Einstein in Bern, but circumstances intervened. Planck sent von Laue, who was shocked by the discovery that Einstein was not a professor, but a patent clerk.

For 4 years after his 1905 papers, Einstein continued to work at the patent office and continued his string of significant contributions to physics. His two main areas of focus were quantum mechanics and relativity. Concerning the former, Einstein's 1905 paper, "On a Heuristic Viewpoint Concerning the Production and Transformation of Light," left many unanswered questions. Einstein pondered the problem of reconciling the wavelike characteristics of light with his particle proposition. As an invited guest to the Salzburg Conference of 1909, Einstein delivered a talk on the subject:

> Light has certain basic properties that can be understood more readily from the standpoint of Newtonian emission theory[28] than the standpoint of the wave theory. I thus believe that the next phase of theoretical physics will bring us a theory of light that can be interpreted as a kind of fusion of the wave and the emission theories

[26]Hitler, *Mein Kampf*, p. 206.
[27]Isaacson, *Einstein*, p. 156.
[28]Recall Newton's theory of light as corpuscles (see Chapter 8). Einstein knocked off Newtonian mechanics but restored Newton's corpuscular theory, for which Newton endured much ridicule.

of light.... These two structural properties simultaneously displayed by radiation should not be considered as mutually incompatible.[29]

With this talk, Einstein not only anticipated the formal development of quantum mechanics; he gave it impetus. In the area of quantum mechanics, the Olympia Academy was ahead of any other research center. Indeed, the force of this patent clerk's ideas was so powerful that the Olympia Academy set the research agenda for the international physics community.

In 1911, the Belgian industrialist, Ernest Solvay, funded a conference intended to bring together the best of the best in the physics community. Among the attendees were Lorentz, Poincaré, Planck, Rutherford, Madame Curie, and Langevin, a veritable who's who of the elite. In addition to these established names, not only did Einstein, by then a professor in Prague, receive an invitation; Lorentz requested that Einstein give a talk on the central issue of the conference, the quantum problem.

Einstein continued his investigation into quantum mechanics with his 1916 paper, "On the Quantum Theory of Radiation." In his paper, Einstein imagines a device that in 1958 Schawlow and Townes of Bell Labs would construct and patent, the laser. Using statistical methods, Einstein analyzes this device, demonstrating its ability to emit photons from a cloud of energized atoms. Writing to Besso, Einstein describes his quandary: "It is a weakness of the theory that it leaves the time and direction of the elementary process to chance."[30] Einstein recognized that further development of quantum mechanics was possible only if one were to veer off the path of determinism and describe nature in probabilistic terms. Determinism, the evolution of every physical process in accordance with a specified law, was Einstein's pillar from which he refused to depart.

There are many famous quotes concerning Einstein's distaste of the abandonment of determinism such as the following communication with Max Born:

> I find the idea quite intolerable that an electron exposed to radiation should choose of its own free will not only its moment to jump off, but also its direction. In that case, I would rather be a cobbler, or even an employee of a gaming house, than a physicist.[31]

In a discussion with Niels Bohr, Einstein proclaimed, "God does not play dice with the universe," to which Bohr retorted, "Einstein, stop telling God what to do."[32]

Einstein was to quantum mechanics what Poincaré and Lorentz were to general relativity. He initiated its development, but left it to others to complete. Einstein saw quantum mechanics as a monster that he battled for the final three decades of his life.

[29] Isaacson, *Einstein*, pp. 156–157.
[30] Ibid.
[31] Ibid. p. 324.
[32] Ibid. p. 326.

BENDING THE LIGHT

In contrast to Einstein's 1905 foray into quantum mechanics, an effort that Einstein not only abandoned but opposed, Einstein enthusiastically continued his research into relativity and considered general relativity to be his most treasured accomplishment. Just as his paper "On a Heuristic Viewpoint Concerning the Production and Transformation of Light" left unanswered questions in quantum mechanics, Einstein's 1905 papers on relativity only raised further issues.

The theory proposed in the 1905 papers is known as special relativity. Yes, it is a special theory, but the adjective in this case indicates a constraint on the theory as in a special case that begs generalization. The special case of special relativity is that all the equations are developed for reference frames in unaccelerated motion. Einstein thought that this was unnatural; he should be able to express the equations of mechanics in any reference frame.

In 1907, Einstein had what he considered to be "the happiest thought of my life." Reflecting on the sensation of a free fall in space under the influence of gravity, Einstein concluded that the cause of the motion was irrelevant. From his reflection, Einstein arrived at the equivalence principle; the sensation and thus the physics involved would be the same regardless of whether gravity or some other cause accelerated a body. The principle allowed Einstein to simplify the analysis of gravity to that of acceleration. Einstein had a vision of what a general theory would look like. Using the equivalence principle, one could incorporate gravity into a general theory of relativity by generalizing special relativity to accommodate accelerating reference frames. It was much easier to set the goal than to achieve it.

As a midpoint to a final result, Einstein wrote his 1911 publication, "On the Influence of Gravitation on the Propagation of Light." This work elegantly exploits the equivalence principle as demonstrated by updating an example that Einstein himself used to illustrate the principle to a general audience. Let us say that you are in outer space aboard a spaceship accelerating in an upward direction, up with respect to the spaceship's floor. You feel the floor of the spaceship pressing against your feet. One wall of the spaceship has a sealed hole that allows a lightbeam from the outside to enter. The lightbeam crosses the spaceship and illuminates a spot on the opposite wall. However, during the time that the light travels between walls, the spaceship has moved upward, so with respect to the spaceship's floor, the illuminated spot on the opposite wall is lower than the beam's entry hole.

Now let's perform the same experiment on a planet whose gravity imputes the identical force on your feet as the spaceship did. You stand in a room that is perfectly still with respect to the planet. Gravity causes you to feel a force pressing on your feet. One wall of the room has a hole that allows a lightbeam from the outside to enter. The lightbeam crosses the room and illuminates a spot on the opposite wall. The question arises as to exactly where the spot of light is on the opposite wall: Einstein's equivalence principle states that you, the observer would not be able to distinguish between the case when you are on the spaceship

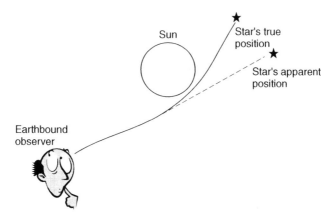

Figure 9.2. Bending of light by the sun's gravity.

and the case when you are on the planet, so the spot of light must fall at the same point with respect to the floor; otherwise you would note a difference. How can this happen? Einstein's conclusion that he states in his 1911 paper is that gravity bends light.

The physics community was abuzz. Could gravity really bend light, or was this an imagined consequence of a thought experiment spun from the mind of a man who was, after all, different? What made the paper all the more intriguing was that Einstein teased the reader by describing an experiment that would settle the controversy. Einstein recognized that the relative position of a star, relative to other stars, would shift when it is aligned or nearly aligned with the sun. This is because the sun would bend the star's light as the light passed in proximity to the sun on its way to an observer on Earth, so the observer would observe the shifted position (see Figure 9.2). Just observe the relative position of a star that is in near alignment with the sun and determine whether that position does, indeed, shift from its normal position. Einstein further provided a calculation that quantifies the shift. There is a catch to this tease; the experiment would have to wait until 1914, because only then would an eclipse of the sun allow an observer to view stars in the direction of the sun. We will return to this experiment later.

AN ERGODIC LIFE

The 1911 paper was one of the first that Einstein produced not as a patent clerk, but as a professor. In retrospect, the eventuality of a professorship was self-evident. However, that's not how it appeared to Einstein. Despite his 1905 contributions, despite having set the agenda for physics at the Salzburg Conference, despite recognition and praise from none other than Planck, by 1909, Einstein still did not have a professorship, although something was in the works. Professor Alfred Kleiner (1849–1916) of the University of Zurich, who was

Einstein's doctoral adviser, lobbied on his behalf. It was not an easy effort. There were other candidates, and it came down to two, a friend of Einsteins, Friedrich Adler, and Einstein.

One has to wonder whether anti-Semitism played a role in Einstein's difficulties. To deny prejudice is foolhardy; we all have prejudices. We'd be empty without them. Our very attitudes are prejudices formed by direct experience and also by the influence of others. Einstein joked about others' attitudes:

> If the theory of relativity turns out to be right, the Germans will call me a great German physicist, the Swiss will call me a great Swiss physicist and the French will call me a great European Physicist. If it turns out to be wrong they will all call me a Jewish physicist.

One individual that showed himself to be of outstanding character was a man molded in Prussia, de Lagarde's hometown and perhaps ground zero of anti-Semetic intellectual thought, Max Planck. Prussian in his bones and damn proud of it Planck showed himself to be first and foremost a scientist. For Planck, good science did not have a national or ethnic character. Science was science. But not everyone felt that way. Einstein's Jewishness did make some uncomfortable and weighed on the professors at Zurich:

> The expressions of our colleague Kleiner, based on several years of personal contact, were all the more valuable for the committee as well as for the faculty as a whole since Herr Einstein is an Israelite and since precisely to the Israelites among scholars are inscribed (in numerous cases not entirely without cause) all kinds of unpleasant peculiarities of character such as intrusiveness, impudence, and a shopkeeper's mentality in the perception of their academic position. It should be said that however among the Israelites that there exist men who do not exhibit a trace of these disagreeable qualities and that it is not proper, therefore, to disqualify a man because he happens to be a Jew....Therefore, neither the committee nor the faculty as a whole considered it compatible with its dignity to adopt anti-Semitism as a matter of policy.[33]

In today's world the very presence of such a document would condemn the faculty to ridicule. At least the document is refreshingly honest in exposing the committee's prejudices. Despite the prejudices that they articulate, the committee followed the policy of not adopting anti-Semitism; they put that policy in writing, and they hired Einstein. The policy was not a one-time exception. The other candidate that they considered, Adler, was Jewish as well, indicating that qualifications were of foremost concern. Yet, if the candidates were not Jewish, no such letter, considerations, or discussions would have been necessary.[34]

[33] Isaacson *Einstein*, p. 152.

[34] Friedrich Adler was the son of Victor Adler, an influential Austrian politician who was once bedmate with of all people the anti-Semite Schonerer. The younger Adler later followed his father into politics. In 1916 opposition to Austria's war efforts motivated Friedrich to assassinate the Austrian prime minister, Karl von Sturgkh. Einstein, anguished by the predicament of his friend, wrote a letter on Friedrich's behalf and perhaps had a role in sparing Adler from the executioner's ax.

While Einstein devoted his intellectual life to relativity, between 1911 and 1914, his personal life was more a study in ergodic systems. After returning from the Salzburg conference, Einstein assumed a position at the University of Zurich. The breakthrough into academia opened more opportunities as other universities started to recruit. It was a short stint in Zurich that proved productive at least outside academia as Mileva gave birth to a second son, Eduard. Shortly after, Einstein took a position in Prague. Neither Einstein nor Mileva felt any attachment to Prague. There, the underlying tension between Mileva and Einstein asserted itself with greater force. They decided to return to Zurich, where Einstein took up a position at his old alma mater, the Polytechnic. The marriage front did not improve in Zurich, and Planck held out an offer in Berlin. By this time Einstein had entered into an affair with his cousin in Berlin. Well, what's a married man to do? Be miserable with his wife in Zurich, or take his family to Berlin, where the presence of a mistress will ensure that everyone is miserable? In 1914 the Einstein family moved to Berlin.

As poison pressed Europe toward war and the pacifist Einstein lent his prestige to the antiwar effort, the house of Einstein followed Europe; it was war. Mileva took the kids and moved into the home of a mutual friend, Fritz Haber. The man who initiated chemical warfare sought to negotiate a truce between the warring couple. It was to no avail. Einstein issued an ultimatum, a detailed code of conduct that would govern their relationship. In essence, Mileva would be tolerated if she assumed the role of an emotionally detached servant. Predictably, the marriage had collapsed. At the Haber home, the terms of separation were set; Einstein was in fact financially generous toward Mileva and his children. Mileva returned to Zurich with the children and remained there for the rest of her life. She was a good mother who did her best to preserve the relationship between the father and the children under difficult circumstances.

VICTORY

The problem of generalizing relativity stymied Einstein because the road passed through mathematical abstractions that he was unaware of. On his return to Zurich in 1912, Einstein turned to his friend Marcel Grossman, also a professor at the Polytechnic. Recall that Grossman specialized in non-Euclidean geometry. After a little digging into the literature, Grossman uncovered the works of Riemann and the Italian mathematician Ricci-Curbastro (1853–1925). Ricci had extended the Riemannian metric, a measurement of distance in curved space, to other constructs known as *tensors* and devised useful notation for performing calculations. Grossman correctly intuited that Riemann and Ricci's work was the vehicle toward the general theory.

Grossman and Einstein worked together. Grossman tutored Einstein in tensor analysis, which ultimately proved to be the gateway to general relativity. But while Einstein and Grossman stood at the gate, and jarred it open, they couldn't pass through. At one point they were so tantalizingly close that one is tortured on discovering that they abandoned the effort. Einstein later explained that he was unable to establish congruity with Newton's law of gravity, a hurdle that the theory must clear. Instead, they arrived at an incorrect theory that they coined *Entwurf*, meaning blueprint, and in 1914, prior to Einstein's departure for Berlin, they jointly published a paper espousing Entwurf.

Einstein knew that his theories were conceptual, developed from principles for which there was evidence, but not certainty. While Newton's mechanics comport with our everyday experience, everyday experience refutes Einstein. The problem is that we observe phenomena in a domain where the effects of relativity are so slight as to be unnoticeable; hence the highly comprehensible and intuitive Newton prevails. Einstein looked for test cases where the effects of relativity would be on display. One such test case, described above, is the bending of light from a star during an eclipse. Another such test case is the orbit of Mercury.

The orbit of Mercury about the sun is not a perfect ellipse. The perihelion, the point where Mercury is closest to the sun, advances with each passage. This is a phenomenon that Newton's laws of motion does not explain. It was a testbed for Entwurf and Einstein put Entwurf to the test.

Assisting Einstein was the one remaining member of the Olympia Academy who had not gone on to better things, Michelle Besso. When the Entwurf calculations of Mercury's orbit did not quite match published data, Einstein shrugged it off. Besso, the patent clerk without a graduate degree, much less a professorship, was less sanguine. He found a mistake that eluded the two distinguished professors, Einstein and Grossmann, but Besso was unable to convince Einstein that there was something amiss with Entwurf.

Einstein continued to enthusiastically promote Entwurf. During the midst of war, in the summer of 1916, David Hilbert, a mathematician who rivals Poincaré in influence, invited Einstein to Gottingen. Hilbert, the Chairman of Mathematics, was worthy of his Gottingen predecessors, Gauss and Riemann. Hilbert wished to learn relativity firsthand from one of its inventors. Einstein obliged. After a week, Einstein was certain that he had converted all of Gottingen to Entwurf disciples. But then his own faith crumbled. It is uncertain what ensued, but in Gottingen Einstein recognized Besso's reservations and admitted that Entwurf was in error.

Despite the flaw with Entwurf, Einstein did convert Hilbert to his cause—a generalization of special relativity that would yield equations of motion in arbitrary reference frames and would treat gravity through spatial relations using the equivalence principle. Hilbert was a convert of such enthusiasm that he decided to embark on the program himself.

In late summer Einstein became aware of Hilbert's efforts. One can imagine something of a sense of rage pounding through Einstein. Later, the public would associate relativity with Einstein and Einstein with relativity. Einstein had already made such an association. Einstein had devoted 11 years of his life to thinking about this problem. He was among the founders of the theory and had advanced it. It was Einstein who explicitly displaced Newtonian mechanics. It was Einstein who teased out the famous formula $E = mc^2$. It was Einstein who proposed the equivalence principle. It was Einstein who, on the basis of the equivalence principle, conjectured that gravity bends light; and it was Einstein who defined the program for generalizing special relativity and incorporating gravity. Through relativity Einstein defined himself as a physicist and as a person, and now David Hilbert (1862–1943), whom Einstein tutored, aimed at kidnapping Einstein's child.

Hilbert, as competitor, was undoubtedly a frightening thought. A list of candidates capable of completing the theory would have included Hilbert, perhaps atop the list. Einstein knew that he could not match Hilbert's mathematical prowess, and Einstein also knew that the final theory would be highly mathematical. Nevertheless, Einstein had some gifts of his own, a keen and unmatched physical intuition. He was all fight. Einstein tuned out the world and retired to his study. There, while World War I raged beyond, Einstein waged his own war.

Einstein returned to his and Grossman's point of departure from Ricci's tensor analysis. He sent letters to Hilbert, informing Hilbert of his progress. In early November Einstein believed that he had conquered the problem. Einstein self-checked his own work by redoing the Mercury calculation that he and Besso performed years earlier. This time, the calculation confirmed the theory. Einstein held a seminar where he announced his result.

On November 15, Einstein received a letter from Hilbert. Hilbert invited Einstein to Gottingen to listen to Hilbert's seminar where Hilbert would announce his result. Enclosed was Hilbert's solution. Einstein responded to Hilbert's letter by declining Hilbert's Gottingen invitation. Einstein also wrote that Hilbert's solution was the same as the one that Einstein had discovered.

Einstein's approach was to generalize Newton's formulation of gravity in a Riemannian framework using tensors. In analogy with Newton's second derivative of the gravitational potential, Einstein used what is known as a contraction on the Riemann–Christoffel tensor. He then set the contraction to zero. It all worked out beautifully as the Mercury calculation confirmed. In Einstein's lecture, he presented this approach—then Hilbert's letter arrived. A subsequent lecture of Einstein's incorporates an additional term known as the "trace term,"[35] one that is necessary for the general theory of relativity, but vanishes when performing the Mercury calculation.

In 1694, a nasty priority debate over the discovery of calculus cast a foul pall over an otherwise celebrated achievement. By contrast, the field equations of general relativity, are known as the *Einstein–Hilbert equations*, and while it may have been otherwise, the protagonists avoided a fight. For his part, Hilbert responded to Einstein's letter with hearty congratulations:

> Cordial congratulations on conquering perihelion motion. If I could calculate as rapidly as you, in my equations the electron would have to capitulate and the hydrogen atom would have to produce its note of apology about why it does not radiate.[36]

[35]The "trace" is a series of terms, but in condensed tensor notation one writes them with a single symbol.
[36]Isaacson, *Einstein*, p. 219.

Hilbert subsequently published his result, stating "The differential equations of gravitation that result are, as it seems to me, in agreement with the magnificent theory of general relativity established by Einstein."[37]

Nearly 100 years after the discovery of the field equations, a debate between Einstein and Hilbert supporters, one that Einstein and Hilbert themselves avoided, continues with each side assiduously claiming that their hero did the deed. It is pointless and misses the real issue of how science gets done. While Einstein may have closeted himself in his study, he took with him the work of Grossman, Riemann, and Ricci. Hilbert may have provided further insight during Einstein's visit to Gottingen. Likewise, Hilbert acquired much of Einstein's knowledge and experience during Einstein's visit to Gottingen. Then and there, Einstein laid out the program that Hilbert pursued. Subsequent letters between the two may have assisted each in their formulation. As if in a sport competition, the audience wants to see a winner, but this was not a sport's competition. This was, in fact, a collaboration, albeit one where the collaborators did not intend to collaborate, resulting in the appropriately named Einstein–Hilbert equations.

So what is the general theory of relativity? It is Einstein's deterministic dream, the cause and effect of motion derived from fundamental laws. General relativity states that an object moves through space along a geodesic, the mathematical term for the shortest path. In conventional Euclidean geometry, a geodesic is a straight line. As Riemann foresaw, general relativity ascribes a geometry to the universe that is non-Euclidean; shortest paths are not straight lines, but curves. The Einstein–Hilbert equations describe how the presence of mass curves space—in the vicinity of mass, shortest paths are curved and the more mass there is, the more curved the shortest path becomes. General relativity views gravity not as a force, but as a geometric curvature of space caused by the presence of mass. The curvature of space is what influences the motion of a small mass in the vicinity of a large mass.

An analogy may prove useful. We have all had the experience of looking at a spoon in a glass filled with water. There appears to be a discontinuity between the handle that lies outside the water and the submerged spoonhead inside the water. What's happening? The water causes the light from the submerged spoon to travel along a different path than the light would if it traveled exclusively through the air. The different paths of light, followed by that emanating from the submerged head and that emanating from the handle, cause the appearance of a discontinuity.

Picture a tank of water with a fish in it. The light from the fish travels along a path that minimizes the time that it takes to arrive at your eye. As water slows down the speed of light, the optimal path is not necessarily a straight line. Instead, as depicted in Figure 9.3, it would take a shorter time if the pathway hastens its departure from the water than if the light were to follow a straight line. (Compare with Figure 9.2, depicting the bending of light by the sun.)

Let's do a thought experiment. Suppose that the presence of water does not slow down the light, but instead speeds up time in the following manner.

---

[37] Ibid. p. 222.

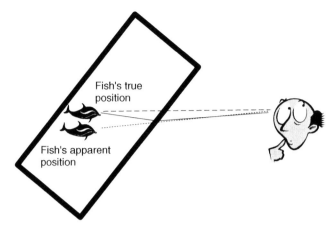

Figure 9.3. The fastest path for light to travel from the fish to the observer is along the solid segments, and the fish is not where it appears to be.

Take two identical clocks and verify that in air they measure time identically. Next, take one clock and place it in the water. Then the presence of water causes the clock to speed up in comparison with the clock that is still in the air. The situation depicted in Figure 9.3 applies. The light hastens its way to the eye, not by following the straight line, but taking the path that skidaddles out of the water because in the water, time gets chewed up at a faster rate.

Let's do one final thought experiment. One could model the phenomenon by imagining that rather than slowing down the speed of light, the presence of water alters spatial distance in the following sense. Take a measuring rod and measure the length of the fishtank wall from the outside. Next submerge the rod inside the tank and measure the same wall. The measurement taken from the submerged rod is longer, meaning that the presence of water has increased the distance between two fixed points. By assuming that the light travels along the path of shortest distance, we arrive at the exact same situation depicted in Figure 9.3. The shortest path is not the direct line, but the alternative path that readily exits the water rather than slogging a long path through the water.

These thought experiments demonstrate that one could explain the phenomenon, the alteration of the shortest path from a straight line to something different, by proposing that the presence of water alters spatial and temporal relations. In fact, the presence of water does neither; the two clocks in the first thought experiment will measure the passage of time identically, and the wall measurement within and outside the tank of the second experiment will likewise be the same. But the presence of matter does alter temporal relations; identical clocks behave differently, slower near a body of matter than distant from the body. Also the presence of matter does change the measurement of distance. How does this occur? Nobody knows, but every experiment performed to date confirms that it does with the resulting curvature of space precisely as described by the Einstein–Hilbert field equations.

THE EXPERIMENT

On July 19, 1914, after surmounting difficulties in arranging financing, the astronomer Erwin Freundlich (1885–1964) departed for Crimea, where the impending solar eclipse would allow him and his colleagues to perform Einstein's experiment. This experiment would confirm or disprove Einstein's theory that gravity bends light. Shortly after crossing the Russian boarder, World War I began, and Russian soldiers intercepted the expedition. Imagine the attempt to explain to Russian officials imbibed with wartime suspicions that the optical equipment, cameras, film, telescopes, and tripods in Freundlich's possession were not for espionage, but for a scientific experiment to determine whether gravity bends light. That conversation did not go so well. The Russians gave little credence to Freundlich and imprisoned the members of the expedition. They remained imprisoned well beyond the date of the eclipse, losing their opportunity to test Einstein's theory.

In 1917, using the Einstein–Hilbert field equations, Einstein once again calculated the shift of a sun-aligned star's apparent position. The updated result yielded a larger shift than that of 1911. On the opposite side of the war's front, across the channel, the English physicist, Arthur Eddington (1882–1944), followed the works of Einstein. The possibilities described by general relativity, a universe based on non-Euclidean geometry in which matter defines space and time while space and time govern the motion of matter (and light), enthralled Eddington. As the war continued, Eddington prepared an expedition to carry out Einstein's experiment for the upcoming 1919 solar eclipse. In the midst of the war, Eddington managed to arrange the necessary funding and procure the necessary equipment for an experiment that would, if successful, glorify a German.

Mercifully, the war ended before Eddington departed from England. Eddington worried about another enemy, clouds. Eddington knew that this was his moment; should clouds obscure the sky, the moment would be lost. To reduce this risk and collect additional data, Eddington arranged for two parties to take observations. One party went to Sobral, Ecuador, and the other, with Eddington himself as the expedition's leader, went to Principe, an island near the equator 150 miles from the coast of Gabon, Africa. On the day of the eclipse, the shadows of clouds cast their ominous presence across the island. During the eclipse, Eddington was far too engaged with changing photographic plates to note whether clouds completely obscured the sky or if the nature would give permission for starlight to pass on to the photographic plates. In the end, nature cooperated and allowed Eddington's camera to glimpse at her. Both the Ecuadoran party and the Principe party obtained the evidence they sought and returned the precious photographic plates to England.

It was not a fastball down the middle of the plate, but a rather difficult call. There were three photographs of the heavens from which to determine the shift, if there was a shift at all. After a belabored effort, the first plate revealed a shift slightly greater than that predicted by Einstein, the second slightly less, and the third much less. Eddington discarded the third as it was not in focus, and then averaged among the remaining two. Perhaps it was a rigged election; Eddington was not an impartial referee but an Einstein fan. Nevertheless, subsequent experiments have verified that he made the right call, Einstein and Hilbert had pitched a strike.

THE FAMOUS AND THE INFAMOUS

A world recovering from war's devastation turned its attention to the heavens. There was a distinctive buzz among physicists as they awaited Eddington's judgment. In London's Burlington House, the home of the Royal Society that Newton presided over two centuries before, the Royal Society and Royal Astronomical Society held a joint meeting. There Sir Frank Dyson, the president of the Royal Astronomical Society, made the announcement that sealed Einstein's triumph over Newton:

> The results of the expeditions to Sobral and Principe leave little doubt that a deflection of light takes place in the neighborhood of the sun and it is the amount demanded by Einstein's generalized theory of relativity.[38]

Then the buzz among the physicists turned into unrestrained whooping among the masses. Overnight, Einstein was an international celebrity.

It is not often that a sport's correspondent specializing in golf reports on the subject of physics. But the *New York Times* wanted coverage of the sensational news and their only correspondent in London was Henry Crouch, a sport's columnist. Unable to get into the mobbed Royal Society meeting, Crouch interviewed Eddington and then sent a dispatch back to New York. The *New York Times* gave the story front-page coverage and followed up for days. When physics makes the front page and golf takes second fiddle, there is, indeed, a strange karma wafting through the cosmos. What did this karma portend?

Could Hollywood have cast a better Einstein than Einstein? The unkempt hair, the exaggerated facial features, the deep eyes with a gaze that was both quizzical and authoritative were all dressing for a sharp wit that the public lapped up. After several failed efforts at explaining relativity to newspaper correspondents hoping for an easy sound byte, Einstein changed tack: "When you are courting a nice girl, an hour seems like a second. When you are sitting on a red hot cinder a second seems like an hour. That's relativity." Undoubtedly, Einstein liked the attention. But he was equally puzzled and on occasion found what he considered to be hero warship unsettling. Einstein's quip "To punish my contempt for authority, fate made me an authority myself" summed up his mixed attitude toward his public role.

Whatever illusions of war accompanied Europe on her way in, reality banished the illusions on the way out. The German surrender halted the killing, but not the anger. The victors acting as judge and jury apportioned blame not to the

[38] Isaacson, *Einstein*, p. 261.

leadership who were calling the shots, but to all Germans. Versailles demanded from Germany wartime reparations that taxed ordinary Germans and crippled their economy. Perhaps if the republic had time to mature on its own, it would have endured. But as with the French Revolution, with intervention from the outside, the republic never had a chance.

The republic had little room to maneuver. The only way to meet the Versailles obligation was to head to the printing press. As excess German marks sloshed around Europe, the currency plummeted and inflation became unbearable. In 1923, prices for everyday necessities like bread would double in 2 days. Einstein himself felt pinched. Mileva complained that Einstein's alimony payment didn't cover expenses. Mileva's complaint rang true as the money that Einstein dutifully forwarded to Mileva became worthless when converted to Swiss francs. But there was nothing Einstein could do.[39] He had no complaints because in comparison with the typical German, Einstein was well off. Old soldiers who survived the war returned home and faced deprivation as hunger and malnourishment prevailed. The year 1923 was the year that Hitler committed treason.

In the fall of 1924 administrative process forwarded Hitler's parole request to Georg Neithardt, the judge who presided over Hitler's initial trial. Hitler had been found guilty of treason for his role in a failed putsch against the Bavarian state government. It was the intention of Hitler and coconspirator General Ludendorff to use Bavaria as a base from which to gain control of Germany. Ancillary to his crime of treason, Hitler had a central role in plotting and executing the putsch; the use of a private paramilitary organization armed with illegal weaponry; the kidnapping of several government officials and private citizens; inciting a confrontation against a police force, resulting in the killing of innocents; and, in addition to the abovementioned crimes, bank robbery. For the crimes, Neithardt sentenced Hitler to a maximum sentence of 5 years with the opportunity of parole. Less than 8 months into his sentence, Neithardt received Hitler's formal request for parole.

Georg Neithardt held sympathies that mirrored Hitler's. The economy was in shambles as the lords of Versailles mugged Germany. The foreigners left Germans not only with an empty belly but also with an empty spirit—this was an affront to German pride. Yet the government did not stand up for its citizens and trembled in front of France and England. Domestically, under the government's nose—indeed, under the guise of democracy—Marxists spread their cancerous propaganda, threatening the law and the established order. Patriotism motivated Hitler's putsch attempt, and while his actions may have been misguided, they were more than understandable. Neithardt signed the papers that released Hitler and set history on an ill-fated course.

How was it that Hitler, a lonely failure incapable of sustaining any social relations and having no prospect prior to the war, came to lead a political movement? The Versailles treaty demanded a hollow German army. Just as mold hides

---

[39] In 1923, in accordance with their divorce agreement, Einstein gave Mileva his entire earnings from the Nobel Prize. The award was a nice sum, enough to purchase a home and retire, and it was paid in Swedish kroner.

itself behind the walls of an otherwise pristine home, a group of German officers wished to hide an illegal military force within Germany. These officers sponsored paramilitary groups disguised as sports clubs. Among the organizers was General Ludendorff, the man that Kaiser Wilhelm ordered to lead Germany's 1918 offensive against the Allies.

Ludendorff's goal was to rid Germany of the republicans and establish a military dictatorship through a rightist political party as a front organization. The SA, Ludendorff's own paramilitary organization, was to be the muscle, while the German Worker's Party (DAP), a rightist political party, would attend to the political sphere. The design required indoctrination of the SA in rightist political philosophy. It is in this endeavor that Hitler showed talent, and his talent came to the attention of Ludendorff. Hitler was a moving speaker with a down-to-earth style. Hitler's delivery so forcefully conveyed his own conviction to the cause that listeners consumed its content. Not only did this ability prove valuable in indoctrinating young SA recruits; it also became invaluable to the DAP. Hitler became the central attraction at staged party rallies; he could draw the crowd. It was this ability to speak that many years later allowed Hitler to attain power.

On December 20, 1924, when Hitler left prison, his rightist party was no more than a fringe group viewed suspiciously by industry, the government, and most of the population. Although Hitler was able to consolidate other rightist parties under his leadership, on October 28, 1929, the political mainstream flowed away from Hitler. On the next day, October 29, the stock market in New York collapsed and precipitated events that would alter the flow of mainstream German politics.

Germany was hard-hit by the financial crisis. Loans from Wall Street banks financed Germany's war reparations as well as an economic recovery that was underway. Overnight, further loans were out of the question; in fact, banks demanded immediate repayment of outstanding loans and Germany was the first to join America in a spiral toward an economic depression. The world's misery was Hitler's nourishment.

Disillusionment with the Social Democrats allowed for gains among the other political parties. A string of elections had ended in a political equilibrium with the electorate split into three somewhat equal camps, a communist and socialist camp, Hitler's Nazi party, and an assortment of conservative parties, the largest of which was Zentrum. Legislation became impossible as no party had the strength to push its agenda through the Reichstag. Hitler never believed in democracy and would use democracy's institutions against itself for his own ambition. With Hitler acting as spoiler, successive governments during the years of 1931 and 1933 were dysfunctional.

Elections in 1932 gave the Nazis the largest share of the vote at 38%. Sizing up Hitler as a divisive figure, the respected president Paul von Hindenberg refused to call on Hitler to form a government. The economic crisis continued and Germany continued, to drift about rudderless. Another election in 1933 resulted in yet another split decision that once again left Germany adrift. While the Nazi party's support decreased to 36% of the electorate, it still had the largest share. Einstein, the deterministic physicist, also sought cause–effect relationships in politics: "An empty stomach is not a good political adviser." Hindenberg softened his stance. Not only did he call on Hitler to lead a government; in addition, recognizing that the Reichstag had become obstructionist, Hindenberg agreed to its temporary suspension:

> With satanic joy in his face, the black-haired Jewish youth lurks in wait for the unsuspecting girl whom he defiles with his blood, thus stealing her from her people. With every means he tries to destroy the racial foundations of the people he has set out to subjugate.
>
> And in politics he begins to replace the idea of democracy with the dictatorship of the proletariat.
>
> In the organized mass of Marxism he has found the weapon which lets him dispense with democracy and in its stead allows him to subjugate and govern the peoples with a dictatorial brutal fist.[40]
>
> Today it is not princes and princes' mistresses who haggle and bargain over state boarders; it is the inexorable Jew who struggles for his domination over nations. No nation can move this hand from its throat except by the sword. Only the assembled and concentrated might of a national passion rearing up in its strength can defy the international enslavement of peoples. Such a process is and remains a bloody one.[41]
>
> As opposed to this, we National Socialists must hold unflinchingly to our aim in foreign policy, namely, to secure for the German people the land and soil to which they are entitled on this earth. And this action is the only one which, before God and our German posterity, would make any sacrifice of blood seem justified...in so far as we have shed no citizen's blood out of which a thousand others are not bequeathed to posterity. The soil on which some day German generations of peasants can beget powerful sons will sanction the investment of the sons of today.[42]
>
> We stop the endless German movement to the south and west, and turn our gaze to the land in the east...and shift to the soil policy of the future.
>
> If we speak of soil in Europe, we can primarily have in mind only Russia and her vassal boarder states.
>
> Here Fate itself seems desirous of giving us a sign. By handing Russia to Bolshevism, it robbed the Russian nation of that intelligentsia which previously brought about and guaranteed its existence as a state. For the organization of a Russian state formation was not the result of the political ability of the Slavs in Russia, but a wonderful example of the state-forming efficacy of the German element in an inferior race....For centuries Russia drew nourishment from this Germanic nucleus to its upper leading strata. Today it can be regarded as totally exterminated and extinguished. It has been replaced by the Jew....The giant Empire in the east is ready for collapse. And the end of Jewish rule in Russia will mean the end of the Russia as a state.

[40]Hitler, *Mein Kampf*, p. 325.
[41]Ibid., p. 651.
[42]Ibid., p. 652.

Our task, the mission of the National Socialist movement, is to bring our own people to such political insight that they will not see their goal for the future in the breath-taking sensation of a new Alexander's conquest, but in the industrious work of the German plow, to which the sword need only give soil.[43]

The man who penned these words assumed his position at Germany's helm and used his position to execute precisely the program that he describes in his book *Mein Kampf*.[44]

EXODUS

In 1933, Max Planck was busy not with science, but with the painful process of executing Hitler's recently enacted law prohibiting the employment of Jews in civil service, and that included all university positions. Academia answered the law with deafening silence and then acquiesced. With great distaste, Planck, too, did what was required of him. He did his best to place the released Jewish faculty in other institutes across Europe. Einstein was one less individual that Planck had to worry about. Einstein was off to Princeton's newly found Institute for Advanced Studies, a position for which Einstein had been negotiating for some time. It was not an easy move; Einstein was comfortable in Berlin and under different circumstances would have continued there. But fate forced the decision, and Einstein resigned himself to fate.

By contrast, fate caused a stirring of internal contradictions within Haber. Haber spent a lifetime in service to Germany. His advances created whole industries that employed thousands and fed the hungry. His inventions yielded the armaments that Germany used to defend itself. For his efforts to win World War I (WWI), much of Europe viewed Haber as a war criminal. After the war at Kaiser Wilhelm Institute, Haber directed the world's most productive research center in the field of chemistry. After all that he had done for Germany, the fatherland spat Haber out of its belly. At the age of 64, Haber would redefine his identity.

Haber returned to his Jewish roots. At a meeting between Haber and his fellow scientist Chaim Weizmann, Weizmann offered Haber a position in Palestine's newly established Hebrew University. Several months later while in transit to Palestine, Haber suffered a fatal heart attack. His final wish before his death was to be buried in Germany aside his dead wife with the words "he served Germany" on his headstone.[45] For all his acumen in science, Haber

---

[43] Ibid., p. 655.

[44] Hitler did not have his father's talent for choosing a name. Hitler's choice for a title was *Four and a Half Years (of Struggle) Against Lies, Stupidity, and Cowardice*. Hitler's publisher persuaded Hitler to change the title to *My Struggle*. The book itself is discombobulated drivel, antics, and rage that evidence does not support. Nevertheless, there was a sector within German society that did support the ideas. Within that sector was the judge presiding over Hitler's trial, Georg Neithardt. The project began while Hitler was imprisoned and prison authorities were aware of it. Some of *Mein Kampf's* contents may well have been known to both prison authorities and Georg Neithardt prior to Hitler's release from prison.

[45] Haber's first wife and later his son committed suicide.

failed to rewire his psyche. Germany rejected Haber, but Haber couldn't reject Germany.

Death spared Haber what would have been a certain nightmare. Had Haber lived, he would have confronted guilt as the inventor of an insecticide that the Nazis used as a gassing agent to kill Jews in concentration camps. Victims included members of Haber's extended family.[46]

Being preeminent in mathematics and physics, Germany attracted Jewish talent from eastern Europe. Among many affected by Hitler's anti-Semitic laws were three Hungarian physicists: Leo Szilard (1898–1964), Eugene Wigner (1902–1995), and Edward Teller (1908–2003). Szilard was the first to envision the harnessing of Einstein's equation, $E = mc^2$, for the purpose of energy production. While in England, he patented the idea of extracting energy from fissioning atoms using neutrons as agents to sustain the fission process. Szilard outlined a chain reaction whereby neutrons from one fissioning atom bombard neighboring atoms, enabling the neighboring atoms to fission and release their neutrons; the released neutrons in turn perpetuate the fission process. Szilard was aware that a successful chain reaction could release enormous amounts of energy in the form of an atomic explosion. Aside from the physics of the bomb, Szilard personally knew some of the German scientists who were working to develop nuclear power. The nightmare he envisioned, should the Nazis succeed in their efforts to build a bomb, stirred Szilard into action.

Szilard and Teller thought it imperative that their newly adopted nation, the United States, begin research into atomic weaponry. As virtual unknowns outside of small academic circles, it was highly improbable that the recent emigrés could exert the required political influence to jump-start an exotic weapon's development program. The emigrés showed they not only had a flare for physics, but for politics as well and found an alternative voice. Who better to warn the U.S. president of the risks in not answering the German research effort than the most famous physicist in the world, the author of the equation $E = mc^2$, Szilard's friend and colleague and himself a German refugee, Albert Einstein.

In the summer of 1939, on the island of Manhattan, Szilard, Wigner, and Teller met Einstein, with whom they discussed the science and politics of the upcoming nuclear age. Concerning applications of science toward weapon development, Wigner harbored a moral conflict and was somewhat ambivalent. The others, while sharing Wigner's moral concerns, did not share his ambivalence. The ugly years in Germany where Einstein saw firsthand the rise of Hitler transformed Einstein the pacifist into Einstein the pragmatist. Szilard and Teller did not have to persuade Einstein of anything, for he readily agreed with their program. Szilard and Einstein authored a letter, signed only by Einstein, and secured a deliverer to Franklin Roosevelt:

---

[46]Death came too late to spare Planck his emotional suffering. After his eldest son's death as a WWI victim, Planck became particularly attached to his second son, Erwin, a surviving prisoner of war. In 1945, the Nazis arrested and in short order executed Erwin after suspecting his involvement in a plot to kill Hitler.

In the course of the last four months it has been made probable —through the work of Joliot in France as well as Fermi and Szilard in America—that it may be possible to set up a chain reaction in a large mass of uranium, by which vast amounts of power and large quantities of new radium-like elements would be generated. Now it appears that this could be achieved in the immediate future.

This new phenomena would lead to the construction of bombs, and it is conceivable—though much less certain—that extremely powerful bombs of a new type may thus be constructed. A single bomb of this type, carried by a boat and exploded in a port, might very well destroy the whole port together with some of the surrounding territory. However, such bombs might prove to be too heavy for transport by air.

This letter served its purpose, and the United States, by virtue of well-founded paranoia, began its quest for the bomb. The code name for this program was the *Manhattan project*. While it took about 2 years to overcome the inertia of bureaucracy, the Manhattan project eventually assembled an international dream-team of physicists and mathematicians. Leading the project was the American-born J. Robert Oppenheimer (1904–1967). Following Hitler's lead, fellow fascist Mussolini enacted anti-Semitic laws prompting Enrico Fermi (1901–1954) to cross the Atlantic in order to protect his Jewish wife. Fermi was an invaluable member of the Manhattan project. Hitler's march in Denmark prompted the half-Jewish Niels Bohr to leave Copenhagen and enlist. Joining Oppenheimer, Szilard, Wigner, Teller, and Niels Bohr were Jon von Neumann and Stanislaw Ulam. The man most responsible for recruiting this unequaled arsenal of talent was Adolf Hitler.

While the American team endeavored to develop fission, American intelligence endeavored to determine the German's nuclear research activities. The most capable of the German scientists and the scientist that American intelligence most feared was the father of the uncertainty principle, Werner Heisenberg. In 1941, Heisenberg's last words with Niels Bohr, prior to Bohr's departure for the United States, concerned the morality of directing scientific research toward war efforts. Heisenberg initiated the topic, as this moral uncertainty weighed measurably on Heisenberg's conscious. The topic so discomfitted Bohr that the conversation never developed. Each went his own separate way, Bohr to work on the bomb for the United States and Heisenberg to lead Germany's nuclear effort.

The German effort did not receive the resources of the Manhattan project, in part because the scientists involved did not believe in the success of the project. Who would advocate for an all-out effort if not the scientists? Perhaps, due to moral discomfort, or perhaps because he did not believe in the feasibility of a chain reaction, Heisenberg was not fully focused on the bomb and remained active in his academic research.

In 1944, as Heisenberg delivered a lecture at Einstein's alma mater, ETH in Switzerland, he was certainly unaware that in the audience, posing as a physics researcher, was an American spy, Morris Berg. Heisenberg calmly delivered his lecture in his area of academic research, S-matrix theory. Heisenberg would not have been so calm had he known that Morris Berg carried a gun and that America's spy agency, the Office of Strategic Services (OSS), authorized Berg to assassinate Heisenberg. In the end the academic nature of Heisenberg's talk as well as Heisenberg's postseminar remarks that he believed the Germans would lose the war, convinced Berg that Heisenberg presented no threat. Seeing no prospective nuclear trigger in Hitler's hand, Berg didn't pull the trigger in his hand.[47]

As for Einstein, other than the famous letter to Roosevelt, he had no role in the development of the bomb and previously didn't foresee his theory of general relativity having any economic or military consequences. Once the bomb exploded and the deed was done, Einstein remarked, "If I had known it would lead to this, I'd have become a plumber."

THE END

In 1940, a movie camera caught for posterity Hitler's joy as he marked Germany's triumph over France by goose-stepping on the Champs de Lysee. The film outlasts the very shortlived Third Reich.[48] By 1944 it was clear to Germany's own officer corp that the nation was en route to defeat. In 1944, those very officers who assisted Hitler in his military buildup and assiduously planned his offensives now plotted against Hitler's life. On July 20, 1944 Colonel Claus von Stauffenberg arrived at Hitler's personal fortress in Bertchessgarten. In his hand was a briefcase containing an explosive. Colonel Stauffenberg proceeded to a bunker where Hitler held meetings with a small inner circle who decided the course of the war. Colonel Stauffenberg placed the briefcase next to Hitler under a conference table and left. From outside the bunker he heard the explosion and assumed that Hitler was gone.

The heavy table buffered the explosion, giving Hitler more time to carry out the design he described in *Mein Kampf*. Was Hitler aware of the little time remaining? Certainly his staff must have informed him of the news on the front. To the west, American and British troops succeeded in landing armies on the continent, breaching German defenses. On the eastern front the situation was more dire.

In July 1942 the Russians ended their initial scorched-earth retreat with a do-or-die defense of Stalingrad. It was a brutal inner-city Stalingrad campaign. Total casualties exceeded 2 million and included a significant number of civilians. Within the city food supplies dwindled, and citizens ate the bark off of trees. Several months after the battle lines were firmly etched across the city of

---

[47]Morris Berg was an interesting character. After graduating from Princeton with a degree in languages, he went on to become a catcher in Major League Baseball. During his career he earned a law degree and passed the bar exams. After his baseball career Berg was able to parley his gift for languages into a successful spy career.

[48]Can anyone recall the first and second (Reichs)? Hitler seeded the subsequent dissolution of the third by basing it on a pure fabrication.

Stalingrad, the Russians were able to cut off the German army's supply route. German troops faced a bitter cold winter where only frozen air and death were plentiful. Frozen soldiers, deprived of adequate food, clothing, shelter, fuel, and munitions, could fight no longer. Under Field Marshal Paulus, the remnants of the German 6th Army, 91,000 out of an initial 250,000 men, surrendered. With this victory, the Russians seized the initiative. Within a year they regained control of the Russian motherland, and pursued the Germans back to the German fatherland.

When Hitler heard of the German surrender, he spoke his mind: "They should have shot themselves with their last bullets." This was not rhetoric, but Hitler's worldview—struggle, struggle, struggle, and if you can no longer struggle, then death is deserving. If Hitler was aware that the sands of time would soon bury the Third Reich, it was irrelevant. He would never compromise but continue his struggle to the bitter end. Hitler summed up the struggle in the concluding paragraphs of *Mein Kampf*:

> A state which in this age of racial poisoning dedicates itself to the care of its best racial elements must some day become the lord of the earth.

> May the adherents of our movement never forget this if ever the magnitude of sacrifices should beguile them to an anxious comparison with the possible results.[49]

After Stalingrad, Hitler continued his efforts to eradicate racial poisoning and prioritized genocide alongside national defense. While Allied armies encircled Germany, Hitler approved the use of precious rail transport and fuel for the purpose of deporting Jews, gypsies (Roma), and other undesirables. The Germans operated Auschwitz until the summer of 1944, when the allied armies locked Germany within their closing grip. Concentration camps were no more than death factories where Nazis introduced assembly-line practices to eliminate their victims. The killing agent of choice was Haber's poison gas, Zyklon B. The European Jewish community did not survive. Only a trace of Jews remained, living among ghosts.

The Russians reached Berlin first. Unable to carry on the Darwinian struggle, Hitler ended it with the only possible ending and shot himself.

RELATIVITY

It is Einstein's 1905 paper "On the Electrodynamics of Moving Bodies" where Einstein makes his relativistic debut. The stage had been set predominantly by Poincaré and Lorentz, but Einstein has his own approach. Einstein first presents an experiment; consider the electrodynamics of the following two cases: (1) a magnet moves over a conductor that is at rest, and (2) the conductor moves over the magnet at rest. The point of the experiment is that the physical results are

---

[49]Hitler, *Mein Kampf*, p. 688.

the same in both cases, but that a classical interpretation of Maxwell's yields asymmetries.

Now that he's identified a problem, in the next paragraph, Einstein gives his fix, which stamps relativity with Einstein's mark:

Examples of this sort together with the unsuccessful attempts to discover any motion of the earth relatively to the light medium suggest that the phenomena of electro-dynamics as well as mechanics possess no properties corresponding to the idea of absolute rest. They suggest rather that, as has already been shown to the first order of small quantities, the same laws of electrodynamics and optics will be valid for all frames of reference for which the equations of mechanics hold good.[50] We will raise this conjecture to the status of a postulate (The purport of which will hereafter be called the "Principle of Relativity"), and also introduce another postulate, which is only apparently irreconcilable with the former, namely, that light is always prop-agated in empty space with a velocity c which is independent of the state of motion of the emitting body. These two postulates suffice for the attainment of a simple and consistent theory of the electrodynamics of moving bodies based on Maxwell's theory for stationary bodies. The introduction of a "luminiferous ether" will prove to be superfluous inasmuch as the view here to be developed will not require an "absolutely stationary space" provided with special properties, nor assign a velocity vector to a point in which the electromagnetic processes take place.

In the remainder of this paper, Einstein carries out his program. Poincaré and Lorentz have already worked out many of the results of this paper. It is Einstein's approach that differs. For Einstein, the results follow from the two postulates, the principle of relativity and the constant speed of light, regardless of the relative motions of the observer and the emitting body. Einstein's approach is not purely pedantic and yields substance of its own. In particular, Einstein gives the frequency and energy content of light in arbitrary reference frames: results that guide Einstein to the equation that defines him in the public eye, $E = mc^2$. In a follow-up paper, "Does the Inertia of a Body Depend upon Its Energy Content," Einstein presents this equation along with supporting arguments.

The postulates by themselves, in particular the principle of relativity, appear quite innocent. But then Einstein jolts the reader's intuition, and one senses the oncoming revolution. The jolt is to our sense of simultaneity, a sense that one needs not invest effort into formally defining because it is instinctive. If we imagine an experiment in which the experimenter releases two identical eggs from the same height at the same time, we easily envision the eggs splattering on the ground at the same time. The simultaneity of their release guarantees that they strike the floor simultaneously. There is no need for a further discussion of simultaneity because it is obvious.

[50]Here Einstein offers a slight peek at what he knew of Lorentz' works. In his 1895 paper, Lorentz had shown that Maxwell's equations to a first-order approximation satisfy the principle of relativity. So Einstein was aware of Lorentz' 1895 work. Poincaré encouraged Lorentz to finish the job and demonstrate not just to the first-order approximation but also that in fact Maxwell's equations do satisfy the principle of relativity. In 1904, Lorentz achieved that result. When Einstein wrote his 1905 paper, Einstein was unaware of Lorentz' 1904 result.

Einstein upsets our instinct. Let us take an example of the American physicist Richard Feynman (1918–1988), who knew how to communicate to the *Star Trek* generation. Imagine a spaceship passing by a planet at very high but constant speed in a constant direction. The spaceship is a disk, like a flying saucer. At the disk's dead center is a laser that, on the push of a button, simultaneously emits light to the front and back of the saucer, where front is the direction of travel. The space traveler pushes the button and observes. Because both lightbeams travel exactly the same distance to reach the front and back walls and light travels at the same speed, the forward and backward beams strike the walls of the spacecraft at the same time (see Figure 9.4). These strikes are simultaneous events. Is that true for the observer on the planet?

For the observer on the planet, the distance traveled by the forward-moving beam and backward-moving beam are not the same. As the beam travels toward the spacecraft walls, the spacecraft moves forward. At the time of the backward strike, the back wall is a bit closer to the original point of emission. Meanwhile, the forward-moving beam has not yet struck the front wall because the front of the spacecraft is farther from the point of emission. Since light travels a shorter distance to get to the back wall, it arrives sooner (see Figure 9.4). The two strikes, which are simultaneous to the space traveler, are not simultaneous to the planet-based observer.

The fact that simultaneity holds for one observer while not for the other has further implications. It confounds the very meaning of time. The observer on the spacecraft could calibrate a timekeeping instrument by the time it takes for the laser to strike the wall. Whichever wall is used is irrelevant. By contrast, the observer on the planet could not use the wall strikes to calibrate a timekeeping instrument. Calibration by the front and back strikes gives a different result.

Protests surface; this is some sort of trickery. Events are either simultaneous or they are not. Events simply cannot be simultaneous to one observer but occurring at different times to another. Furthermore, time has only one measurement, not

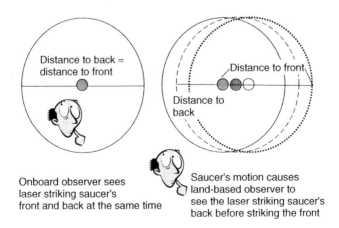

Figure 9.4. Simultaneity is not universal.

more than one. One of these observers must have got it wrong. There must be some preferred eye from which to calibrate time and space.

Nature leaves no hint of a supreme eye; there is not a preferred point of observation. As described in Einstein's words above, "The phenomena of electrodynamics as well as mechanics possess no properties corresponding to the idea of absolute rest." If there is a preferred point, where is it? Two spacecraft in deep space pass each other in uniform motion. Observers on both space craft conclude that their spaceship is at rest while the other is moving. Who determines which observer is correct? There is no universal dictator. Rather than designating a supreme eye, nature is democratic and leaves everything to the eye of the beholder. The measure of time and space, as we will see below, truly differs among different observers.

But nature has structure. Beyond the beauty of structures that we observe in nature, there is utility in structure, without which we couldn't function. If there is no supreme eye, how does nature structure itself? While there is no universal eye, there are universal relations between observers, and these relations are the underlying structure. General relativity gives the most general form of the relations as we understand them. Special relativity gives relations among observers under special conditions that we give below.

Let's give further meaning to the notion that nature has no supreme eye. Suppose that there are two space travelers on separate spaceships. It will be helpful to give them names, so we'll call one O and the other O'. Each has identical equipment for measuring distance and time, and the equipment has been tested at a common location. After running through tests to verify their measurements, each embarks on their own mission, after which they are to meet at specified space–time coordinates. If O and O' know nothing about relativity, and each proceeds to the rendezvous coordinates, there is a good chance that they never meet, even though both have in fact followed instructions.

What has happened? O and O''s perceptions of space and time are different. Even though they have identical equipment for measuring space and time, their instruments perform differently. If O and O' are to meet, they must convert their measurements to a common reference frame. General relativity gives the means to perform this conversion, while special relativity allows for the conversion in a special case.

SPECIAL RELATIVITY

The Principle of Relativity and Symmetry

Special relativity applies to observers in inertial reference frames. Recall that this means that the observers' relative motion with respect to one another is fixed; that is, each perceives the other moving at constant speed and direction. This case is natural because it is the setting for the principle of relativity; two observers moving in inertial frames describe physical phenomena within their inertial frame identically. Let O and O' be the observers. If O describes an experiment to O', one that is totally housed within O's inertial frame, then both perform the experiment they will obtain identical results.

For example, if each nails one end of a spring (assume the properties of the spring to be identical) to a point on their inertial reference frame and applies the same force to the spring, their reported measurements of length will be identical. Alternatively, if one is not in inertial frames, let us say, if O′ accelerates in the direction of the spring's length, then their reported measurements are not the same. O′'s spring is subject to not only the force applied by O′ but also the force on O′'s accelerating reference frame; accordingly, O′'s measurement differs from O's.

As the principle of relativity applies only to inertial reference frames, it is natural to constrain the analysis to inertial reference frames. General relativity extends the analysis to noninertial reference frames and requires much more sophisticated mathematics.

There is another restriction on special relativity applied for similar reasons, namely, homogeneity of space and time. By this it is meant that the properties of space do not alter from point to point and time to time and are also independent of orientation. Under these conditions, identical experiments yield identical results at different locations and orientations. O and O′ are at different locations and orientations (O′ points the spring in a different direction). Yet they perform the spring experiment and obtain identical results.

Homogeneity forces the assumption that the observers and any of their apparatus do not affect the properties of space and time around them. It explicitly excludes gravity, which, we all observe, affects space nonhomogeneously. The spring experiment would yield different results by location and orientation under the influence of gravity.

As a side remark, Einstein recognized that through the equivalence principle (equivalence between acceleration and gravity), if he could mange to extend the analysis of special relativity to the case of accelerating inertial frames, he would be able to incorporate gravity as well. The result is general relativity.

Returning to homogeneity, the assumption allows one to partition any experiment into pieces, and sum the pieces together to get a final result. An example gives meaning to this concept and shows its utility. Let us consider the spring experiment in which we wish to determine as a final result the time that the spring is a predesignated length with a constant force acting on it. We can arbitrarily partition the experiment into breakpoints where we make new time measurements by resetting the clock at each breakpoint as the spring expands toward the final target. The final result may be obtained by adding up all the results at each breakpoint. Homogeneity of time allows us to add the times.

Next, let O have a partner for each breakpoint. Each partner performs the entire experiment at a different location and different time. Each partner reports results for a unique breakpoint, and the results are summed. The results concur with the final result as though performed solely by O without breakpoints. Not only that; the sum of positions and times up to any arbitrary breakpoint is exactly what O would observe at that breakpoint.

Homogeneity allows one to consider any sequence of events as a sum of subevents in which one arbitrarily reinitializes the time and location of the subevents. The additivity of subevents is mathematically very useful.

The Lorentz Transformations

Enough probing about the body of relativity; let's dig to its heart. Let our two space travelers travel in inertial frames. Suppose that O communicates to O' a certain event that he plans in the future, perhaps a fireworks display, and requests that O' observe the event firsthand through his telescope. O passes on the coordinates of the event so that O' knows when and where to look. This time, using the Lorentz transformations, O' converts O's coordinates to those that are commensurate with the inertial frame in which O' takes measurements. When O' peers through the telescope, there are fireworks at the exact point and time that the Lorentzian calculations indicate. Here we derive the Lorentz transformations, the formulas that convert the space–time coordinates of one inertial reference frame to another.

Following Einstein, the derivation is a result of the two postulates, the principle of relativity and a fixed measurement of the speed of light among all observers regardless of their relative motion with respect to the emitting source. The derivation also uses homogeneity. We note that Poincaré also derived the Lorentz transformations; in fact, Poincaré was the first to name the transformations after Lorentz.[51] Poincaré's derivation is more mathematical and more general than Einstein's. But while Poincaré reveals an elegant mathematical structure, Einstein reveals a simple physical concept. Einstein motivates the derivation as described below.

If there is not a common coordinate system to which both space travelers refer, how can one discuss a space–time structure? We need a starting point, something that is common to both observers. What is available are events. Each observer coordinatizes an event differently, but both can take observations of events, and then compare coordinates. It is this comparison of events and their coordinates that leads to the Lorentz transformations.

We indicate the coordinates of O's reference by (x,y,z,t) and the coordinates of O''s by (x', y', z', t'), in which t and t' indicate time and the remaining variables are spatial variables. We may make further assumptions as follows:

1. Both O and O' can coordinate experiments to make measurements of time and space.
2. The coordinate axes along which they make spatial measurements are always parallel, and the $x:x'$ axes are aligned with one another, and the positive direction for each is the same.
3. O and O''s relative motion is along the $x:x'$ axis.

[51] Poincaré, "On the Dynamics of the Electron".

4. Both O and O′ consider their reference frame to be at rest; O observes O′ moving at a speed of *s* while O′ observes O moving at a speed of −*s*.

5. Both use identical measuring devices for time and space, and if they were taking measurements together with zero relative motion, they would return identical results.

6. The observers calibrate their instruments so that $t = t' = 0$ and $x = x' = 0$ at the very moment when they pass one another along the $x:x'$ axis.

7. The observers are able to orient themselves in any manner necessary to perform the experiments.

8. As noted above, the space and time through which O and O′ travel are homogeneous.

First O and O′ perform an experiment to determine whether their measurements along the $y:y'$ axis are the same. Once again, we follow Feynman, who proposes that each observer compare the length of their unit-measuring stick. as they pass one another at $t = t' = 0$ and $x = x' = 0$. Figure 9.5 depicts the experiment. Both observers hold their measuring sticks along their respective $y:y'$ axes and await the arrival of their counterpart. The event that they observe is the passing of the measuring sticks, and their particular point of interest is whether the measuring sticks are of the same height. As both observers view the same phenomenon, they both agree on its outcome; if O observes that O's measuring rod is longer, O′ observes the same outcome. But the experiment contains a symmetry between the observers; if O observes that O's measuring stick is longer, then, by symmetry, O′ must observe that O′'s measuring stick is longer, contradicting an agreed-on outcome. In fact, one sees that the only possible result without contradiction is that both observers agree that their measuring sticks are at the same height. Since both observers agree that their measuring sticks are the same when aligned with the $y:y'$ axis, their coordinatization of the $y:y'$ axis are the same; that is, if O reports that the y coordinate of some event is h, then O′ reports the same value, h, for the y' coordinate of the event. The same experiment and conclusion applies to

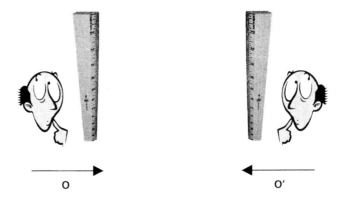

Figure 9.5. Measuring length orthogonal to direction of travel.

the $z:z'$ axis. The $z:z'$ relation is unnecessary for deriving the remaining Lorentz transformations. Accordingly, these coordinates are not mentioned again until summarization of the final result.

We next want the relationship between the time and $x:x'$ coordinate measures. If, for example, O reports the coordinates of an event (x,t) so that the coordinates are known, we wish to determine O''s report for the coordinates of the same event (x',t'). There are two known coordinates, x and t, and two unknown coordinates, x' and t'. Observations of two events yielding two relationships are necessary to determine the full relationships. However, using symmetry, we will find it possible to obtain two relationships from the observations of only one event.

As each observer agrees that light travels at the same speed, distance can be used to measure time. Simply take the distance and divide it by the speed of light, yielding the time. For this experiment, O agrees to set a target at a designated height y along the y axis. O points a laser at the target and switches on the laser at the exact moment when O' passes by. We calculate each observer's perception of the time that the light strikes the target. O perceives that the distance between the point of emission and the target is y; therefore the strike time is $t = \frac{y}{c}$ (see Figure 9.6).

To calculate O''s perception of the distance between the point of emission and the point where the target lies at strike time, one uses the Pythagorean theorem. Figure 9.6 depicts the geometry. From the preceding result, O''s y' coordinate of the target is the same as O's y coordinate, $y' = y$. Since the target moves with velocity $-s$ in O''s rest frame, we have $x' = -st'$.

Using the Pythagorean theorem, the calculated distance from emission point to strike point is $d'^2 = y^2 + (st')^2$. We also have $t' = \frac{d'}{c}$, from which one arrives at the following:

$$t'^2 = \frac{y^2 + (st')^2}{c^2}$$

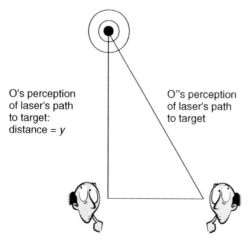

O's perception of laser's path to target: distance = y

O''s perception of laser's path to target

Figure 9.6. Path of light to a target.

Substituting the result that $y = ct$ gives the following:

$$t'^2 = \frac{(ct)^2 + (st')^2}{c^2}$$

Solving for t' and simplifying yields the following result:

$$t' = \frac{t}{\sqrt{1 - \frac{s^2}{c^2}}} \tag{9.1}$$

Substituting the value of t' into the equation $x' = -st'$ yields

$$x' = \frac{-st}{\sqrt{1 - \frac{s^2}{c^2}}} \tag{9.2}$$

There are several points of note. First, the equations do not contain the variable y. Both x' and t' are expressed as functions of t alone. Like a catalyst, the y coordinate assists with the calculation, but once the desired variables are in place, y exits the result. As we already have the relationship between y and y' coordinates, they are the same, so we will not explicitly refer to the relationship unless necessary.

Another point of note is that the relationships hold for both positive and negative values of t, a point that we make use of below. How do we think of the result with a negative time? The time $t = t' = 0$ corresponds to an event, O and O' passing one another. For both O and O', any event occurring prior to the passing must occur at a negative value of time.

We have one pair of observations corresponding to the same event. Table 9.1 gives the coordinates of the event as determined by O and O'. The final column gives O''s determination in terms of O's coordinates.

Continuing on, we need another pair of corresponding observations. Next, let us imagine that the symmetric experiment is performed; that is, instead of placing the laser gun and target on O's inertial frame, we place it on O''s inertial frame and do everything as before. We've already done the calculation. In the corresponding observations of equations (9.1) and (9.2), simply switch the roles of the the unprimed and primed coordinates and shift the sign on the s variable. The sign shift follows from O''s observation that O's velocity is $-s$.

TABLE 9.1. Event Coordinates

O's Coordinate	O's Observation	O''s Coordinate	O''s Observation
x	0	x'	$\dfrac{-st}{\sqrt{1-\frac{s^2}{c^2}}}$
t	t	t'	$\dfrac{t}{\sqrt{1-\frac{s^2}{c^2}}}$

The shift results in the following equations:

$$t = \frac{t'}{\sqrt{1 - \frac{s^2}{c^2}}} \qquad (9.3)$$

$$x = \frac{st'}{\sqrt{1 - \frac{s^2}{c^2}}} \qquad (9.4)$$

Solving for t' in terms of t yields the following:

$$t' = \sqrt{1 - \frac{s^2}{c^2}}\,t \qquad (9.5)$$

Once again there are several points of note: (1) it is important to explicitly recall $x' = 0$ for this event; (2) as before, the equations hold for positive as well as negative values of t; and (3) if we substitute equation (9.5) into (9.4), the result is the correct relation between x and t for this event, $x = st$ or equivalently $t = \frac{x}{s}$. In order to obtain the Lorentz transformations in their final forms, we need to substitute for t in the right-hand side of equation (9.5):

$$t' = \sqrt{1 - \frac{s^2}{c^2}\frac{x}{s}} \qquad (9.6)$$

We now have a second pair of observations as recorded in Table 9.2.

For reasons that are evident below, it is of interest to find O″'s corresponding coordinates for an event that O coordinatizes at an arbitrary value of x and $t = 0$. Using homogeneity, we are able to accomplish this with the results at hand. O' observes the final event of interest at $(x,0)$. This is the sum of coordinates that occur at two subevents, one at (x,t) and one at $(0,-t)$. Lucky for us for the case $t = \frac{x}{s}$, we have performed the subevents and have O″'s corresponding coordinates in Tables 9.1 and 9.2. The O' coordinates of the final event must be the sum of the corresponding subevents. Accordingly, we find x' by summing its value in Table 9.2 with the negative of its value (we substitute $-t$ for t) in Table 9.1:

$$x' = 0 + \frac{st}{\sqrt{1 - \frac{s^2}{c^2}}} \qquad (9.7)$$

TABLE 9.2. Second Pair of Event Observations

O's Coordinate	O's Observation	O″'s Coordinate	O″'s Observation
x	x	x'	0
t	$\frac{x}{s}$	t'	$\sqrt{1 - \frac{s^2}{c^2}\frac{x}{s}}$

We note that $st = x$ gives the final result:

$$x' = \frac{x}{\sqrt{1 - \frac{s^2}{c^2}}} \tag{9.8}$$

Similarly, we add the appropriate values in the tables to arrive at the final value for t':

$$t' = \sqrt{1 - \frac{s^2}{c^2}\frac{x}{s}} - \frac{t}{\sqrt{1 - \frac{s^2}{c^2}}} \tag{9.9}$$

Simplifying equation (9.9) gives the following result:

$$
\begin{aligned}
t' &= \sqrt{1 - \frac{s^2}{c^2}\frac{x}{s}} - \frac{t}{\sqrt{1 - \frac{s^2}{c^2}}}\\
&= \sqrt{1 - \frac{s^2}{c^2}\frac{x}{s}} - \frac{1}{\sqrt{1 - \frac{s^2}{c^2}}}\frac{x}{s}\\
&= \frac{-sx}{c^2\sqrt{1 - \frac{s^2}{c^2}}}
\end{aligned} \tag{9.10}
$$

We have applied a summation of two results as allowed by homogeneity to construct a result for a third corresponding event. Table 9.3 summarizes the result for the third event.

We are now ready for the final assault on the Lorentz transformations. Let O mark the event of the coordinates at (x,t). O's coordinates of this event are equivalent to the sum of O's coordinates of two subevents occurring at $(x,0)$ and $(0,t)$. O''s coordinates for the final event must be the sum of O's corresponding coordinates for the same subevents (the results are available from Tables 9.1 and 9.3):

$$x' = \frac{x}{\sqrt{1 - \frac{s^2}{c^2}}} - \frac{st}{\sqrt{1 - \frac{s^2}{c^2}}} \tag{9.11}$$

TABLE 9.3. Results for Third Event

O's Coordinate	O's Observation	O''s Coordinate	O''s Observation
x	x	x'	$\dfrac{x}{\sqrt{1-\frac{s^2}{c^2}}}$
t	0	t'	$\dfrac{-sx}{c^2\sqrt{1-\frac{s^2}{c^2}}}$

$$t' = \frac{-sx}{c^2\sqrt{1 - \frac{s^2}{c^2}}} + \frac{t}{\sqrt{1 - \frac{s^2}{c^2}}} \tag{9.12}$$

To get the full set of equations, one includes the relations for the y' and z' coordinates:

$$y' = y \tag{9.13}$$

$$z' = z \tag{9.14}$$

Equations 9.11–9.14 are the Lorentz transformations for the case of relative motion along the $x:x'$ axis. Let's consider the results a bit more.

Consider the calculation of an observer viewing a space probe launched from a spaceship. In the observer's inertial frame the speed of the spaceship at $0.8c$, and in the spaceship's inertial frame the probe's speed is also $0.8c$. Let's use the Lorentz transforms to determine the speed of the space probe from the initial observer's reference frame. Let O launch the probe and O' be the observer. O' views O's spaceship moving at $s' = -0.8c$. O launches a probe from his spaceship at $s = -0.8c$. How fast is O''s perception of the probe's speed? (see Figure 9.7)

For each observer, O launches the probe at the same coordinates, $x = x' = t = t' = 0$. In O's reference frame, at time $t = 1$, the probe is at the position $x = -0.8c$. Using equations (9.11) and (9.12), O''s coordinates for the rocket are $t' = 2.733$ and $x' = -2.667c$. Dividing O''s perceived distance by O''s perceived time gives O''s perceived velocity of the rocket: $s' = \frac{-2.667c}{2.733} = -0.976c$. According to O', the speed of the probe is $0.976c$.

This disagrees with the classical method of calculating the probe's speed in O''s reference frame. Classically one adds the speed of the probe in O's reference frame to the speed of the spaceship in O''s with the result that the probe in O''s reference frame moves at $1.6c$. By Einstein's second postulate, it is impossible for any observer to perceive motion faster than the speed of light, and the Lorentz transforms comport with this postulate. Indeed, we can up the ante. Let's say that

Figure 9.7. Probe speed measurement by O and O'.

at time $t = 0$, O shoots off a laser gun in the negative direction. Going through the same calculation gives $t = 1$, $x = c$, $t' = 2.733$, $x' = -2.733c$. If both observers had instruments to measure the speed of the laser, both measurements would indicate a speed of c.

Note that when the value s is small in comparison with c, when the relative velocity of the observers is small in comparison with the speed of light, the results are very close to Newton's classical results framed around Euclidean geometry. In such a case, the denominators of equations (9.11) and (9.12) are very close to one and the coordinates of both observers are nearly indistinguishable. Let us consider an Earthbound example. The speed of a jet plane is on the order of 500 miles per hour, or around 0.14 mile per second, whereas the speed of light is on the order of 186,000 miles per second. The term $\sqrt{1 - \frac{s^2}{c^2}} = 0.99999999999972$, which is nearly 1. The relativistic effects are there, but we do not notice them. For most everyday purposes, classical geometry and Newtonian mechanics work fine, so that the pilot and the air-traffic controller can communicate coordinates as though they exist in Euclid's universe without any adverse effects.

The Twin Paradox

Einstein considered the ramifications of the Lorentz transformations to have real consequences. Otherwise, identical clocks really do measure time in accordance with their relative motion. Otherwise identical measuring sticks do not assess space identically, but judge distance from their own perspectives. A most famous illustration of this is known as the *twin paradox*. Here we use the Lorentz equations to demonstrate the paradox.

Let's imagine two identical twins, O and O', initially on the same reference frame. At an arbitrary time $t = 0$, a spaceship moving at a speed $0.95c$ in O's reference frame passes the twins by and O' hops aboard. The spaceship carries O' for 10 years in O's reference frame before another spaceship moving back toward O comes along and O' once again hops on board for the return journey. In O's reference frame, the speed of the return spaceship is also $0.95c$. We know that when O' returns, O has aged 20 years. How old is O's twin brother, O'?

Using the Lorentz transform, we can answer this question. Equation (9.12) informs us of the time that O' ages when O' hops aboard the return spaceship. First, it is necessary to determine x, O's measurement of O''s position at the turnaround point. Since $s = 0.95c$, and $t = 10$, it follows that $x = 9.5c$.[52] Also, the term $\sqrt{1 - \frac{s^2}{c^2}} = 0.31$. This places the value of the first and second terms at $\frac{-sx}{c^2\sqrt{1 - \frac{s^2}{c^2}}} = -28.90$ and $\frac{t}{\sqrt{1 - \frac{s^2}{c^2}}} = 32.02$. When O' turns around, O' has aged 3.12 years. By symmetry, O' ages another 3.12 years during the return flight, for a total aging time of 6.24 years. When the twins O and O' reunite, O is around 13 years and 9 months older.

---

[52]Excuse the absence of units here. We will be more specific below.

How far did O' travel? That depends on who answers the question. O considers that O' traveled at a speed of $0.95c$ for 20 years. The total distance of travel is then $20 \times 0.95c = 1.17 \times 10^{14}$ miles. Conversely, O' observes O recede into the distance at a speed $0.95c$ followed by an approach at the same speed. However, O' concedes O''s own motion because O' is the one who initially jumped on to the moving spaceship and then once more jumped onto the return flight. As O''s flight time is 6.24 years, O' measures the distance at $6.24 \times 0.95c = 3.66 \times 10^{13}$ miles.

The reader may wonder how it is possible for O' to hop aboard a spaceship traveling by at a speed of $0.95c$. It is a fair point. Einstein first introduced the paradox without the twins, and in his 1905 paper, he considered the more realistic notion of traveling along a smooth curved path. Using the notation v as the velocity in place of s above, Einstein relates the paradox as follows:

> From this ensues the following peculiar consequences. If at the points A and B of K there are stationary clocks which viewed in the stationary frame are synchronous; and if the clock at A is moved with velocity v along the line AB to B, then on its arrival at B the two clocks no longer synchronize, but the clock from A to B lags behind the other which has remained at B by $\frac{1}{2}tv^2/c^2$ (up to magnitudes of fourth and higher order),[53] t being the time occupied in the journey from A to B.

> It is at once apparent that this result still holds good if the clock moves from A to B in a polygonal line, and also when the points A and B coincide.

> If we assume that the result proved for a polygonal line is also valid for a continuously curved line, we arrive at this result: If one of two synchronous clocks at A is moved in a closed curve with constant velocity until it returns to A, the journey lasting t seconds, then by the clock which has remained at rest the traveled clock on its arrival at A will be $\frac{1}{2}tv^2/c^2$ seconds slow. Thence we conclude that a balance clock at the equator must go more slowly by a very small amount, than a precisely similar clock situated under the poles under otherwise identical conditions.

Relativistic Velocity

The speed of the space probe investigated following the Lorentz transforms indicates that since time and distance of an event depend on the observer's inertial frame, so does the velocity of a moving object. Below we generalize the example and specify the relationship between the velocities of a body as viewed by O and O'. We describe the velocity of the body by the vectors (u,v,w) and (u',v',w') where u and u' refer to the corresponding $x:x'$ components of the velocity and there is the same correspondence between $v:v',w:w'$ and $y:y',z:z'$. The question we address is given as (u,v,w); determine (u',v',w').

[53]Einstein approximates the value $\dfrac{1}{\sqrt{1-\frac{v^2}{c^2}}}$ by its second-order Taylor series approximation, $\dfrac{1}{\sqrt{1-\frac{v^2}{c^2}}} \approx 1 + \frac{1}{2} + \left(\frac{v^2}{c^2}\right)$. Alternatively one can apply Newton's binomial theorem $(1+z)^p = 1 + pz + \frac{p(p-1)}{2!}z^2 + \frac{p(p-1)(p-2)}{3!}z^3 + \cdots$. Setting $z = -\frac{v^2}{c^2}$ and $p = -\frac{1}{2}$ gives the approximation, up to the first term, $(1 - \frac{v^2}{c^2})^{-(1/2)} \approx 1 + \frac{1}{2}\left(\frac{v^2}{c^2}\right)$.

The velocity of a body in uniform motion is determined by dividing the difference of the position of the body at two separate events by the time lapse between the events. Let (x_0, y_0, z_0, t_0) and (x_1, y_1, z_1, t_1) be O's coordinatization of the body at two events. Then we have the following relations:

$$u = \frac{x_1 - x_0}{t_1 - t_0} \tag{9.15}$$

$$v = \frac{y_1 - y_0}{t_1 - t_0} \tag{9.16}$$

$$w = \frac{z_1 - z_0}{t_1 - t_0} \tag{9.17}$$

The same relations hold for O':

$$u' = \frac{x_1' - x_0'}{t_1' - t_0'} \tag{9.18}$$

$$v' = \frac{y_1' - y_0'}{t_1' - t_0'} \tag{9.19}$$

$$w' = \frac{z_1' - z_0'}{t_1 - t_0} \tag{9.20}$$

Applying the Lorentz transforms to equations 9.15–9.17 yields the transformation from O's measurment of the body's velocity to O'''s:

$$u' = \frac{x_1' - x_0'}{t_1' - t_0'}$$

$$= \frac{\dfrac{x_1 - t_1 s}{\sqrt{1 - \frac{s^2}{c^2}}} - \dfrac{x_0 - t_0 s}{\sqrt{1 - \frac{s^2}{c^2}}}}{\dfrac{t_1 - \frac{x_1 s}{c^2}}{\sqrt{1 - \frac{s^2}{c^2}}} - \dfrac{t_0 - \frac{x_0 s}{c^2}}{\sqrt{1 - \frac{s^2}{c^2}}}}$$

$$= \frac{(x_1 - x_0) - (t_1 - t_0) s}{(t_1 - t_0) - (x_1 - x_0) \frac{s}{c^2}} \tag{9.21}$$

$$= \frac{(x_1 - x_0) - (t_1 - t_0) s}{(t_1 - t_0) - (x_1 - x_0) \frac{s}{c^2}} \times \frac{\frac{1}{t_1 - t_0}}{\frac{1}{t_1 - t_0}}$$

$$= \frac{u - s}{1 - \frac{us}{c^2}}$$

A similar calculation for v' yields the following result:

$$v' = \frac{y_1' - y_0'}{t_1' - t_0'}$$

$$= \frac{y_1 - y_0}{\dfrac{t_1 - \frac{x_1 s}{c^2}}{\sqrt{1 - \frac{s^2}{c^2}}} - \dfrac{t_0 - \frac{x_0 s}{c^2}}{\sqrt{1 - \frac{s^2}{c^2}}}}$$

$$= \frac{(y_1 - y_0)\sqrt{1 - \frac{s^2}{c^2}}}{(t_1 - t_0) - (x_1 - x_0)\frac{s}{c^2}} \qquad (9.22)$$

$$= \frac{(y_1 - y_0)\sqrt{1 - \frac{s^2}{c^2}}}{(t_1 - t_0) - (x_1 - x_0)\frac{s}{c^2}} \times \frac{\frac{1}{t_1 - t_0}}{\frac{1}{t_1 - t_0}}$$

$$= \frac{\sqrt{1 - \frac{s^2}{c^2}}\, v}{1 - \frac{us}{c^2}}$$

Similarly

$$w' = \frac{z_1' - z_0'}{t_1' - t_0'}$$

$$= \frac{\sqrt{1 - \frac{s^2}{c^2}}\, w}{1 - \frac{us}{c^2}} \qquad (9.23)$$

It is worth noting that while the y and z coordinates of the body are the same for both O and O', the speeds in these directions differ. Different speeds in these directions occur as a result of different time measurements, which, in turn, depend on the relative speed of the inertial frames along the $x:x'$ direction.

Let us propose an extreme circumstance. Consider the case when O observes the body moving at the speed of light and O''s inertial frame also moves at the speed of light, but in the opposite direction so that $u = c$, $s = -c$. How does O' measure the body's speed? Placing the values of u and s into equation (9.21) yields the following:

$$u' = \frac{u - s}{1 - \frac{us}{c^2}}$$

$$= \frac{c - (-c)}{1 + \frac{c^2}{c^2}}$$

$$= \frac{2c}{2}$$

$$= c$$

The result matches that of the required proposition: that movement at the speed of light is perceived as a constant in all inertial reference frames.

Relativistic Mass

Let us add an additional postulate to Einstein's original two: conservation of momentum. We accept the classical definition of momentum as a vector; it is the product of mass and the velocity vector, $\mathbf{p} = m\mathbf{v}$. Below, we consider motion only along the $x{:}x'$ direction with corresponding momentum $p = mu$. What results from the postulate, conservation of momentum?

Let us consider the following thought experiment. In a given inertial frame, a body with mass m moves at 25% of the speed of light. Another body is in pursuit with a mass of $2m$ and a speed of $0.5c$. The total momentum of both bodies is $0.25cm + cm = 1.25cm$. When the pursuing body catches up with the target body, there is an entanglement that leaves the pursuer at rest, which, according to the proposition, means that the entanglement transferred all the momentum of the pursuing body to the target body.[54] The momentum of the target body is now $1.25cm$. If we allow our classical experience to guide our interpretation of the outcome, we predict that the final velocity of the target body is $v = \frac{p}{m} = 1.25c$. We have arrived at a result that is in conflict with the Einstein's second proposition that no body can travel faster than the speed of light. Fortunately, this does not mean that the additional proposition is inconsistent with Einstein's second proposition. It means that our classical instincts have led us astray and we must determine how to proceed in a relativistic setting.

If we want to incorporate the additional proposition, we must accept the fact that the final momentum of the target body is, indeed, $1.25cm$. The only way to proceed is to find a manner in which the individual masses of the bodies differ before and after the entanglement. How do we envision that mass changes? Homogeneity of space and time does not allow mass to change as a result of differences in location and time. Accordingly, we seek a relationship between a body's speed and its mass. Let m_0 be the rest mass of a body, its mass when at rest with respect to the inertial frame of the observer. We consider the mass of the body when it is in motion to be as follows.

$$m = m_0 f(\xi) \tag{9.24}$$

where ξ is the speed of the body.[55] We note that by definition, $f(0) = 1$ and seek the function $f(\xi)$.

---

[54] Such an entanglement is allowed because the momentum is conserved. One might consider, for example, that the pursuer uncorks a spring-loaded punch arm once the target is in reach. The punch brings the pursuer to a standstill while the target flies off at high speed.

[55] We cannot consider the mass to be a function of the velocity, since the velocity is dependent on the orientation of the axes, but homogeneity requires a mass that is independent of orientation. The speed, which is independent of orientation, is all that remains.

Toward this end, let us apply relativity to an elastic collision between two masses. Assume that in O's inertial frame, there are twin bodies of identical mass, that move with the same speed but in opposite directions along the x axis and then collide. Assume also that after collision, they recoil along the x axis. As the collision is elastic, both the total momenta and the classical kinetic energy of the system remain the same. After the collision, the direction of the bodies reverses but their speed remains the same. The velocities are $u_{1-} = u_{2+} = u$ and $u_{1-} = u_{2+} = -u$, in which the number of the subscript denotes the body, body 1 or body 2, and the sign of the subscript $(-)$ denotes the precollision parameter, while the sign $(+)$ denotes the postcollision parameter. Figure 9.8 illustrates the collision. For convenience, we take the collision point to be $(x,t) = (0,0)$

Next let's examine the collision from the perspective of O' and consider the case when O sees O' moving with the speed $s = -u$ in the negative direction. From the perspective of O', body 1 is initially in motion and body 2 is initially at rest, while after the collision, body 1 is at rest and body 2 is in motion. Figure 9.9 illustrates the collision from the perspective of O'.

Applying equation (9.24), we note that O considers the pre- and postcollision masses of the bodies to be the same:

$$m_{1-} = m_{1+} = m_{2-} = m_{2+} = m_0 f(u)$$

Equality follows from symmetry. Both bodies have the same rest mass and move with the same speed. O''s view of the pre- and postcollision masses is the following:

$$m'_{1-} = m'_{2+} = m_0 f(u') \tag{9.25}$$

$$m'_{1+} = m'_{2-} = m_0 \tag{9.26}$$

Precollision configuration: both bodies
move toward each other at the same speed

Precollision configuration: both bodies
move away from each other at the same speed

Figure 9.8. Collision as viewed by O.

Precollision configuration: body 1 moves toward body 2

Postcollision configuration: body 2 moves away from body 1

Figure 9.9. Collision as viewed by O'.

Note that the second equation follows because the masses are at rest and by definition $f(0) = 1$. Also note that while the masses of the individual bodies change, the total pre- and postcollision summed mass of the bodies is the same: $m'_{1-} + m'_{2-} = m'_{1+} + m'_{2+}$. If one determines the pre- and postcollision masses, one can also determine the function f. This is precisely our objective.

Conservation of momentum is necessary for the calculation. The centroid of the collision is the weighted average of the positions, and conservation of momentum and mass requires that the speed of the centroid be the same before and after the collisions.[56] It is this property that yields the masses.

Let k denote the velocity of the centroid as viewed by O. By design the centroid stays at rest at O's origin, $k = 0$. Applying equation (9.21) gives the velocity of the centroid as measured by O':

$$k' = \frac{k - s}{1 - \frac{ks}{c^2}}$$

$$= \frac{0 - (-u)}{1 - \frac{0}{c^2}} \tag{9.27}$$

$$= u$$

The value k' is also the weighted average of the velocities of the individual bodies:[57]

$$k' = \frac{m'_{1+}u'_{1+} + m'_{2+}u'_{2+}}{m'_{1+} + m'_{2+}} \tag{9.28}$$

[56]If x_c is the position of the centroid, $x_c(t) = \frac{m_1 x_1(t) + m_2 x_2(t)}{m_1 + m_2}$. The numerator is the momentum and the denominator is the mass. Differentiating with respect to time yields the velocity of the centroid. $k = \frac{m_1 u_1 + m_2 u_2}{m_1 + m_2}$, where k is the velocity of the centroid. The collision alters neither the total momentum nor the total mass, so velocity of the centroid is also unaltered by the collision.
[57]While equation (9.28) contains postcollision parameters, precollision parameters may also be used. Above we arbitrarily choose the postcollision parameters.

By design, $u'_{1+} = 0$, giving the following result:

$$k' = \frac{m'_{2+}u'_{2+}}{m'_{1+} + m'_{2+}} \tag{9.29}$$

Applying equation (9.21) gives the following result for u'_{2+}:

$$
\begin{aligned}
u'_{2+} &= \frac{u_{2+} - s}{1 - \dfrac{u_{2+}s}{c^2}} \\
&= \frac{u - (-u)}{1 + \dfrac{u^2}{c^2}} \tag{9.30} \\
&= \frac{2u}{1 + \dfrac{u^2}{c^2}}
\end{aligned}
$$

There is a point to note. While the center of mass in O's reference frame remains exactly midway between the two bodies, this does not hold in O''s reference frame. For O', body 1 is at rest and body 2 moves less than twice the speed of the centroid. The centroid is closer to body 2 than body 1, and so the mass of body 2 must be greater than that of body 1. Equating equations (9.27) and (9.29), making the substitution equation (9.24), and simplifying yields the value of the function f:

$$\frac{m'_{2+}u'_{2+}}{m'_{1+} + m'_{2+}} = u \qquad [\text{eq.}(9.29) = \text{eq.}(9.27)]$$

$$\frac{m_0 f(u'_{2+})u'_{2+}}{m_0 + m_0 f(u'_{2+})} = u \qquad [\text{eq.}(9.24)]$$

$$m_0 f(u'_{2+})u'_{2+} = \left[m_0 + m_0 f(u'_{2+})\right]u$$

$$f(u'_{2+})u'_{2+} = \left[1 + f(u'_{2+})\right]u$$

$$\left(u'_{2+} - u\right)f(u'_{2+}) = u$$

$$f(u'_{2+}) = \frac{u}{u'_{2+} - u} \tag{9.31}$$

The only remaining step is to express u in terms of u'_{2+} and simplify. Equation (9.30) is the starting point:

$$u'_{2+} = \frac{2u}{1 + \dfrac{u^2}{c^2}}$$

$$\left(1 + \frac{u^2}{c^2}\right) u'_{2+} = 2u$$

$$\left(1 + \frac{u^2}{c^2}\right) u'_{2+} - 2u = 0$$

$$\frac{u'_{2+}}{c^2} u^2 - 2u + u'_{2+} = 0$$

Applying the quadratic formula to the final equation yields the following result:

$$u = \frac{1 \pm \sqrt{1 - \left(\frac{u'_{2+}}{c}\right)^2}}{\frac{u'_{2+}}{c^2}} \tag{9.32}$$

There is a choice to make. Should we take the positive or negative root? The positive root is problematic. As u'_{2+} approaches zero, u becomes unbounded, and as u'_{2+} approaches c, u approaches $2c$. These solutions imply that body 1's initial speed as seen by O may be greater than the speed of light and are not feasible. The negative root maintains the value of u within its constrained limits so long as u'_{2+} is within its constrained limits, $| u'_{2+} | < c$.[58] As such, choose the negative root. Placing the solution of u in equation (9.32) into equation (9.31) and simplifying gives the following result:

$$f(u'_{2+}) = \frac{u}{u'_{2+} - u}$$

$$= \frac{\frac{1 - \sqrt{1 - \frac{(u'_{2+})^2}{c^2}}}{\frac{u'_{2+}}{c^2}}}{u'_{2+} - \frac{1 - \sqrt{1 - \frac{(u'_{2+})^2}{c^2}}}{\frac{u'_{2+}}{c^2}}} \tag{9.33}$$

At this point, the notation u'_{2+} is not meaningful. It is a speed dependent on the arbitrary initial state of the bodies. We may choose any value and adopt simpler notation: $u'_{2+} = \xi$.

$$f(\xi) = \frac{\frac{1 - \sqrt{1 - \frac{\xi^2}{c^2}}}{\frac{\xi}{c^2}}}{\xi - \frac{1 - \sqrt{1 - \frac{\xi^2}{c^2}}}{\frac{\xi}{c^2}}}$$

[58]Limiting arguments are necessary to arrive at a value when $u'_{2+} = 0$. Those familiar with L'Hôpital's rule find the value $u = 0$ when $u'_{2+} = 0$.

$$= \frac{1 - \sqrt{1 - \frac{\xi^2}{c^2}}}{\frac{\xi^2}{c^2} - \left(1 - \sqrt{1 - \frac{\xi^2}{c^2}}\right)}$$

$$= \frac{1 - \sqrt{1 - \frac{\xi^2}{c^2}}}{\frac{\xi^2}{c^2} - \left(1 - \sqrt{1 - \frac{\xi^2}{c^2}}\right)} \times \frac{1 + \sqrt{1 - \frac{\xi^2}{c^2}}}{1 + \sqrt{1 - \frac{\xi^2}{c^2}}}$$

$$= \frac{\frac{\xi^2}{c^2}}{\frac{\xi^2}{c^2}\sqrt{1 - \frac{\xi^2}{c^2}}}$$

$$= \frac{1}{\sqrt{1 - \frac{\xi^2}{c^2}}} \tag{9.34}$$

The relativistic mass is

$$m = \frac{m_0}{\sqrt{1 - \frac{\xi^2}{c^2}}} \tag{9.35}$$

If one adopts the conventional notation that v is the speed of the body, then one has the following equation:

$$m = \frac{m_0}{\sqrt{1 - \frac{v^2}{c^2}}} \tag{9.36}$$

For the remainder of the chapter, we use the conventional notation and v represents the speed of the body.

Energy

Let us further analyze the collision described above. In O'''s inertial frame, body 2 is initially at rest and then moves with speed v. The mass is initially m_0 and then becomes $\frac{m_0}{\sqrt{1 - \frac{v^2}{c^2}}}$.

The following expression gives the difference between the final and initial masses:

$$m - m_0 = m_0 \left(\frac{1}{\sqrt{1 - \frac{v^2}{c^2}}} - 1\right) \tag{9.37}$$

One may apply Newton's binomial theorem to the term $\frac{1}{\sqrt{1 - \frac{v^2}{c^2}}} = \left(1 - \frac{v^2}{c^2}\right)^{(-1/2)}$ or use a Taylor expansion to express equation (9.37) as a series. Taking the

first two terms of the binomial expansion gives the approximation $\dfrac{1}{\sqrt{1-\frac{v^2}{c^2}}} \approx$
$1 + \dfrac{v^2}{2c^2}$. The approximation works well when the term $\dfrac{v^2}{c^2}$ is small. Placing the approximation into equation (9.37) results in the following:

$$m - m_0 = m_0 \left(\frac{1}{\sqrt{1 - \frac{v^2}{c^2}}} - 1 \right)$$

$$\approx \frac{m_0 v^2}{2} \frac{1}{c^2}$$

The expression $\dfrac{m_0 v^2}{2}$ is the classical post-collision kinetic energy, which ought to be close to the relativistic kinetic energy when the term $\dfrac{v^2}{c^2}$ is small. Multiplying both sides by c^2 gives the following result:

$$m - m_0 = m_0 \left(\frac{1}{\sqrt{1 - \frac{v^2}{c^2}}} - 1 \right)$$

$$m - m_0 \approx m_0 \left(1 + \frac{v^2}{2c^2} - 1 \right)$$

$$(m - m_0)\, c^2 \approx \frac{m_0 v^2}{2} \tag{9.38}$$

The interpretation of equation (9.38) is that the product of body 2's mass gain and the square of the speed of light is equivalent to its energy gain. Einstein then proposed the equivalence between mass and energy, which is now expressed in the famous formula $E = mc^2$.

Using calculus along with the relations between force, momentum and energy, one may eliminate the approximation. Consider F to be the force on body 2 during the collision; E, the energy of body 2; and, as above, m, its mass and v, its speed. Using differential notation we arrive at the following equations:

$$dE = F\,dx \tag{9.39}$$

$$d(mv) = F\,dt \tag{9.40}$$

Equation (9.39) demonstrates that the differential energy of a body dE is equal to the product of the force over the differential distance dx. Equation (9.40) gives the relation between the differential momentum and the product of the force over the differential time dt. These equations appear the same as their Newtonian counterparts, but there is a difference. Whereas Newton considers the mass to be a constant, in relativity, it is not.

We integrate the first equation over the collision interval. A subscript denotes the beginning and the end of the collision interval. The spatial collision interval ranges from $x = x_0 = 0$ to $x = x_1$, and the temporal collision interval ranges from $t = t_0 = 0$ to $t = t_1$. Note that at $t = 0$, body 2 is at rest, $v_0 = 0$:

$$E_1 - E_0 = \int_0^{x_1} F\,dx \qquad \text{[integration of eq.(9.39)]}$$

$$= \int_0^{t_1} F\frac{dx}{dt}dt \qquad \text{(change of variables)}$$

$$= \int_0^{t_1} Fv\,dt \qquad \left(\text{definition of velocity } v = \frac{dx}{dt}\right)$$

$$= \int_0^{m_1 v_1} v\,d(mv) \qquad \text{[eq.(9.40): } Fdt = d(mv)] \qquad (9.41)$$

Next, use the relation between mass and speed [eq. (9.36)] to simplify the term $d(mv)$:

$$d(mv) = m_0 d\left(\frac{v}{\sqrt{1 - \frac{v^2}{c^2}}}\right)$$

$$= m_0 \frac{1}{\left(1 - \frac{v^2}{c^2}\right)^{3/2}}dv \qquad (9.42)$$

With this simplification, it is possible to evaluate the integral in (9.41):

$$E_1 - E_0 = \int_0^{m_1 v_1} v\,d(mv)$$

$$= m_0 \int_0^{v_1} \frac{v}{\left(1 - \frac{v^2}{c^2}\right)^{3/2}}dv \qquad \text{[eq.(9.42)]}$$

$$= m_0 \frac{c^2}{\sqrt{1 - \frac{v^2}{c^2}}}\Big|_0^{v_1} \qquad (9.43)$$

$$= \frac{m_0 c^2}{\sqrt{1 - \frac{v_1^2}{c^2}}} - m_0 c^2$$

$$= (m_1 - m_0)c^2$$

The final result equates the change in a body's energy to the change in a body's mass through the constant c^2 from which we deduce $E = mc^2$.

A Strange Universe

Newton guided Europe out of the Counter-Reformation toward the Enlightenment. It was a zone where common sense and reason triumphed. God orchestrated the universe's motion using sensible laws that revealed themselves through everyday experiences; an apple flung by a branch flexing with the wind makes its way to the ground on a parabolic trajectory; a planet orbits the sun along an elliptic pathway. It was certain that every problem would cede to Newtonian reasoning, and that the reasoning would reveal the simple elegance of God's design.

Einstein jolted humanity's complacency. Disregarding the circle, nature does not structure herself around human notions of perfection, disregarding Euclidean geometry; nature does not structure herself around humanity's notion of common sense. Instead, nature structures herself around her own imperatives for which humanity's common sense is utterly inadequate. The path that the apple takes is not the same for all observers, and the planets do not orbit in perfect ellipses. Indeed, commonly held beliefs of such basic notions as time and distance, beliefs that held sway for many millennium, were all false. Einstein's equations yield strange phenomena. Mass can collapse in a black hole from which nothing, including light, can escape. General relativity inspired the big bang theory; a belief that the whole universe sprang from a single dimensionless point. Like Shakespeare's Puck watching mortals from beyond, nature conspires to perform in ways beyond our imagination and perhaps amuses herself while watching human attempts to fathom her. The attempts are not in vain. With each generation we probe further, understand more, and then, nature unleashes yet another surprise.

EPILOGUE

Let us return to the mirror of our preface and imagine the view it affords on both our present and past. Suppose that Saint Augustine stands in front of that mirror. It might well be an "I told you so" moment:

> Did I not warn you to remain ignorant of the workings of God? God provides for your needs and asks in return only that you offer your thanks. God has the wisdom to use his creation in a tempered manner; humankind does not. Instead, your greed drove you to seek his workings and convert them to your own use. What has this wrought? It has wrought the Reformation and loss of Church authority followed by chaos in Europe; the Thirty Years' War; the French wars under Louis XIV, the French Revolution and Napoleonic Wars; social revolutions of the midnineteenth century; the Franco-Germanic War; World War I; economic depression; and World War II. It has wrought injustice committed by those with technical prowess against defenseless populations on a scale unimaginable when humans constrained their knowledge to the word of the Bible, the subjugation of native populations in the Americas; the enslavement of Africans; European colonization across Africa and Asia, along with its incumbent exploitation. It has wrought the replacement of the spear with its limited capacity to kill by weapons undreamed of when humans knew their role beneath God, rifles; machine guns; tanks; conventional explosives delivered by artillery, aircraft, and short-range missiles; chemical weapons; deadly biological agents; the harnessing of fission in the atom bomb; the harnessing of fission to set off yet an even more powerful fusion-based explosion, the hydrogen bomb; along with the capacity to deliver this terrifying assault against not only

Shifting the Earth: The Mathematical Quest to Understand the Motion of the Universe,
First Edition. Arthur Mazer.
© 2011 John Wiley & Sons, Inc. Published 2011 by John Wiley & Sons, Inc.

human life, but all God's creations, using a shower of intercontinental ballistic missiles. Now all of humanity is at risk of self-annihilation.

While I believe that the counterargument speaking on behalf of the advances is stronger, I have to concede some merit to those who align with Augustinian philosophy. Perhaps Augustine was right. But can we go back? Would Augustine himself return to his world if he knew ours? That for over a century there has been an influx of immigrants from nations that have not fully transformed through the industrial revolution to nations that have, and not the other way around, indicates a preference for the benefits that the transition has bestowed. So we have grown and become more knowledgeable. We look in the mirror and see more possibilities than before, possibilities with both positive and negative consequences, and know that there is no turning back. It is the scientists who, peering at the heavens, brought about the earthly industrial revolution, and it is the scientists who will lead us to the next transformational revolution, whatever that may be. It is our collective self-image, our wisdom or lack of wisdom, that determines the ends to which we commit scientific knowledge.

BIBLIOGRAPHY

Alder, Ken, *The Measure of All Things* Free Press, New York, 2003.

Al-Khwarizmi, *Algebra* (translated by Frank Rosen; publisher unknown), London, 1831.

Apollonius, *Conics* (translated by William Donahue), Green Lion Press, Santa Fe, 1997.

Appelbaum, Wilbur, *Encyclopedia of the Scientific Revolution: From Copernicus to Newton*, Garland, London, 2008.

Archimedes, *The Works of Archimedes* (translated by Thomas Heath), Dover, New York, 2002.

Aristotle, *The Basic Works of Aristotle* (edited by Richard McKeon), Modern Library, New York, 2001.

Arnold, Vladimir, *Geometrical Methods in the Theory of Ordinary Differential Equations*, Springer, New York, 1983.

Artmann, Benno, *Euclid: The Creation of Mathematics*, Springer, New York, 1999.

Asprey, Robert, *The Reign of Napoleon*, Basic Books, New York, 2001.

Bagnall, Nigel, *The Punic Wars*, St. Martin's Press, New York, 1990.

Bardi, Jason, *The Calculus Wars*, Thunder's Mouth Press, New York, 2006.

Beckmann, Petr, *A History of Pi*, St. Martin's Press, New York, 1976.

Blom, Philipp, *The Vertigo Years*, Basic Books, New York, 2008.

Blond, Anthony, *A Scandalous History of the Roman Emperors*, Carroll & Graf Publishers, New York, 2000.

Boorstin, Daniel, *The Discoverers*, Vintage Books, New York, 1985.

Bottazini, Umberto and Van Egmond, Warren, *The Higher Calculus: A History of Real and Complex Analysis from Euler to Weierstrass*, Springer-Verlag, New York, 1986.

Boyer, Carl, *A History of Mathematics* (revised by Uta Merzbach), Wiley, Hoboken, 1991.

Shifting the Earth: The Mathematical Quest to Understand the Motion of the Universe,
First Edition. Arthur Mazer.
© 2011 John Wiley & Sons, Inc. Published 2011 by John Wiley & Sons, Inc.

Brecht, Bertolt, *Life of Galileo*, Penguin, New York, 2008.

Burman, Edward, *The Inquisition: The Hammer of Heresy*, Dorset, London, 1984.

Calaprice, Alice and Lipscombe, Trevor, *Albert Einstein, a Biography*, Greenwood Press, Westport, 2005.

Cardano, Girolamo *Ars Magna* (translated by Richard Witmer), Dover, New York, 2007.

Christianson, Gale, Isaac Newton, Oxford University Press, Oxford, 2005.

Cicero, Marcus Tulio, *Selected Works* (translated by Michael Grant), Penguin Classics, London, 1971.

Connor, James, *Kepler's Witch*, HarperCollins, New York, 2005.

Copernicus, Nicolaus, *On the Revolutions of Heavenly Spheres* (translated by Charles Wallis), Prometheus Books, New York, 1995.

_____, Einstein, Albert; Hawking, Stephen; Galileo, Galilee; Kepler, Johanes; and Newton, Isaac, *On the Shoulders of Giants*, Running Press, Philadelphia, 2002.

Crowe, Michael, *Theories of the World from Antiquity to the Copernican Revolution*, Dover, New York, 1990.

_____, *Mechanics from Aristotle to Einstein*, Green Lion Press, Santa Fe, 2007.

Cullen, Christopher, *Astronomy and Mathematics in Ancient China*, Cambridge University Press, Cambridge, 2008.

Davison, James, *Cortesans and Fishcakes*, St. Martin's Press, New York, 1998.

Descartes, René, *The Geometry of Rene Descartes*, Dover, New York, 1954.

Einstein, Albert, *The Collected Papers of Albert Einstein* (translated by Anna Beck), Princeton University Press, Princeton, 1987.

_____, *Letters to Solovine*, Carol Publishing Group, New York, 1993.

_____, *Relativity: The Special and General Theory*, Forgotten Books, New York, 2010.

Einstein, Albert; Lorentz, Hendrik; Minkowski, Hermann; and Weyl, Hermann, *The Principle of Relativity*, Dover, New York, 1952.

Estep, William, *Renaissance and Reformation*, Eerdmans, Grand Rapids, 1986.

Euclid, *The Elements* (translated by Thomas Heath), Green Lion Press, Santa Fe, 2002.

Evans, James, *The History and Practice of Ancient Astronomy*, Oxford University Press, Oxford, 1998.

Ferguson, Kitty, *Tycho and Kepler*, Walker, New York, 2002.

Feynman, Richard, *Six Not So Easy Pieces*, Basic Books, Cambridge, 1997.

Feynman, Richard; Leighton, Robert; and Sands, Mathhew, *The Feynman Lectures on Physics*, Addison-Wesley Longman, New York, 1970.

Fildes, Alan and Fletcher, Joann, *Alexander the Great: Son of the Gods*, J. Paul Getty Museum Press, Los Angeles, 2004.

Freeman, Charles, *The Closing of the Western Mind*, Vintage Books, New York, 2005.

Galileo, Galilei, *Dialogue Concerning the Two Chief World Systems* (translated by Stillman Drake), University of California Press, Berkeley, 2001.

Galison, Peter, *Einstein's Clocks, Poincare's Maps*, Norton, New York, 2004.

Gingerich, Owen, *The Eye of Heaven: Copernicus, Ptolemy, Kepler*, American Institute of Physics, New York, 1993.

Ginzburg, Carlo, *The Cheese and the Worms: The Cosmos of a Sixteenth-Century Miller* (translated by John and Anne Tedeschi), Johns Hopkins University Press, Baltimore, 1992.

Gleick, James, *Isaac Newton*, Pantheon, New York, 2003.

Gottleib, Anthony, *The Dream of Reason: A History of Philosophy from the Greeks to the Renaissance*, Norton, New York, 2000.

Grafton, Anthony, *The Thirty Years War*, The New York Review of Books, New York, 2005.

Grant, Michael, *Constantine the Great*, History Book Club, New York, 2000.

Green, Peter, *The Hellenistic Age*, Random House, New York, 2007.

———, *Alexander of Macedon 356–323 B.C.: A Historical Biography*, University of California Press, Berkeley, 1992.

Greenberg, Marvin, *Euclidean and Non-Euclidean Geometries: Development and History*, Freeman Publishers, New York, 2007.

Greene, Brian, *The Fabric of the Cosmos*, Vintage Books, New York, 2004.

Gullberg, Jan, *Mathematics: From the Birth of Numbers*, Norton, New York, 1997.

Hager, Thomas, *The Air of Alchemy: A Jewish Genius, a Doomed Tycoon, and the Scientific Discovery that Fed the World but Fueled the Rise of Hitler*, Three Rivers Press, New York, 2008.

Hawking, Stephen, *A Brief History of Time*, Bantam Books, New York, 1988.

Heath, Thomas, *A History of Greek Mathematics*, Vols. I and II, Dover, New York, 1921.

———, *The Copernicus of Antiquity*, Macmillan, New York, 1920.

Hibbert, Christopher, *The Days of the French Revolution*, The Penguin Group, London, 1980.

Hitler, Adolf, *Mein Kampf* (translated by Ralph Abraham), Mariner Books, New York, 1999.

Hofstadter, Douglas, *Godel, Escher, Bach: An Eternal Golden Braid*, Basic Books, New York, 1979.

Isaacson, Walter, *Einstein: His Life and Universe*, Simon & Schuster, New York, 2007.

Johnson, Marion, *The Borgias*, The Penguin Group, London, 2001.

Kann, Robert, *A History of the Habsburg Empire, 1526–1918*, University of California Press, Berkeley, 1974.

Kepler, Johannes, *New Astronomy* (translated by William Donahue), Cambridge University Press, Cambridge, 1993.

———, *The Optical Part of Astronomy* (translated by William Donahue), Green Lion Press, Santa Fe, 2000.

———, *Epitome of Copernican Astronomy and Harmonies of the World* (translated by Charles Wallis), Prometheus Books, New York, 1995.

Kershaw, Ian, *Hitler, 1889–1936: Hubris; 1936–1945: Nemesis*, Norton (two-volume set), New York, 1998, 2000.

King, David, *Vienna 1814*, Three Rivers Press, New York, 2008.

Koestler, Arthur, *The Sleepwalkers: A History of Man's Changing Vision of the Universe* the Penguin Group, New York, 1990.

Kramnick, Isaac, *The Portable Enlightement Reader*, The Penguin Group, New York, 1995.

Kuhn, Thomas, *The Copernican Revolution*, Shambhala, Boston, 1991.

Kumar, Manjit, *Quantum: Einstein, Bohr and the Great Debate about the Nature of Relativity*, Norton, New York, 2010.

Laubenbacher, Reinhard and Pangelley, David, *Mathematical Expeditions: Chronicles by the Explorers*, Springer, New York, 2000.

Leibniz, Gottfried, *The Early Mathematical Manuscripts of Leibniz* (translated by J. M. Child), Cosimo Classics, New York, 2008.

Levi, Anthony, *Louis XIV*, Carroll & Graf, New York, 2004.

Linton, C. M., *From Eudoxus to Einstein: A History of Mathematical Astronomy*, Cambridge University Press, Cambridge, 2004.

Litvinoff, Barnet, *1492*, Little, Brown Book Group, New York, 1991.

Livy, *Hannibal's War: Books Twenty-One to Thirty* (translated by J. C. Yardley), Oxford University Press, Oxford, 2006.

Longfellow, Ki, *Flow Down Like Silver: Hypatia of Alexandria*, Eio Books, Belvedere, 2009.

MacCulloch, Diarmaid, *The Reformation*, The Penguin Group, New York, 2003.

Marsden, Jerrold and Ratiu, Tudor, *Introduction to Mechanics and Symmetry*, Springer, New York, 2002.

Martin, Thomas, *Ancient Greece*, Yale University Press, New Haven, 1996.

Mazer, Arthur, *The Ellipse: A Historical and Mathematical Journey*, Wiley, Hoboken, 2010.

Murdoch, Adrian, *The Last Pagan: Julian the Apostate and the Death of the Ancient World*, Inner Traditions, Rochester, 2008.

Newton, Isaac, *Principia* (translated by Andrew Motte), Prometheus Books, New York, 1995.

Oberman, Heiko, *Luther: Man Between God and the Devil*, Yale University Press, New Haven, 1989.

O'Connel, Marvin, *The Counter Reformation, 1550–1610*, Harper & Row, New York, 1974.

O'Malley, John, *The First Jesuits*, Harvard University Press, Cambridge, 1995.

Penrose, Roger, *The Road to Reality: A Complete Guide to the Laws of the Universe*, Knopf, New York, 2005.

Planck, Max, *Eight Lectures on Theoretical Physics* (translated by A. P. Wills), Dover Publications, New York, 1992.

Plato, *The Republic* (translated by Desmond Lee), Penguin Classics, London, 1987.

―――, *Apology, Crito, and Phaedo of Socrates* (translated by Henry Cary), Forgotten Books, Philadelphia, 2010.

Plutarch, *Plutarch's Lives*, Vol. I (translated by John Dreyden; edited by Arthur Clough), Random House, New York, 2001.

Poincaré, Henri, *Science and Hypothesis*, Dover Publications, New York, 1952.

Poincaré, Henri, "On the Dynamics of the Electron," *Rendiconti del Circolo Matematico di Palermo*, vol. 21, pp. 129–176, 1905.

Pollard, Justin and Reid, Howard, *The Rise and Fall of Alexandria: Birthplace of the Modern World*, The Penguin Group, New York, 2006.

Ptolemy, *The Almagest* (translated by G. J. Toomer), Princeton University Press, Princeton, 1998.

Roper, Lyndell, *Witch Craze: Terror and Fantasy in Baroque Germany*, Yale University Press, New Haven, 2006.

Rowland, Ingrid, *Giordano Bruno: Philosopher/Heretic*, University of Chicago Press, Chicago, 2008.

_____, *From Heaven to Arcadia: The Sacred and the Profane in the Renaissance*, The New York Review of Books, New York, 2008.

Rudin, Walter, *Principles of Mathematical Analysis*, McGraw-Hill, New York, 1976.

Saliba, George, *Islamic Science and the Making of the European Renaissance*, MIT Press, Boston, 2007.

Shuckburgh, Evelyn *A History of Rome to the Battle of Actium*, Macmillan, New York, 1894.

Sobel, Dava, *Galileo's Daughter*, Walker, New York, 1999.

_____, *Longitude*, Walker, New York, 1995.

Spencer, Charles, *Blenheim: Battle for Europe*, Phoenix, London, 2007.

Spivak, Michael, *Calculus*, Cambridge University Press, Cambridge, 2006.

_____, *Differential Geometry*, Vols. I–V, Publish or Perish, Houston, 1999.

Terrall, Mary, *The Man Who Flattened the Earth: Maupertuis and the Sciences of the Enlightenment*, University of Chicago Press, Chicago, 2006.

Thorne, Kip, *Black Holes and Time Warps: Einstein's Outrageous Legacy*, Norton, New York, 1994.

Van Helden, Albert, *Measuring the Universe: Cosmic Dimensions from Aristarchus to Halley*, University of Chicago Press, Chicago, 1986.

Vivante, Bella, *Events that Changed Ancient Greece*, Greenwood Press, Westport, 2002.

Voelkel, James, *The Composition of Kepler's Astronomia Nova*, Princeton University Press, Princeton, 2001.

Wheatcroft, Andrew, *The Habsburgs*, The Penguin Group, London, 1996.

INDEX

acceleration, 186, 190–191, 203, 235, 245, 251, 272

achronycal observation, 139

Adler, Friedrich, 253

Africa, 26, 40–41, 96, 259, 293

Aix-la-Chapelle, Treaty of, 169

alchemy, 176

Alexander of Macedon, 24–26, 42

Alexandria, 27, 41–42, 65, 67–68, 70

Almagest, 68–71, 73, 75, 77, 79, 81, 83, 85–87, 89, 92, 150

Ambrose, Bishop of Milan, 64–65

angular momentum, 143–145, 181, 191–193, 197, 206–211, 214, 216, 223–224

angular momentum, conservation of, 143–145, 181, 191–192, 206–209

Antigonus, 41

anti-indulgence sentiment, 99

aphelion, 73, 76, 148–149, 151–153, 163, 219

apogee, 73, 81–82, 215

Apollonius, 42–47, 50, 52–53, 62, 67–68, 70, 74, 97, 105, 128, 137, 139, 141, 150, 154–155, 162, 221, 223, 235

apsides, 141

Aquarius, 10, 12, 138, 146, 150

aqueducts, 67

Aquinas, Thomas, 94

Archimedes, 38–39, 41–45, 66, 68, 71, 97, 128, 141, 144, 150, 154–159, 161, 176

Aristarchus, 29–37, 42, 68, 97, 100, 103, 105, 134, 137

Aristophanes, 5

Aristotle, 5, 14, 25, 43, 68–69, 94, 186, 232

assimilation, 227, 239

astrology, 93, 123

astronomer, 29, 42, 70, 104, 123, 136, 138, 147, 181–182, 259

Astronomers, 182

astronomy, 29, 93, 98, 101, 103–104, 120, 123, 131–134, 136, 140

Astynomoi, 6–7

Athenians, 5–8, 27

atmosphere, 3, 27, 100, 148

atom, 245, 256, 265, 293

atomic bomb, 265

Attilid Kingdom, 27

Augsburg, 100

Augustus, 65–66

Auschwitz, 268

Averroism, 94

axioms, 28, 43, 183, 185

Ayscough, Hannah, 173

Shifting the Earth: The Mathematical Quest to Understand the Motion of the Universe,
First Edition. Arthur Mazer.
© 2011 John Wiley & Sons, Inc. Published 2011 by John Wiley & Sons, Inc.